Nanoformulations for Sustainable Agriculture and Environmental Risk Mitigation

Nanoformulations for Sustainable Agriculture and Environmental Risk Mitigation

Edited by

Zeba Khan

Center of Agricultural Education (CAE), Faculty of Agricultural Sciences, Aligarh Muslim University, 202 002 Aligarh, India

Nicoleta Anca Șuțan

Department of Natural Sciences, University of Pitesti, Târgul din Vale 1, 110040 Pitesti, Romania

CABI is a trading name of CAB International

CABI
Nosworthy Way
Wallingford
Oxfordshire OX10 8DE
UK

CABI
200 Portland Street
Boston
MA 02114
USA

Tel: +44 (0)1491 832111
E-mail: info@cabi.org
Website: www.cabi.org

T: +1 (617)682-9015
E-mail: cabi-nao@cabi.org

The views expressed in this publication are those of the author(s) and do not necessarily represent those of, and should not be attributed to, CAB International (CABI). Any images, figures and tables not otherwise attributed are the author(s)' own. References to internet websites (URLs) were accurate at the time of writing.

CAB International and, where different, the copyright owner shall not be liable for technical or other errors or omissions contained herein. The information is supplied without obligation and on the understanding that any person who acts upon it, or otherwise changes their position in reliance thereon, does so entirely at their own risk. Information supplied is neither intended nor implied to be a substitute for professional advice. The reader/user accepts all risks and responsibility for losses, damages, costs and other consequences resulting directly or indirectly from using this information.

CABI's Terms and Conditions, including its full disclaimer, may be found at https://www.cabi.org/terms-and-conditions/.

A catalogue record for this book is available from the British Library, London, UK.

ISBN-13: 9781800623071 (hardback)
        9781800623088 (ePDF)
        9781800623095 (ePub)

DOI: 10.1079/9781800623095.0000

Commissioning Editor: Ward Cooper
Editorial Assistant: Lauren Davies
Production Editor: James Bishop

Typeset by Exeter Premedia Services Pvt Ltd, Chennai, India
Printed and bound in the USA by Integrated Books International, Dulles, Virginia

Contents

Contributors

Baroi, Anda Maria, National Institute for Research & Development in Chemistry and Petrochemistry—ICECHIM Bucharest, Bucharest, Romania; University of Agronomic Sciences and Veterinary Medicine of Bucharest, Bucharest, Romania. Email: anda.baroi@icechim.ro

Călinescu, Mirela Florina, Research Institute for Fruit Growing Pitesti—Maracineni, 402, Marului Strada, Arges, 110006, Romania. Email: elacalinescu@yahoo.com

Cîrstea, Georgiana, Regional Center of Research and Development for Materials, Processes and Innovative Products Dedicated to the Automotive Industry (CRC&D-AUTO), University of Pitesti, Doaga Street, No. 11, 110440 Pitesti, Romania. Email: georgiana.cirstea@upit.ro

Dobrescu, Codruța Mihaela, Department of Natural Sciences, University of Pitesti, Targul din Vale 1, 110040 Pitesti, Romania. Email: codrutza_dobrescu@yahoo.com

Dorobăț, Leonard Magdalin, Department of Natural Sciences, University of Pitesti, Targul din Vale1, 110040 Pitesti, Romania. Email: coltanabe@yahoo.com

Drăghiceanu, Oana Alexandra, Department of Natural Sciences, University of Pitesti, Târgul din Vale 1, 110040 Pitesti, Romania. Email: oana.draghiceanu@upit.ro

Eltanahy, Eladl, Botany Department, Faculty of Science, Mansoura University, 35516, Egypt. Email: eladl@mans.edu.eg

Fierascu, Irina, National Institute for Research & Development in Chemistry and Petrochemistry—ICECHIM Bucharest, Bucharest, Romania; University of Agronomic Sciences and Veterinary Medicine of Bucharest, Bucharest, Romania. Email: irina.fierascu@icechim.ro

Fierascu, Radu Claudiu, National Institute for Research & Development in Chemistry and Petrochemistry—ICECHIM Bucharest, Bucharest, Romania; University Politehnica of Bucharest, 313 Splaiul Independentei Str., 060042, Bucharest, Romania. Email: fierascu.radu@icechim.ro

Hashim, Rauzah, Centre for Fundamental and Frontier Sciences in Nanostructure Self Assembly, Department of Chemistry, Faculty of Science, University Malaya, 50603 Kuala Lumpur, Malaysia. Email: rauzah@um.edu.my

Hegazy, Amany Saad, Botany Department, Faculty of Science, Mansoura University, 35516, Egypt. Email: amanysaad@mans.edu.eg

Ishak, Khairul Anwar, Centre for Fundamental and Frontier Sciences in Nanostructure Self-Assembly, Department of Chemistry, Faculty of Science, University of Malaya, 50603 Kuala Lumpur, Malaysia; Institute of Biological Sciences, Faculty of Science, University of Malaya, 50603 Kuala Lumpur, Malaysia. Email: ka.ishak@um.edu.my

Jayathilaka, Dilki, Sri Lanka Institute of Nanotechnology, Mahenwatta, Pitipana, Homagama 10206, Sri Lanka. Email: dilkiJ@slintec.lk

Kaushani, Kotuwegoda Guruge, Department of Materials and Mechanical Technology, Faculty of Technology, University of Sri Jayewardenepura, Homagama, Sri Lanka. Email: kaushi.kguruge@gmail.com

Kottegoda, Nilwala, Department of Chemistry/Center for Advanced Materials Research, Faculty of Applied Sciences, University of Sri Jayewardenepura, Sri Soratha Mawatha, Nugegoda 10250, Sri Lanka. Email: nilwala@sjp.ac.lk

Neblea, Monica Angela, Department of Natural Sciences, University of Pitesti, Targul din Vale 1, 110040 Pitesti, Romania. Email: monica_neb@yahoo.com

Nicuță, Daniela, Department of Biology, Ecology and Environmental Protection, Faculty of Sciences, "Vasile Alecsandri" University of Bacău, Bacău, Romania. ORCID ID: 0000-0002-0547-5908. Email: daniela.nicuta@ub.ro

Ortan, Alina, University of Agronomic Sciences and Veterinary Medicine of Bucharest, Bucharest, Romania. Email: alina_ortan@hotmail.com

Osman, Yehia A., Botany Department, Faculty of Science, Mansoura University, 35516, Egypt. Email: yaolazeik@mans.edu.eg

Popescu, Gheorghe Cristian, Department of Applied Sciences and Environmental Engineering, University of Pitesti, Targul din Vale 1, 110040, Pitesti, Romania. ORCID ID: https://orcid.org/0000-0001-5432-4607. Email: christian_popescu2000@yahoo.com

Popescu, Monica, Department of Natural Sciences, University of Pitesti, Targul din Vale 1, 110040, Pitesti, Romania. ORCID ID: https://orcid.org/0000-0002-0757-2867. Email: monica_26_10@yahoo.com

Priyadarshana, Gayan, Department of Materials and Mechanical Technology, Faculty of Technology, University of Sri Jayewardenepura, Homagama, Sri Lanka. Email: gayanp@sjp.ac.lk

Razak, Ahmed A., Botany Department, Faculty of Science, Mansoura University, 35516, Egypt. Email: ahmed_bt@mans.edu.eg

Roșu, Ana-Maria, Department of Chemical and Food Engineering, Faculty of Engineering, "Vasile Alecsandri" University of Bacău, Bacău, Romania. ORCID ID: 0000-0001-5551-4752. Email: University of Bacau ana.rosu@ub.ro

Sandaruwan, Chanaka, Sri Lanka Institute of Nanotechnology, Mahenwatta, Pitipana, Homagama 10206, Sri Lanka. Email: chanakas@slintec.lk

Sărdărescu, Ionela Daniela, University Politehnica of Bucharest, 313 Splaiul Independentei Str., 060042, Bucharest, Romania; National Research and Development Institute for Biotechnology in Horticulture Stefanesti Arges, 37 Bucharest-Pitesti Strada, 117715, Arges, Romania. Email: ionela.toma93@yahoo.com

Sayed-Ahmed, Khaled, Department of Agricultural Biotechnology and The Center for Excellence in Research of Advanced Agricultural Sciences, Faculty of Agriculture, Damietta University, New Damietta 34517, Egypt. Email: drkhaled_1@du.edu.eg

Shabana, Yasser M., Plant Pathology Department, Faculty of Agriculture, Mansoura University, El-Mansoura 3556, Egypt. Email: yassershabana2@yahoo.com

Soare, Liliana Cristina, Department of Natural Sciences, University of Pitesti, Targul din Vale 1, 110040 Pitesti, Romania. Email: cristina.soare@upit.ro

Șuțan, Nicoleta Anca, Department of Natural Sciences, University of Pitesti, Targul din Vale 1, 110040 Pitesti, Romania. ORCID ID: http://orcid.org/0000-0001-7459-628X. Email: anca.sutan@upit.ro

Thiranagama, Gayanath, Sri Lanka Institute of Nanotechnology, Mahenwatta, Pitipana, Homagama 10206, Sri Lanka. Email: gayanathT@slintec.lk

Ungureanu, Camelia, University Politehnica of Bucharest, 313 Splaiul Independentei Str., 060042, Bucharest, Romania. Email: ungureanucamelia@gmail.com

Vîlcoci, Denisa Ștefania, Regional Center of Research and Development for Materials, Processes and Innovative Products Dedicated to the Automotive Industry (CRC&D-AUTO), University of Pitesti, Doaga Street, No. 11, 110440 Pitesti, Romania. Email: stefania.vilcoci@upit.ro

Vizitiu, Diana, National Research and Development Institute for Biotechnology in Horticulture Stefanesti Arges, 37 Bucharest-Pitesti Strada, 117715, Arges, Romania. Email: vizitiud@yahoo.com

Voicu, Roxana-Elena, Department of Biology, Ecology and Environmental Protection, Faculty of Sciences, "Vasile Alecsandri" University of Bacau, University of Bacău, Bacău, Romania. ORCID ID: 0000-0001-9860-9671. Email: roxana.voicu@ub.ro

Foreword

_____

Dear Reader,

This is a book that takes the implications of nanotechnologies for food security seriously. It moves quickly from the diseases and pests that affect agricultural crops to the most cutting-edge techniques and technologies for combating and treating them using nanotechnology. It also delves into the dark corners of genetic information manipulation using nanotechnology to boost agricultural productivity. Because the interaction of nanomaterials with biological systems is implied from the start of the book, nanomaterial management to reduce risks and ensure safe use is a defining feature of this book.

A book to read, learn from, and think about!

Prof. Dr. Telat Yanik

Preface

For decades, the scientific world has not stopped researching, deepening, discovering and applying the results of the most profound research in the field of nanotechnologies. Since Richard Feynman launched the hypothesis of the use of particles with the smallest dimensions, hundreds of nanomaterials have been obtained and used for a wide variety of purposes and in the most diverse fields, and research continues for probably several hundred or perhaps thousands of such products.

Humans have always successfully applied biotechnologies and nanobiotechnologies to develop competitive products, for his own benefit, but also to improve the quality of life. Ensuring food security, global warming, diminishing natural resources, and a slowing down of agricultural productivity and profitability are some of the challenges that the world is facing today, and these threats have become increasingly fierce over time.

This book highlights the urgent needs to be fulfilled, such as increasing the quantity and quality of agricultural products, the safe control of pests and diseases, the development and implementation of efficient agricultural technologies, the development of sustainable agriculture, and the reduction of environmental risks, and the use of nanotechnologies as an exceptional source of feasible alternatives, some already discovered, others unidentified or insufficiently explored, that can be used to address these issues.

Nanotechnology is an emerging and expanding interdisciplinary research field with relevant results in the agri-tech revolution. Thus, the concentration of studies in a concise approach, in a book that presents the latest nanoformulations for sustainable agriculture and environmental risk mitigation, represented a real challenge. After hundreds of thousands of scientific articles and books have addressed these topics, after scientists and the general public have been well informed about nanotechnologies and nanomaterials, is there still interest in this topic?

The answer to this question is the main purpose of the book—to provide up-to-date information regarding the application of nanotechnologies in agriculture and to perpetuate interest in this topic until a sustainable agriculture and a green environment are achieved.

Introduction

In the past decade, nanotechnology in agriculture has advanced significantly. Agriculture can benefit from the development of more efficient and less contaminated agrochemicals (nanoformulations), smart delivery systems that can detect biotic or abiotic challenges before they impair productivity (nanosensors), and novel genetic modification techniques that allow greater efficiency during crop improvement. Nanoformulations are being used more frequently in agriculture to improve food quality, lower agricultural inputs, improve nutrient contents, and extend shelf life. To limit the usage of harmful chemicals, many nano-agricultural products are now being developed. Numerous areas of food security, disease treatment, novel methods for pathogen detection, efficient delivery systems, and packaging materials are all impacted by nanotechnology. This book explores the crucial and cutting-edge roles that nanotechnology can play in promoting sustainable agriculture. It contains the most recent information on nanoformulation and its implication in sustainable agricultural applications.

The chapters will highlight uses for these nanoforms that can significantly improve the current situation of a worldwide food shortage and address the negative consequences caused by the widespread use of chemical agrochemicals, which are claimed to promote biomagnification in an ecosystem. The use of nanoformulations and nano-enabled goods in reducing the danger associated with the existing usage of pesticides will also be covered.

1 Agricultural Nanotechnologies: Current Scenario and Future Implications for Global Food Security

Gheorghe Cristian Popescu[1]* and Monica Popescu[2]*

[1]*Department of Applied Sciences and Environmental Engineering, University of Pitesti, Pitesti, Romania;* [2]*Department of Natural Sciences, University of Pitesti, Pitesti, Romania*

Abstract

Agriculture is one of the oldest production activities on the planet with a fundamental role in food security and the supply of raw materials for other industries, such as bioenergy, clothing, construction, pharmaceutical, and food. Sustainable agriculture is an increasingly promoted concept that ensures a beneficial correlation between economic, social, and environmental objectives in agricultural production. Nanotechnology is a relatively new cutting-edge field with implications in many fields of activity that could provide sustainable solutions for agricultural production systems. Nanotechnology is expected to improve the level of precision farming and food security. Nanoparticles prepared in various formulations, such as nanofertilizers, nanopesticides, and nanosensors, have potential benefits for biomass accumulation, crop yields and quality, and crop protection. There are still many challenges for the application of nanotechnology and the future implications for agriculture and food security. Nanotechnologies are alternative solutions for improving agricultural performance and form a part of new technologies.

Keywords: agroecological risk, crop yields and quality, food security, agri-nanotechnology, sustainability

1.1 General Considerations Regarding the Role of Nanotechnologies in the Development of Sustainable Agriculture

Sustainable management of agroecosystems involves identification and exploitation of eco-friendly approaches for crop production. The most important technological tools that can ensure a sustainable approach to agricultural production are fertilization and plant protection against pests and diseases. New sustainable formulas for the production of fertilizers and plant-protection products can improve sustainability in agricultural production systems. The new concept of agricultural nanoproducts is safer and more effective than conventional methods (Muhammad *et al.*, 2020).

In this context, nanotechnologies that use nanomaterials (NMs) prepared by various chemical, physical, or biological processes and techniques could solve some of the problems related to environmental sustainability issues

---

*Corresponding authors: christian_popescu2000@yahoo.com and monica_26_10@yahoo.com

© CAB International 2023. *Nanoformulations for Sustainable Agriculture and Environmental Risk Mitigation* (eds Z. Khan and N.A. Șuțan)
DOI: 10.1079/9781800623095.0001

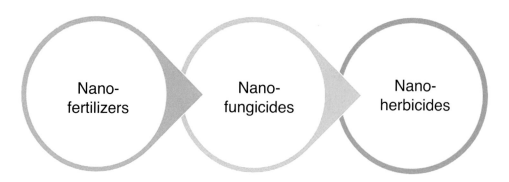

Fig. 1.1. Types of nano-agricultural products used in sustainable agriculture.

in agricultural systems. NMs are eco-friendly products that can be used as nanofertilizers, nanosensors or nanopesticides in agriculture. These agricultural nanoproducts can change the conventional ways of agro-practices into more sustainable practices (Neme *et al.*, 2021). Green synthesis of NMs using natural resources represents an innovative way to achieve new eco-friendly products for sustainable agriculture. The main route of producing green nanoparticles (NPs) from natural resources is by the use of enzymes, plant extracts, or bacteria (Al-Asfar *et al.*, 2018). Agricultural waste could be an essential resource for manufacturing of NMs (AboDalam *et al.*, 2022; Babu *et al.*, 2022). New concepts for pesticides and fertilizers such as nanofertilizers, nanofungicides and nanoherbicides (Fig. 1.1) should generate new sustainable solutions for agricultural systems and food security.

Agro-nanotech innovations could have a beneficial contribution toward achieving sustainable development goals in agriculture and food security (Sarkar *et al.*, 2022). Nanotechnologies can fulfil a number of important objectives for agriculture, including sustainability and food security (Fig. 1.2).

1.2 Nanotechnology Applications in Agriculture for Crop Yields and Quality

Productivity of the agriculture sector and quality crops are essential to ensure food security and safety. Agriculture is the main component for ensuring food security. In many areas of the world, there has been an increase in the degradation of natural resources, but also reduced productivity of agricultural production systems. Access to natural resources and their quality directly and significantly influence crop yields and quality. The generation of alternative solutions and the use of new technologies are increasingly necessary to meet the requirements of modern and efficient agricultural systems. Nanotechnology is a modern scientific and practical approach that uses various types of NMs and nanoproducts for the agricultural sector, with promising results in terms of quantity and quality of plant yields (Mali *et al.*, 2020). Nanotechnology, based on NPs prepared in various formulations such as nanofertilizers, nanopesticides, and nanoherbicides, is a new way to enhance agricultural crop production.

NPs have become one of the most important tools for agricultural purposes due to their physico-chemical and biological properties. The most common types of NPs used in many applications for agricultural or environmental purposes are made using cerium dioxide (CeO_2NPs), copper (CuNPs), iron (FeNPs), titanium dioxide (TiO_2NPs), zinc oxide (ZnONPs), silicon dioxide (SiO_2NPs) or silver (AgNPs) (Fig. 1.3). Many types of NMs from different resources have been studied for agricultural purposes. Currently, a multitude of studies are focused on applications of nanotechnology in cereal crops (Wang *et al.*, 2021), industrial crops, orchards (Abdelmigid *et al.*, 2022), vegetables (Faizan *et al.*, 2018), vineyards (Gohari *et al.*, 2021), medicinal and aromatic plants (Elsayed *et al.*, 2022),

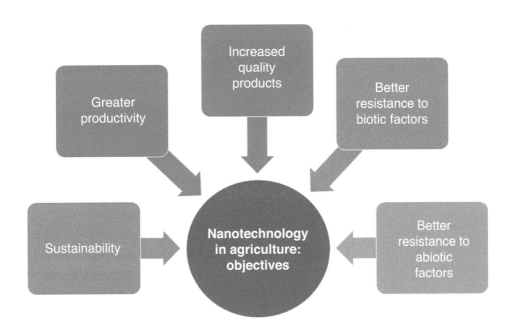

Fig. 1.2. The main objectives of agricultural nanotechnology to increase the quality of products.

ornamental plant and cut flowers (Salachna *et al.*, 2019; Tofighi Alikhani *et al.*, 2021; Phong *et al.*, 2022) (Fig. 1.4).

The productivity and quality of agricultural ecosystems is influenced by a number of factors, among which plant nutrition is essential. Fertilizers have an important role in agricultural technology by providing essential additional nutrients for plant growth and development requirements. Nanofertilizers, as a part of nanotechnology applications in agriculture, have shown great potential for increased crop production (Jakhar *et al.*, 2022). These innovative products are eco-friendly NMs and can be used as a sustainable tool for soil management fertility (Milani *et al.*, 2015). Many research and development studies are focused on improving the performance of agricultural technologies. Enhancement of various physiological, biochemical, and morphological parameters of plants using nanoformulations has been reported for many species (Table 1.1.).

AgNPs are the most used NPs obtained by physical, chemical, or biological processes. These NPs are used in many agricultural studies to improve technologies. Silver NMs can be used as plant growth stimulants in agriculture and horticulture sectors. AgNPs have been used in the production of potted lilies. Different concentrations of AgNPs (25–150 ppm) were applied to evaluate the growth and flowering features of *Lilium* plants. *Lilium* plants treated with AgNPs had higher plant biomass, accelerated flowering, formation of more flowers, and flowered for longer. A concentration of 100 ppm AgNPs stimulated the accumulation of assimilatory pigments, and treated plants had the highest amount of calcium and potassium in their leaves (Salachna *et al.*, 2019). At a concentration of $100\,mg\,l^{-1}$, use of AgNPs for tulip culture showed a positive effect was for flowers, biomass and bulbs (Byczyńska *et al.*, 2019). AgNPs stimulated the flowering and fruiting parameters for *in vitro* propagation of *Passiflora edulis* L.; plants treated with AgNPs at $7\,mg\,l^{-1}$ formed the most passion fruits and the fruits had the largest diameter (Phong *et al.*, 2022). The effect of AgNPs (10–20 nm in size) on seed germination and yield of wheat were evaluated for various concentrations from 25 to 50 ppm. Any plants treated with AgNPs applied to soil showed improvements in dry weight,

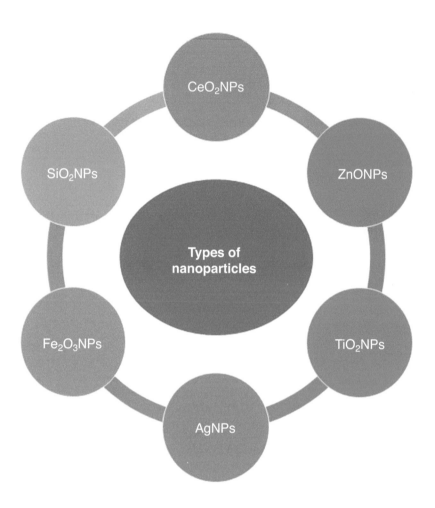

Fig. 1.3. The most commonly used nanoparticles (NPs) in agricultural production systems for crop performance.

chlorophyll content and fresh weight, with all of these biochemical and morphological parameters having better values than those recorded for the control variant (Razzaq *et al.*, 2016). The best response for AgNP application to wheat crop was observed at a concentration of 25 ppm. AgNPs can be used in chrysanthemum breeding (Tymoszuk and Kulus, 2020) or passion fruit breeding (Phong *et al.*, 2022). ZnONPs was applied at various concentrations to evaluate the effect on the growth and photosynthetic capacity of tomato plants. ZnONPs treatment at 8 mg l^{-1} was found to be the most effective, and at this concentration, ZnONPs significantly improved

the growth processes and photosynthetic efficiency of tomato plants (Faizan *et al.*, 2018).

Green synthesis of NPs from natural resources such as enzymes, plant extracts and bacteria is the new route to achieve new eco-friendly products for sustainable agricultural systems. In recent years, green processes using NPs have been documented to promote environmental sustainability in the agricultural sector. Foliar application of green-synthesized ZnONPs using plant extract from *Eucalyptus lanceolatus* enhanced the stem and leaf surface area in maize plants. Green-synthesized ZnONPs also improved the germination rate of maize seeds,

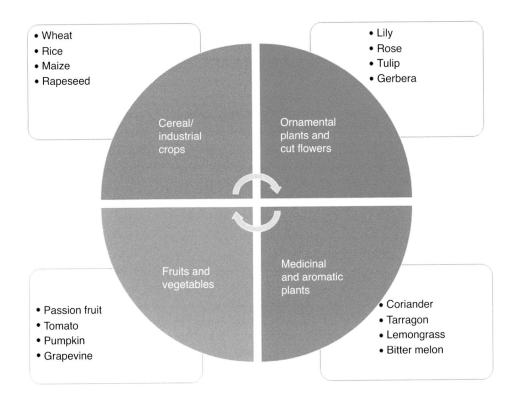

- Wheat
- Rice
- Maize
- Rapeseed

- Lily
- Rose
- Tulip
- Gerbera

Cereal/
industrial
crops

Ornamental
plants and
cut flowers

Fruits and
vegetables

Medicinal
and aromatic
plants

- Passion fruit
- Tomato
- Pumpkin
- Grapevine

- Coriander
- Tarragon
- Lemongrass
- Bitter melon

Fig. 1.4. Applications of nanotechnologies in agriculture sectors.

and the fresh biomass and root system of the plants (Sharma *et al.*, 2022). Other application of nanotechnology in agriculture used silicon-based NPs (SiNPs). SiNPs are a current silicon source that can be applied to improve the ability of plants to withstand biotic and abiotic stress conditions (Rajput *et al.*, 2021). A concentration of 150 mg l^{-1} was found to be optimal for SiNPs for lemongrass plants in order to have good values of agronomic parameters (Mukarram *et al.*, 2021). The growth process and antioxidant enzyme activities of coriander cultivated under different lead concentrations were improved by foliar application of SiNPs. Therefore, SiNPs could be an interesting approach in minimizing the level of lead toxicity in plants (Fatemi *et al.*, 2021).

SiO$_2$NPs have also been used in experiments with plants in hydroponic conditions. The effect of four concentrations of SiO$_2$NPs on *Gerbera jamesonii* L. in hydroponic conditions was investigated and several biochemical and morphological parameters were measured. The treatment of 80 mg l^{-1} of SiO$_2$NPs increased the number of flowers and longevity of gerbera flowers compared with the control experimental variant. The same treatment also generated the lowest bud abortion rate for gerbera plants (Tofighi Alikhani *et al.*, 2021). Application of SiO$_2$NPs resulted in higher development of *Triticum aestivum* L. in terms of shoot and root length or fresh and dry weight of plants (Akhtar and Ilyas, 2022). CeO$_2$NPs were applied at different concentrations (100, 200, and 500 mg l^{-1}) to evaluate the photosynthesis rate of peas (*Pisum sativum* L.). The lowest concentration induced a better photosynthesis rate and thus the vegetative growth processes were stimulated (Skiba *et al.*, 2020). At the same time, CeO$_2$NPs at a concentration of 500 mg l^{-1} promoted the biomass production of wheat plants, while higher concentration of NPs negatively affected the growth processes and plant photosynthesis (Abbas *et al.*, 2020).

Table 1.1. Nanomaterials application in agriculture with implications for crop yields.

Type of nanoparticle (NP)/nanoproduct	Crop	Reference
Cerium dioxide NPs	*Pisum sativum* L. (pea)	Skiba *et al.*, 2020
	Vitis vinifera L. (grapevine)	Gohari *et al.*, 2021
	Lactuca sativa L. (lettuce)	Gui *et al.*, 2015
	Triticum aestivum L. (wheat)	Abbas *et al.*, 2020
	Brassica napus L. (rapeseed)	Li *et al.*, 2022
	Raphanus sativus L. (radish)	Gui *et al.*, 2017
Titanium dioxide, zinc oxide, ferric oxide and silicon dioxide NPs	*Linum usitatissimum* L. (flax)	Singh *et al.*, 2021
Selenium and zinc oxide NPs	*Brassica napus* L. (rapeseed)	El-Badri *et al.*, 2021
Silicon dioxide NPs	*Zea mays* L. (maize)	Suriyaprabha *et al.*, 2014
	Cymbopogon flexuosus (Steud.) Wats (lemongrass)	Mukarram *et al.*, 2021
	Cucurbita pepo L. (Squash)	Siddiqui *et al.*, 2014
	Coriandrum sativum L. (coriander)	Fatemi *et al.*, 2021
Zinc oxide NPs	*Zea mays* L. (maize)	Sharma *et al.*, 2022
	Lycopersicon esculentum Miller (tomato)	Faizan *et al.*, 2018
	Artemisia dracunculus L. (tarragon)	Hassanpouraghdam *et al.*, 2022
Nano-iron (magnetic iron)	*Artemisia dracunculus* L. (tarragon)	Hassanpouraghdam *et al.*, 2022
Silver NPs	*Triticum aestivum* L. (wheat)	Mohamed *et al.*, 2017
Cobalt NPs	*Glycine max* (L.) Merr. (soybean)	Hoe *et al.*, 2018

Under salt stress conditions, a positive effect of the presence of selenium NPs (SeNPs) and ZnONPs in the growing medium for two rapeseed cultivars was observed on germination rate. The role of SeNPs and ZnONPs in enhancing salinity tolerance in rapeseed plants was also highlighted (El-Badri *et al.*, 2021). Foliar application of CeO$_2$NPs improved rapeseed salt stress tolerance and increased chlorophyll content, leaf length and carbon assimilation rate. The contribution of NMs in plant physiology may thus become an important tool to enhance salinity toxicity tolerance. One experiment evaluated the effect of TiO$_2$NPs, ZnONPs, Fe$_2$O$_3$NPs, and SiO$_2$NPs on the morphological features and physiological processes of linseed (*Linum usitatissimum* L.), a flowering plant cultivated as a fibre and food crop, under salinity stress conditions (Singh *et al.*, 2021). The presence of NPs had a positive impact on all treated plants and on growth processes, as well as improving the assimilation of

carbon and nutrients. Application of NPs had a stimulating action of the antioxidant enzymatic system in *L. usitatissimum*.

Drought is an abiotic stress that negatively affects crop production. Different levels of NPs have been tested to evaluate the growth processes in wheat plants. Ikram *et al.* (2020) reported that SeNPs stimulate the growth processes of wheat plants under drought stress conditions. The positive response of the application of NPs is influenced by the applied dose. SeNPs applied at concentration of 30 mg l^{-1} seem to be optimal to enhance morphological parameters (plant height, root length, and leaf number) in wheat under normal and drought conditions.

Foliar application of CeO$_2$NPs at 10 mg l^{-1} improved the salt tolerance of maize and the photosynthetic capacity in salt-stressed maize leaves. This was indicated by improved values for physiological, morphological, and biochemical traits of plants treated with NPs under salinity stress conditions (Liu *et al.*, 2022). These results provide a beneficial alternative method for alleviating abiotic stress factors such as salt soil stress for crops. Gohari *et al.* (2021) explored the benefits of CeO2NPs at various concentrations (25, 50, and 100 mg l^{-1}) in combating induced salt stress (25 and 75 mM NaCl) in grapevine cuttings by evaluating the effect of NPs and salinity on some biochemical, viticultural, and physiological attributes. The growth parameters of grapevine such as leaf number and plant height demonstrated higher values compared with the control. The findings of this study demonstrated that CeO$_2$NPs can improve the growth and physiological parameters of grapevine under salinity stress conditions.

NPs can reduce oxidative stress and cadmium (Cd) accumulation in wheat. Different concentrations of ZnONPs (25–100 mg l^{-1}) and iron oxide NPs (Fe$_2$O$_3$NPs; 5–20 mg l^{-1}) have been tested in wheat plants under cadmium contamination. Plants treated with NPs registered a lower concentration of Cd in shoots, grains, and roots, and demonstrated enhanced plant biomass (Rizwan *et al.*, 2019). Cobalt (Co) stress negatively influences the physiological processes of plants, and is an important contaminant in agricultural systems. ZnONPs enhanced the plant growth, biomass, photosynthetic activity, seed zinc contents and nutrient uptake under Co stress. The results showed that ZnONPs enhanced the plant antioxidant defense system in maize under Co stress and allowed crops to be cultivated in Co-contaminated areas (Salam *et al.*, 2022). Fe$_2$O$_3$NPs stimulated photosynthetic capacity and plant growth of soybean (*Glycine max* L.) and alleviated arsenic toxicity (Bhat *et al.*, 2022). Co, Cd, and arsenic negatively affect plant biomass accumulation and plant developmental processes and can generate substantial harvest losses. NPs can thus play an essential role in alleviating Co, Cd, and arsenic stress and other phytotoxicities in plants.

1.3 Nanotechnology Applications in Agriculture for Crop Protection

A crop protection management plan is an essential tool of modern agricultural technologies in order to ensure a good crop yield. Agricultural crops are usually attacked by insects and various pathogens. In performance agriculture, pest management relies heavily on pesticide application. Considering the new requirements related to environmental standards, the management of pests in agricultural crops must have a minimal impact on the environment and not affect the quality of the crops. Nanotechnology is developing innovative products based on nanopesticides to enhance crop resistance to pests and diseases (Rajwade *et al.*, 2020; Tran *et al.*, 2022). Nanopesticides can enhance crop protection and agriculture productivity, with low environmental impact (Scott-Fordsmand *et al.*, 2022). The main innovative nanoformulations against phytopathogens are nanoinsecticides, nanofungicides and nanoherbicides (Fig. 1.5). Numerous research studies indicate that products based on NMs will be essential instruments for agricultural systems to prevent and control crop pests and pathogens (Rajwade *et al.*, 2020; Zhao *et al.*, 2022). The development of alternative plant-protection products such as nanopesticides represents a key for efficient plant-protection management (Table 1.2.).

Foliar application of nanofungicides based on Cu, Ag, and Zn NMs is an eco-friendly approach for traditional synthetic fungicides against a number of plant pathogens including: *Verticillium dahliae*, *Alternaria alternata*, *Fusarium oxysporum f. sp. radices-lycopersici*, *Botrytis cinerea*, *Monilia fructicola*, *Colletotrichum gloeosporioides*, and *Fusarium solani* (Malandrakis *et al.*, 2019). CoNPs and AgNPs did not show an antifungal effect for *Phytophthora cactorum* or *Sparassis crispa*, while both types showed good antifungal activity for *Rhizoctonia solani*, *Fusarium redolens*, and *Meripilus giganteus* (Aleksandrowicz-Trzcińska *et al.*, 2018). TiO$_2$NPs and ZnONPs, as nano-insecticides, were tested to evaluate the insecticidal effect on *Bactericera cockerelli* under laboratory and greenhouse conditions in tomato. Both NP products showed good insecticidal activity (Gutiérrez-Ramírez *et al.*, 2021). Green-synthesized ZnONPs were applied at various concentrations (100, 200, 300, 400, and 500 mg l^{-1}) against *Macrosiphum euphorbiae* (potato aphid) and *Spodoptera litura* (tobacco cutworm) associated with tomato crops (Thakur *et al.*, 2022). The findings showed that nanoformulations of ZnONPs at higher concentrations induced the best insecticidal efficacy.

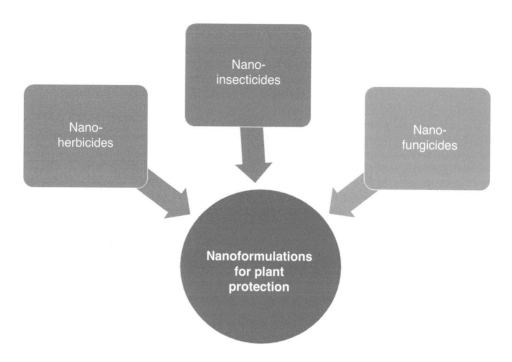

Fig. 1.5. Nanoformulations for crop protection: nanoproducts/nano-agrochemicals.

Green-synthesized AgNPs from the poly-saccharide extract of the red alga *Ceratocystis fimbriata* Ellis & Halst. proved to have a good antibacterial action against infection by *Xanthomonas oryzae* pv. *oryzae*. The biosynthe-sized AgNPs proved to be an effective substrate for nanopesticide production (Roseline *et al.*, 2021). Carbon 60 NPs (C$_{60}$NPs), CuONPs, and TiO$_2$NPs were selected to evaluate the antifungal efficiency against *Rhizopus stolonifer* infections (Pang *et al.*, 2021). The best antipathogenic effect was observed for treatment with 50 mg CuONPs l^{-1}, followed by TiO$_2$NPs and C$_{60}$NPs.

CuNPs showed an efficient antifungal activity against *Phytophthora infestans* isolated from the tomato field. The findings demon-strated that CuNP products can be more efficient against *P. infestans* than traditional pesticides (Giannousi *et al.*, 2013). A field experiment using chitosan-based nanopesticides showed an efficient control management of plumeria rust caused by *Coleosporium plumeria*. ZnNPs have shown a strong potential to control phytopathogens (Zhou *et al.*, 2022). Foliar application of ZnONPs at a concentration of

100 mM was good for controlling of *Fusarium graminearum* in wheat plant experiments (Savi *et al.*, 2015). Green-synthesized ZnONPs using a leaf extract of *Sida rhombifolia* Linn. were evaluated against growth of *Bacillus subtilis* and showed antibacterial efficacy (Kavya *et al.*, 2020). SiO$_2$NPs were tested against two phytopathogens, Colorado potato beetle (*Leptinotarsa decemlineata*) and cabbage beetles (*Phyllotreta* spp.). The results showed that NPs had an entomocidal effect on these pests and could be included for crop pest management (Shatalova *et al.*, 2022).

Agricultural productivity is significantly reduced due to plant pests and pathogens. It is essential to fight against plant phytopathogens to achieve better yields and quality products. There is a continuous need to identify new and effective methods to prevent and combat diseases and pests in agricultural crops. The recent studies and field experiments described demonstrate that nano-agrochemicals can prevent and control plant diseases and will play an essential role in future plant-protection management.

Table 1.2. Applications of nanoformulations for plant protection.

Type of nanoparticle (NP)/ nanopesticide	Phytopathogens	Reference(s)
Nano-chlorine (chlorantraniliprole) and nano-sulfur (thiocyclam)	*Agrotis ipsilon* (Hufnagel)	Awad *et al*., 2022
Selenium NPs	*Agrotis ipsilon* (Hufnagel)	Amin *et al*., 2021
Chitosan-based nanopesticides	*Coleosporium plumera*	Zhou *et al*., 2022
Zinc oxide NPs	*Fusarium graminearum* Schwabe, *Macrosiphum euphorbiae* (Thomas), *Spodoptera litura* (Fabricius), *Bacillus subtilis* Cohn, *Spodoptera frugiperda* (J.E. Smith)	Savi *et al*., 2015; Kavya *et al*., 2020; Pittarate *et al*., 2021; Thakur *et al*., 2022
Copper oxide NPs	*Gibberella fujikuroi* (Sawada) Wollenw, *Fusarium oxysporum* f. sp. *Lactucae* (J.C. Hubb. & Gerik), *Phytophthora infestans* (Mont) de Bary	Shang *et al*., 2020; Shang *et al*., 2021; Giannousi *et al*., 2013
Carbon 60, copper oxide, titanium dioxide	*Rhizopus stolonifer* (Ehrenb.)	Pang *et al*., 2021
Silver NPs	*Xanthomonas oryzae* (Ishiyama)	Roseline *et al*., 2021
Calcium oxide NPs	*Colletotrichum brevisporum* (Noireung, Phouliv., L. Cai & K.D. Hyde)	Maringgal *et al*., 2020
Copper, silver and zinc NPs	*Botrytis cinerea* (Pers.), *Monilia fructicola* (G. Winter) Honey, *Fusarium oxysporum* f. sp. *radicis* Jarvis & Shoemaker, *Fusarium oxysporum* f. sp. *radicis-lycopersici* (Jarvis & Shoemaker), *Colletotrichum gloeosporioides* *Alternaria alternata*, *Fusarium solani*, *Verticillium dahliae* (Kleb.)	Malandrakis *et al*., 2019
Copper and zinc NPs	*Grifola frondosa* (Dicks.: Fr.) *Sparassis crispa* (Wulfen Fr.), *Fistulina hepatica* (Huds) Fr., *Fusarium oxysporum*, *Phytophthora cactorum* (Lebert & Cohn) J. Schröt., *Meripilus giganteus* (Pers.) P. Karst., *Rhizoctonia solani* (Kuhn), *Fusarium redolens* (Wollenw.)	Aleksandrowicz-Trzcińska *et al*., 2018
MgO NPs	*Fusarium oxysporum*, *Fusarium solani* Schlechtendahl	Abdallah *et al*., 2022
Titanium dioxide and zinc oxide NPs	*Bactericera cockerelli* (Sulc)	Gutiérrez-Ramírez *et al*., 2021
Silicon dioxide NPs	*Phyllotreta cruciferae* (Goeze), *Leptinotarsa decemlineata* (Say), *Spodoptera littoralis* (Boisduval), *Aphis craccivora* Koch, *Liriomyza trifolii* (Burgess in Comstock)	Thabet *et al*., 2021; Shatalova *et al*., 2022

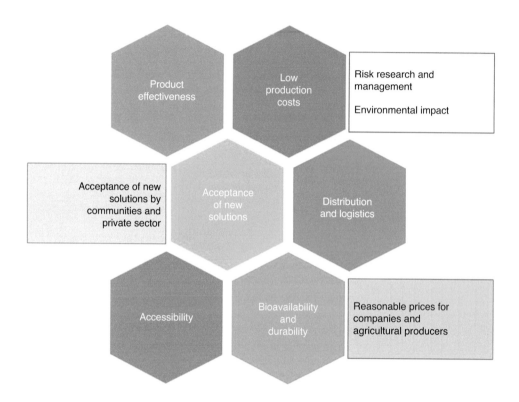

Fig. 1.6. The main challenges of agricultural nanotechnology.

1.4 Opportunities, Challenges and Future Implications of Nanotechnology Applications in Agriculture and Food Security

Nanotechnologies, based on particles at a nano-level scale, are a relatively recently researched and developed field. This fact generates a series of challenges, opportunities and future implications for the research and development sector. At the same time, nanotechnologies bring a series of challenges and implications for companies and agricultural producers. Nanotechnologies have many applications in the agricultural sector and form part of the new technologies available for agriculture. New technologies and innovations provide key solutions for environmental protection, efficient agriculture and food security (Popescu *et al.*, 2022). The nanoindustry offers a series of products with an essential role in agricultural production systems, such as nanofertilizers, nanofungicides, and nanoherbicides.

The main challenges of nanotechnologies are presented in Fig. 1.6. The large-scale production of nanoproducts, the acceptance of new nanotechnological solutions by farmers and agricultural companies, the effectiveness of the products, the production costs, and an affordable purchase price are concerns and challenges of the nanoproducts industry with applications in agriculture. Agriculture requires large quantities of products for fertilization and protection against diseases and pests. Therefore, production of nano-agricultural products on a large scale is a key need to cover the agricultural systems requirements. Nanotechnology products bring a series of opportunities to the following fields of interest for agriculture: the production sector, the waste sector, the environmental sector, and green technology (Fig. 1.7).

The main future implications of agricultural nanotechnologies will be improvements and beneficial changes in regulatory systems in agriculture, agricultural policies and the

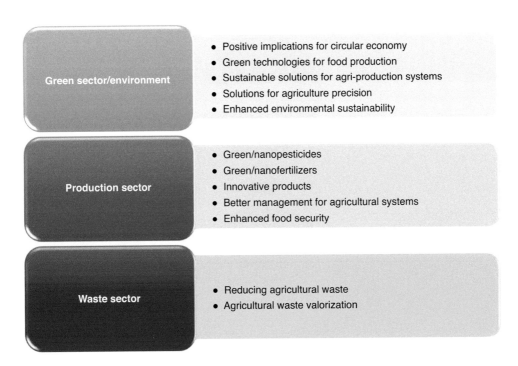

Fig. 1.7. The main opportunities for agricultural nanotechnology.

performance of agricultural systems (Fig. 1.8). Agricultural inputs at the nanoscale will bring new approaches and implications for agricultural technologies.

Future implications of nanotechnologies suppose greater sustainability and green solutions for agroecosystems but also implications for agricultural technologies, regulatory systems, and agricultural policies (Fig. 1.8).

Nanotechnologies can provide viable instruments for agricultural issues to enhance agricultural production and food security. This new approach could resolve a number of environmental issues, aid in plant protection and crop improvement, and enhance productivity. Nanotechnologies are also involved in preventing soil degradation and in circular economies by agricultural waste valorization. Agricultural waste could be an essential resource for the production of NMs (AboDalam *et al.*, 2022; Babu *et al.*, 2022). In recent years, NP products have become a beneficial alternative instrument in alleviating abiotic stress factors, such as salt or drought soil stress (Gohari *et al.*, 2021; Li *et al.*, 2022; Rajput *et al.*, 2022).

Plant-protection management is another positive application of nanotechnology. Agricultural crops are susceptible to attack by numerous diseases and pests. Further studies with the aim of identifying nano-agrochemicals with strong effects against phytopathogens under field conditions are needed. Agricultural crops require permanent monitoring of pathogens and effective plans to prevent and combat diseases and pests. Frequent applications of synthetic pesticides can have an environmental impact (Awad *et al.*, 2022), and increasing pest resistance to traditional plant-protection products requires novel pest control methods. Identifying sustainable and effective solutions for plant protection is a challenge but also an opportunity for nanotechnologies.

The agricultural sector produces large amounts of waste. Agricultural waste management could be improved by the application of nanotechnologies. Agricultural waste can be exploited as a beneficial resource for nanoscience products at the nanoscale. This application is an important eco-friendly opportunity for nanotechnologies to reduce environmental

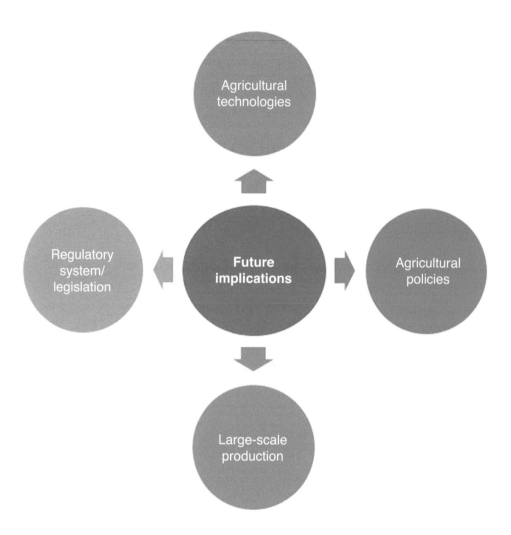

Fig. 1.8. The main future implications of agricultural nanotechnology.

pollution. Nanobiotechnology provides new instruments and products for the agricultural sector such as nano-insecticides, nanofertilizers, nanopesticides and nano-emulsions. Currently, one of the main purposes for agricultural and food security research is to develop eco-friendly production technology based on these innovative tools. Green synthesis of NPs from natural resources, such as plant extracts, bacteria and enzymes, is a new opportunity to develop innovative nanoproducts for eco-friendly agricultural technologies.

Many NM-based products are being developed for commercial and technological purposes in the agricultural sector, but several studies also show some effects on the environment or health. Nanotechnologies in terms of efficiency, costs, risks for humans and animals, and environmental safety are still subjects of further study by researchers and companies. There are concerns regarding the interaction between these NP-based products with the environment, animals or humans. Identification of green sources based on natural resources is an important and useful option for the safety of the environment, animals, ecosystems and human health (Neme *et al.*, 2021). The accumulation of NPs in plants and the impact on the health of animals or people must be a permanent concern of companies and scientists. Toxicity to animals and humans, and cytotoxic and genotoxic effects in plants caused by NPs have been reported (Chaud *et al.*, 2021; Neme *et al.*, 2021; Padilla-Camberos *et al.*, 2022; Rajput *et al.*, 2022; Sibiya *et al.*, 2022). There are some concerns about the uptake of NPs by plants, humans, animals, and aquatic and terrestrial ecosystems. Thus, a complete environmental assessment must be carried out for each type of nanoproduct. Specific regulatory frameworks for environmental and human health risk assessments must be applicable to nanoformulations and proper guidelines for each type of nanoformulation are needed. All potential risks of NP applications should be evaluated and ensured before approval for agricultural purposes and food production.

Farmers and agri-business owners should focus more on applying new technologies intended to improve quality and agricultural production, in an effort to respond to the current and future global challenges. At the same time, farmers and agricultural companies want affordable and efficient products. The economic impact of nano-agrochemicals must be evaluated carefully, with the cost of agricultural technologies being an essential factor for the sustainability of agri-businesses. In order for nanoproducts to be made on an industrial scale, it is necessary that the production cost and the selling price be reasonable. Another challenge for nanotechnologies is the availability of raw materials. The need for plant protection and fertilization is immense in agriculture. In order to meet these needs in agriculture, the raw material for nanoproducts must be accessible and found in abundance. As a new domain for the research and development sector, the delivery, packaging, storage, and durability of components, as well as safe application techniques of nanoproducts are other essential socio-economic variables for further research.

Research activities related to nanoscience should address research, testing, and production of NMs and green NPs, such as methods of synthesis, characterization of physico-chemical properties, environmental issues, and field studies. The size of the NPs, concentration, and method of application are other factors that must be addressed in depth for each type of nanoproduct. In addition, the research must be extended to other species and cultures located in different climatic zones. Appropriate regulatory frameworks for dealing with nanotechnologies are essential to increase confidence in these new nanoscale products and to minimize potential risks. Every country that produces nanoproducts must have a strict legislative framework, concrete guidelines and procedures for the use of nanoproducts, and a nanoproducts certification system.

In conclusion, nanotechnology is an advanced technology that could resolve a number of agricultural and environmental issues, from plant protection and resistance to abiotic stress factors to productivity and quality. Nanotechnologies thus provide an opportunity and potential solutions for future agricultural sustainability and food security.

References

Abbas, Q., Liu, G., Yousaf, B., Ali, M.U., Ullah, H. *et al.* (2020) Biochar-assisted transformation of engineered-cerium oxide nanoparticles: effect on wheat growth, photosynthetic traits and cerium accumulation. *Ecotoxicology and Environmental Safety* 187, 109845. DOI: 10.1016/j.ecoenv.2019.109845.

Abdallah, Y., Hussien, M., Omar, M.O.A., Elashmony, R.M.S., Alkhalifah, D.H.M. *et al.* (2022) Mung bean (*Vigna radiata*) treated with magnesium nanoparticles and its impact on soilborne *Fusarium solani* and *Fusarium oxysporum* in clay soil. *Plants* 11(11), 1514. DOI: 10.3390/plants11111514.

Abdelmigid, H.M., Morsi, M.M., Hussien, N.A., Alyamani, A.A., Alhuthal, N.A. *et al.* (2022) Green synthesis of phosphorous-containing hydroxyapatite nanoparticles (nHAP) as a novel nano-fertilizer: preliminary assessment on pomegranate (*Punica granatum* L.). *Nanomaterials* 12(9), 1527. DOI: 10.3390/nano12091527.

AboDalam, H., Devra, V., Ahmed, F.K., Li, B. and Abd-Elsalam, K.A. (2022) Rice wastes for green production and sustainable nanomaterials: an overview. *Agri-Waste and Microbes for Production of Sustainable Nanomaterials* 2022, 707–728. DOI: 10.1016/B978-0-12-823575-1.00009-3.

Akhtar, N. and Ilyas, N. (2022) Role of nanosilicab to boost the activities of metabolites in *Triticum aestivum* facing drought stress. *Plant and Soil* 477(1–2), 99–115. DOI: 10.1007/s11104-021-05285-1.

Al-Asfar, A., Zaheer, Z. and Aazam, E.S. (2018) Eco-friendly green synthesis of Ag@Fe bimetallic nanoparticles: antioxidant, antimicrobial and photocatalytic degradation of bromothymol blue. *Journal of Photochemistry and Photobiology B: Biology* 185, 143–152. DOI: 10.1016/j.jphotobiol.2018.05.028.

Aleksandrowicz-Trzcińska, M., Szaniawski, A., Olchowik, J. and Drozdowski, S. (2018) Effects of copper and silver nanoparticles on growth of selected species of pathogenic and wood-decay fungi *in vitro*. *The Forestry Chronicle* 94(2), 109–116. DOI: 10.5558/tfc2018-017.

Amin, M.A., Ismail, M.A., Badawy, A.A., Awad, M.A., Hamza, M.F. *et al.* (2021) The potency of fungal-fabricated selenium nanoparticles to improve the growth performance of *Helianthus annuus* L. and control of cutworm *Agrotis ipsilon*. *Catalysts* 11(12), 1551. DOI: 10.3390/catal11121551.

Awad, M., Ibrahim, E.-D.S., Osman, E.I., Elmenofy, W.H., Mahmoud, A.W.M. *et al.* (2022) Nano-insecticides against the black cutworm *Agrotis ipsilon* (Lepidoptera: Noctuidae): toxicity, development, enzyme activity, and DNA mutagenicity. *PloS One* 17(2), e0254285. DOI: 10.1371/journal.pone.0254285.

Babu, S., Rathore, S.S., Singh, R., Kumar, S., Singh, V.K. *et al.* (2022) Exploring agricultural waste biomass for energy, food and feed production and pollution mitigation: a review. *Bioresource Technology* 360, 127566. DOI: 10.1016/j.biortech.2022.127566.

Bhat, J.A., Bhat, M.A., Abdalmegeed, D., Yu, D., Chen, J. *et al.* (2022) Newly-synthesized iron-oxide nanoparticles showed synergetic effect with citric acid for alleviating arsenic phytotoxicity in soybean. *Environmental Pollution* 295, 118693. DOI: 10.1016/j.envpol.2021.118693.

Byczyńska, A., Zawadzińska, A. and Salachna, P. (2019) Silver nanoparticles preplant bulb soaking affects tulip production. *Acta Agriculturae Scandinavica, Section B — Soil & Plant Science* 69(3), 250–256. DOI: 10.1080/09064710.2018.1545863.

Chaud, M., Souto, E.B., Zielinska, A., Severino, P., Batain, F. *et al.* (2021) Nanopesticides in agriculture: benefits and challenge in agricultural productivity, toxicological risks to human health and environment. *Toxics* 9(6), 131. DOI: 10.3390/toxics9060131.

El-Badri, A.M., Batool, M., Wang, C., Hashem, A.M., Tabl, K.M. *et al.* (2021) Selenium and zinc oxide nanoparticles modulate the molecular and morpho-physiological processes during seed germination of *Brassica napus* under salt stress. *Ecotoxicology and Environmental Safety* 225, 112695. DOI: 10.1016/j.ecoenv.2021.112695.

Elsayed, A.A., Ahmed, E.G, Taha, Z.K., Farag, H.M., Hussein, M.S. *et al.* (2022) Hydroxyapatite nanoparticles as novel nano-fertilizer for production of rosemary plants. *Scientia Horticulturae* 295, 110851. DOI: 10.1016/j.scienta.2021.110851.

Faizan, M., Faraz, A., Yusuf, M., Khan, S.T. and Hayat, S. (2018) Zinc oxide nanoparticle-mediated changes in photosynthetic efficiency and antioxidant system of tomato plants. *Photosynthetica* 56(2), 678–686. DOI: 10.1007/s11099-017-0717-0.

Fatemi, H., Pour, B. and Rizwan, M. (2021) Foliar application of silicon nanoparticles affected the growth, vitamin C, flavonoid, and antioxidant enzyme activities of coriander (*Coriandrum sativum* L.) plants grown in lead (Pb)-spiked soil. *Environmental Science and Pollution Research International* 28(2), 1417–1425. DOI: 10.1007/s11356-020-10549-x.

Giannousi, K., Avramidis, I. and Dendrinou-Samara, C. (2013) Synthesis, characterization and evaluation of copper based nanoparticles as agrochemicals against *Phytophthora infestans*. *RSC Advances* 3(44), 21743–21752. DOI: 10.1039/c3ra42118j.

Gohari, G., Zareei, E., Rostami, H., Panahirad, S., Kulak, M. *et al.* (2021) Protective effects of cerium oxide nanoparticles in grapevine (*Vitis vinifera* L.) cv. flame seedless under salt stress conditions. *Ecotoxicology and Environmental Safety* 220, 112402. DOI: 10.1016/j.ecoenv.2021.112402.

Gui, X., Rui, M., Song, Y., Ma, Y., Rui, Y., *et al.* (2017) Phytotoxicity of CeO_2 nanoparticles on radish plant (*Raphanus sativus*). *Environmental Science and Pollution Research International* 24(15), 13775–13781. DOI: 10.1007/s11356-017-8880-1.

Gui, X., Zhang, Z., Liu, S., Ma, Y., Zhang, P. *et al.* (2015) Fate and phytotoxicity of CeO_2 nanoparticles on lettuce cultured in the potting soil environment. *PLoS One* 10(8), e0134261. DOI: 10.1371/journal.pone.0134261.

Gutiérrez-Ramírez, J.A., Betancourt-Galindo, R., Aguirre-Uribe, L.A., Cerna-Chávez, E., Sandoval-Rangel, A. *et al.* (2021) Insecticidal effect of zinc oxide and titanium dioxide nanoparticles against *Bactericera Cockerelli* Sulc. (Hemiptera: Triozidae) on tomato *Solanum lycopersicum*. *Agronomy* 11(8), 1460. DOI: 10.3390/agronomy11081460.

Hassanpouraghdam, M.B., Mehrabani, L.V., Kheirollahi, N., Soltanbeigi, A. and Khoshmaram, L. (2022) Foliar application of graphene oxide, Fe, and Zn on *Artemisia dracunculus* L. under salinity. *Scientia Agricola* 80, e20210202. DOI: 10.1590/1678-992x-2021-0202.

Hoe, P.T., Linh, T.M., Van, N.T., Buu, N.Q., Mai, N.C. *et al.* (2018) Effects of nanoparticles zinc oxide and nano cobalt on the germination of soybean (*Glycine max* (L.) Merr). *Vietnam Journal of Biotechnology* 16(3), 501–508. DOI: 10.15625/1811-4989/16/3/13472.

Ikram, M., Raja, N.I., Javed, B., Mashwani, Z. Hussain, M. *et al.* (2020) Foliar applications of bio-fabricated selenium nanoparticles to improve the growth of wheat plants under drought stress. *Green Processing and Synthesis* 9(1), 706–714. DOI: 10.1515/gps-2020-0067.

Jakhar, A.M., Aziz, I., Kaleri, A.R., Hasnain, M., Haider, G. *et al.* (2022) Nano-fertilizers: a sustainable technology for improving crop nutrition and food security. *NanoImpact* 27, 100411. DOI: 10.1016/j.impact.2022.100411.

Kavya, J.B., Murali, M., Manjula, S., Basavaraj, G.L., Prathibha, M. *et al.* (2020) Genotoxic and antibacterial nature of biofabricated zinc oxide nanoparticles from *Sida rhombifolia* Linn. *Journal of Drug Delivery Science and Technology* 60, 101982. DOI: 10.1016/j.jddst.2020.101982.

Li, Y., Liu, J., Fu, C., Khan, M.N., Hu, J. *et al.* (2022) CeO_2 nanoparticles modulate Cu–Zn superoxide dismutase and lipoxygenase-IV isozyme activities to alleviate membrane oxidative damage to improve rapeseed salt tolerance. *Environmental Science: Nano* 9(3), 1116–1132. DOI: 10.1039/D1EN00845E.

Liu, Y., Cao, X., Yue, L., Wang, C., Tao, M. *et al.* (2022) Foliar-applied cerium oxide nanomaterials improve maize yield under salinity stress: reactive oxygen species homeostasis and rhizobacteria regulation. *Environmental Pollution* 299, 118900. DOI: 10.1016/j.envpol.2022.118900.

Malandrakis, A.A., Kavroulakis, N. and Chrysikopoulos, C.V. (2019) Use of copper, silver and zinc nanoparticles against foliar and soil-borne plant pathogens. *The Science of the Total Environment* 670, 292–299. DOI: 10.1016/j.scitotenv.2019.03.210.

Mali, S.C, Raj, S. and Trivedi, R. (2020) Nanotechnology a novel approach to enhance crop productivity. *Biochemistry and Biophysics Reports* 24, 100821. DOI: 10.1016/j.bbrep.2020.100821.

Maringgal, B., Hashim, N., Tawakkal, I.S.M.A., Hamzah, M.H. and Mohamed, M.T.M. (2020) Biosynthesis of CaO nanoparticles using *Trigona* sp. honey: physicochemical characterization, antifungal activity, and cytotoxicity properties. *Journal of Materials Research and Technology* 9(5), 11756–11768. DOI: 10.1016/j.jmrt.2020.08.054.

Milani, N., Hettiarachchi, G.M., Kirby, J.K., Beak, D.G., Stacey, S.P. *et al.* (2015) Fate of zinc oxide nanoparticles coated onto macronutrient fertilizers in an alkaline calcareous soil. *PloS One* 10(5), e0126275. DOI: 10.1371/journal.pone.0126275.

Mohamed, A.K.S., Qayyum, M.F., Abdel-Hadi, A.M., Rehman, R.A., Ali, S. *et al.* (2017) Interactive effect of salinity and silver nanoparticles on photosynthetic and biochemical parameters of wheat. *Archives of Agronomy and Soil Science* 63(12), 1736–1747. DOI: 10.1080/03650340.2017.1300256.

Muhammad, Z., Inayat, N., and Majeed, A. (2020) Application of nanoparticles in agriculture as fertilizers and pesticides: challenges and opportunities. In: Rakshit, A., Singh, H., Singh, A., Singh, U., and Fraceto, L. (eds) *New Frontiers in Stress Management for Durable Agriculture*. Springer, Singapore, pp. 281–293.

Mukarram, M., Khan, M.M.A. and Corpas, F.J. (2021) Silicon nanoparticles elicit an increase in lemongrass (*Cymbopogon flexuosus* (Steud.) Wats) agronomic parameters with a higher essential oil yield. *Journal of Hazardous Materials* 412, 125254. DOI: 10.1016/j.jhazmat.2021.125254.

Neme, K., Nafady, A., Uddin, S. and Tola, Y.B. (2021) Application of nanotechnology in agriculture, postharvest loss reduction and food processing: food security implication and challenges. *Heliyon* 7(12), e08539. DOI: 10.1016/j.heliyon.2021.e08539.

Padilla-Camberos, E., Juárez-Navarro, K.J., Sanchez-Hernandez, I.M., Torres-Gonzalez, O.R. and Flores-Fernandez, J.M. (2022) Toxicological evaluation of silver nanoparticles synthesized with peel extract of *Stenocereus queretaroensis*. *Materials* 15(16), 5700. DOI: 10.3390/ma15165700.

Pang, L.J., Adeel, M., Shakoor, N., Guo, K.-R., Ma, D.-F. *et al.* (2021) Engineered nanomaterials suppress the soft rot disease (*Rhizopus stolonifer*) and slow down the loss of nutrient in sweet potato. *Nanomaterials* 11(10), 2572. DOI: 10.3390/nano11102572.

Phong, T.H., Hieu, T., Tung, H.T., Mai, N.T.N., Khai, H.D. *et al.* (2022) Silver nanoparticles: a positive factor for *in vitro* flowering and fruiting of purple passion fruit (*Passiflora edulis* Sim f. *edulis*). *Plant Cell, Tissue and Organ Culture* 151(2), 401–412. DOI: 10.1007/s11240-022-02361-x.

Pittarate, S., Rajula, J., Rahman, A., Vivekanandhan, P., Thungrabeab, M. *et al.* (2021) Insecticidal effect of zinc oxide nanoparticles against *Spodoptera frugiperda* under laboratory conditions. *Insects* 12(11), 1017. DOI: 10.3390/insects12111017.

Popescu, G.C., Popescu, M., Khondker, M., Clay, D.E., Pampana, S. *et al.* (2022) Agricultural sciences and the environment: reviewing recent technologies and innovations to combat the challenges of climate change, environmental protection, and food security. *Agronomy Journal* 114(4), 1895–1901. DOI: 10.1002/agj2.21164.

Rajput, V.D., Faizan, M., Upadhyay, S.K., Kumari, A., Ranjan, A. *et al.* (2022) Influence of nanoparticles on the plant rhizosphere microbiome. In: Rajput, V.D., Verma, K.K., Sharma, N. and Minkina, T. (eds) *The Role of Nanoparticles in Plant Nutrition under Soil Pollution*. Springer, Cham, Switzerland, pp. 83–102. DOI: 10.1007/978-3-030-97389-6.

Rajput, V.D., Minkina, T., Feizi, M., Kumari, A., Khan, M., *et al.* (2021) Effects of silicon and silicon-based nanoparticles on rhizosphere microbiome, plant stress and growth. *Biology* 10(8), 791. DOI: 10.3390/biology10080791.

Rajwade, J.M., Chikte, R.G. and Paknikar, K.M. (2020) Nanomaterials: new weapons in a crusade against phytopathogens. *Applied Microbiology and Biotechnology* 104(4), 1437–1461. DOI: 10.1007/s00253-019-10334-y.

Razzaq, A., Ammara, R., Jhanzab, H.M., Mahmood, T., Hafeez, A., *et al.* (2016) A novel nanomaterial to enhance growth and yield of wheat. *Journal of Nanoscience and Technology* 2, 55–58.

Rizwan, M., Ali, S., Ali, B., Adrees, M., Arshad, M. *et al.* (2019) Zinc and iron oxide nanoparticles improved the plant growth and reduced the oxidative stress and cadmium concentration in wheat. *Chemosphere* 214, 269–277. DOI: 10.1016/j.chemosphere.2018.09.120.

Roseline, T.A., Sudhakar, M.P. and Kulanthaiyesu, A. (2021) Synthesis of silver nanoparticle composites using *Calliblepharis fimbriata* aqueous extract, phytochemical stimulation, and controlling bacterial blight disease in rice. *ACS Agricultural Science & Technology* 1(6), 702–718. DOI: 10.1021/acsagscitech.1c00189.

Salachna, P., Byczyńska, A., Zawadzińska, A., Piechocki, R. and Mizielińska, M. (2019) Stimulatory effect of silver nanoparticles on the growth and flowering of potted oriental lilies. *Agronomy* 9(10), 610. DOI: 10.3390/agronomy9100610.

Salam, A., Khan, A.R., Liu, L., Yang, S., Azhar, W. *et al.* (2022) Seed priming with zinc oxide nanoparticles downplayed ultrastructural damage and improved photosynthetic apparatus in maize under cobalt stress. *Journal of Hazardous Materials* 423(Pt A), 127021. DOI: 10.1016/j.jhazmat.2021.127021.

Sarkar, M.R., Rashid, Md.H., Rahman, A., Kafi, Md.A., Hosen, Md.I. *et al.* (2022) Recent advances in nanomaterials based sustainable agriculture: an overview. *Environmental Nanotechnology, Monitoring & Management* 18, 100687. DOI: 10.1016/j.enmm.2022.100687.

Savi, G.D., Piacentini, K.C., de Souza, S.R., Costa, M.E.B., Santos, C.M.R. *et al.* (2015) Efficacy of zinc compounds in controlling Fusarium head blight and deoxynivalenol formation in wheat (*Triticum aestivum* L.). *International Journal of Food Microbiology* 205, 98–104. DOI: 10.1016/j.ijfoodmicro.2015.04.001.

Scott-Fordsmand, J.J., Fraceto, L.F. and Amorim, M.J.B. (2022) Nano-pesticides: the lunch-box principle-deadly goodies (semio-chemical functionalised nanoparticles that deliver pesticide only to target species). *Journal of Nanobiotechnology* 20(1), 13. DOI: 10.1186/s12951-021-01216-5.

Shang, H., Ma, C., Li, C., White, J.C., Polubesova, T. *et al.* (2020) Copper sulfide nanoparticles suppress *Gibberella fujikuroi* infection in rice (*Oryza sativa* L.) by multiple mechanisms: contact-mortality, nutritional modulation and phytohormone regulation. *Environmental Science* 7(9), 2632–2643. DOI: 10.1039/D0EN00535E.

Shang, H., Ma, C., Li, C., Zhao, J., Elmer, W. *et al.* (2021) Copper oxide nanoparticle-embedded hydrogels enhance nutrient supply and growth of lettuce (*lactuca sativa*) infected with *fusarium oxysporum* f. sp. *lactucae*. *Environmental Science & Technology* 55, 13432–13442. DOI: 10.1021/acs.est.1c00777.

Sharma, P., Urfan, M., Anand, R., Sangral, M., Hakla, H.R. *et al.* (2022) Green synthesis of zinc oxide nanoparticles using *Eucalyptus lanceolata* leaf litter: characterization, antimicrobial and agricultural efficacy in maize. *Physiology and Molecular Biology of Plants* 28(2), 363–381. DOI: 10.1007/s12298-022-01136-0.

Shatalova, E.I., Grizanova, E.V. and Dubovskiy, I.M. (2022) The effect of silicon dioxide nanoparticles combined with entomopathogenic bacteria or fungus on the survival of colorado potato beetle and cabbage beetles. *Nanomaterials* 12(9), 1558. DOI: 10.3390/nano12091558.

Sibiya, A., Jeyavani, J., Santhanam, P., Preetham, E., Freitas, R. *et al.* (2022) Comparative evaluation on the toxic effect of silver (Ag) and zinc oxide (ZnO) nanoparticles on different trophic levels in aquatic ecosystems: a review. *Journal of Applied Toxicology* 42(12), 1890–1900. DOI: 10.1002/jat.4310.

Siddiqui, M.H., Al-Whaibi, M.H., Faisal, M. and Al Sahli, A.A. (2014) Nano-silicon dioxide mitigates the adverse effects of salt stress on *Cucurbita pepo L. Environmental Toxicology and Chemistry* 33(11), 2429–2437. DOI: 10.1002/etc.2697.

Singh, P., Arif, Y., Siddiqui, H., Sami, F., Zaidi, R. *et al.* (2021) Nanoparticles enhance the salinity toxicity tolerance in *Linum usitatissimum* L. by modulating the antioxidative enzymes, photosynthetic efficiency, redox status and cellular damage. *Ecotoxicology and Environmental Safety* 213, 112020. DOI: 10.1016/j.ecoenv.2021.112020.

Skiba, E., Pietrzak, M., Gapińska, M. and Wolf, W.M. (2020) Metal homeostasis and gas exchange dynamics in *Pisum sativum* L. exposed to cerium oxide nanoparticles. *International Journal of Molecular Sciences* 21(22), 8497. DOI: 10.3390/ijms21228497.

Suriyaprabha, R., Karunakaran, G., Yuvakkumar, R., Rajendran, V. and Kannan, N. (2014) Foliar application of silica nanoparticles on the phytochemical responses of maize (*Zea mays* L.) and its toxicological behavior . *Synthesis and Reactivity in Inorganic, Metal-Organic, and Nano-Metal Chemistry* 44(8), 1128–1131. DOI: 10.1080/15533174.2013.799197.

Thabet, A.F., Boraei, H.A., Galal, O.A., El-Samahy, M.F.M., Mousa, K.M. *et al.* (2021) Silica nanoparticles as pesticide against insects of different feeding types and their non-target attraction of predators. *Scientific Reports* 11(1), 14484. DOI: 10.1038/s41598-021-93518-9.

Thakur, P., Thakur, S., Kumari, P., Shandilya, M., Sharma, S. *et al.* (2022) Nano-insecticide: synthesis, characterization, and evaluation of insecticidal activity of ZnO NPs against *Spodoptera litura* and *Macrosiphum euphorbiae*. *Applied Nanoscience* 12(12), 3835–3850. DOI: 10.1007/s13204-022-02530-6.

Tofighi Alikhani, T., Tabatabaei, S.J., Mohammadi Torkashvand, A., Khalighi, A. and Talei, D. (2021) Effects of silica nanoparticles and calcium chelate on the morphological, physiological and biochemical characteristics of gerbera (*Gerbera jamesonii* L.) under hydroponic condition. *Journal of Plant Nutrition* 44(7), 1039–1053. DOI: 10.1080/01904167.2020.1867578.

Tran, N.N., Le, T.N.Q., Pho, H.Q., Tran, T.T. and Hessel, V. (2022) Nanofertilizers and nanopesticides for crop growth. In: Chen, J.T. (ed.) *Plant and Nanoparticles*. Springer, Singapore, pp. 367–394. DOI: 10.1007/978-981-19-2503-0.

Tymoszuk, A. and Kulus, D. (2020) Silver nanoparticles induce genetic, biochemical, and phenotype variation in chrysanthemum. *Plant Cell, Tissue and Organ Culture* 143(2), 331–344. DOI: 10.1007/s11240-020-01920-4.

Wang, Y., Deng, C., Rawat, S., Cota-Ruiz, K., Medina-Velo, I. *et al.* (2021) Evaluation of the effects of nanomaterials on rice (*Oryza sativa* L.) responses: underlining the benefits of nanotechnology for agricultural applications. *ACS Agricultural Science & Technology* 1(2), 44–54. DOI: 10.1021/acsagscitech.1c00030.

Zhao, W., Liu, Y., Zhang, P., Zhou, P., Wu, Z. *et al.* (2022) Engineered Zn-based nano-pesticides as an opportunity for treatment of phytopathogens in agriculture. *NanoImpact* 28, 100420. DOI: 10.1016/j.impact.2022.100420.

Zhou, Y., Wu, J., Zhou, J., Lin, S. and Cheng, D. (2022) pH-responsive release and washout resistance of chitosan-based nano-pesticides for sustainable control of plumeria rust. *International Journal of Biological Macromolecules* 222(Pt A), 188–197. DOI: 10.1016/j.ijbiomac.2022.09.144.

2 Nanoformulation Synthesis and Mechanisms of Interactions with Biological Systems

Denisa Ştefania Vîlcoci[1], Nicoleta Anca Sutan[2]*, Oana Alexandra Drăghiceanu[2]*, Liliana Cristina Soare[2]* and Georgiana Cîrstea[1]

[1]*Center of Research and Development for Materials, Processes and Innovative Products Dedicated to the Automotive Industry (CRC&D-AUTO), University of Pitesti, Pitesti, Romania;* [2]*Department of Natural Sciences, University of Pitesti, Pitesti, Romania*

Abstract

Nanoparticles have produced a real revolution in many fields of science in terms of their special properties, having applicability in medicine, agriculture, horticulture, the food industry, and environmental protection. The growing interest of the scientific community in this field is also visible through the increase in the number of publications in the last decade. In 2012, in the Scopus database there were 52 records with the keyword nano-formulation, while in 2021, the number increased 8.2-fold (427), while on the Web of Science the increase was 10.5-fold. Nanoparticle synthesis uses dry or wet methods, high-temperature synthesis, biosynthesis or waste-material biosynthesis, and the methods of preparing and manufacturing nanoparticles/nanoformulations with specific properties and designed to meet the application requirements are critical in their successful production and use. In addition, the environmental safe use of nanoformulations requires *in vitro* and *in vivo* risk assessment of their impact at the cellular and subcellular levels, as well as their interaction with the various groups of living organisms characteristic of natural, agricultural and horticultural ecosystems.

Keywords: polymeric nanoparticles, nano-emulsions, nanoformulations, bioavailability, toxicity

2.1 Introduction

With humanity currently facing numerous challenges, it is essential to ensure technologies that support high yield, increased and cost-effective crop production, and ecological and sustainable protection of crops (Javed *et al.*, 2019). With food consumption increasing every year and one-third of crops affected by decreased soil fertility, natural disasters and stressful conditions, innovative approaches are the only solution to this problem (Amna *et al.*, 2023).

As stated by the European Commission, nanotechnology is the key to sustainable development (Baker *et al.*, 2019). The use of nanotechnology in agriculture has led to favourable results in terms of increased production due to the properties of nanoformulations (NFs) (Zhang and Goss, 2022). Various nanomaterials (NMs), such as liposomes, micelles, carbon nanotubes,

*Corresponding authors: anca.sutan@upit.ro, oana.draghiceanu@upit.ro and cristina.soare@upit.ro

© CAB International 2023. *Nanoformulations for Sustainable Agriculture and Environmental Risk Mitigation* (eds Z. Khan and N.A. Şuţan)
DOI: 10.1079/9781800623095.0002

quantum dots, organic dendrimers, polymer nanocapsules, nanoshells, nanospheres, and metalloid/metal oxide nanoparticles (NPs) have created new horizons for covering people's needs (Tripathi and Prakash, 2022). Globally, there are many studies in the field of applying NPs in agriculture.

As the application of NMs improves plant nutrition without damage to the soil and ensures protection against pathogens and diseases, the integration of nanotechnologies in agriculture is of particular importance because it is considered to achieve sustainable agricultural production (Jhanzab *et al.*, 2022). NFs can increase agricultural yield, and at the same time reduce ecosystems pollution. NMs can be used to diagnose diseases in field crops, monitor the quality of packaged foods, and determine the quality and health of soils (Srivastava *et al.*, 2021).

The abusive use of pesticides and fertilizers has led to their accumulation in the environment, which can have negative repercussions on quality of life. Dendrimer-type NFs have been tested for the rapid removal of pesticides from the environment (Guo *et al.*, 2019). In the same context of enhancing food-related quality of life, simultaneously with the reduction of chemical inputs, metal and metal oxide NPs have been integrated into plant disease diagnosis and disease management strategies (Elmer and White, 2018), into the technologies for increasing soil fertility and efficiency of nutrient uptake, and into improving agronomic yields (Kalia *et al.*, 2019; Mejias *et al.*, 2021). NPs and NFs have also been evaluated for their potential as nanofertilizers and controlled-release systems (Beig *et al.*, 2022), plant breeding, and postharvest quality control.

In this chapter, we focus on the synthesis of NPs and NFs, and their applications as nano-based agri-inputs, as well as their interaction with biological systems.

2.2 Synthesis and Applications of Nanoformulations

In the last decade, several NFs (e.g. nanopesticides, nanofertilizers, nanosensors) have gained momentum as they have shown promising applications in the agri-food sector (e.g. sustainability of agricultural practices and waste reduction) due to their unique physico-chemical characteristics (Nile *et al.*, 2020; Amna *et al.*, 2023). A number of types of NMs (dendrimers, liposomes, micelles, quantum dots, carbon nanotubes and metallic NPs) under different formulations are known, and a diverse range of methods for their synthesis have been developed.

2.2.1 Synthesis of polymeric NPs

Some of the most important types of NPs have been shown to be polymeric NPs, which are dispersions of solid particles (Mallakpour and Behranvand, 2016). Polymeric NPs can be nanospheres or nanocapsules. Nanospheres consist of polymer networks that adsorb compounds of interest on to their surface or within the network. Nanocapsules consist of a polymeric shell that has the property of controlled release of the compounds dissolved in their oily center (Zielińska *et al.*, 2020). A number of polymeric NP synthesis methods are known, including dispersion of preformed polymers, monomer polymerization and supercritical fluid technology. To obtain polymeric NPs by dispersion of preformed polymers, it is necessary to dissolve them in certain solvents, followed by formation of emulsions, which can be obtained by ultrasonication or very high-speed homogenization, or by magnetic stirring. The current most commonly used solvent is ethyl acetate, which has been shown to be much less toxic than the solvents used in the past, namely dichloromethane and chloroform. Subsequently, the polymer NP suspension is formed by evaporating the solvent used (Anton *et al.*, 2008). Polymeric NPs can be collected from the suspension by ultracentrifugation and washing, followed by freeze-drying to obtain the NPs as a powder.

Emulsion polymerization of monomers is the fastest and most readily scalable method of obtaining NPs and is of greater interest to the scientific community because it allows *de novo* polymers with well-established features to be produced. This process is based on emulsion polymerization techniques, in particular in mini- and micro-emulsions. This process can be carried out using either an organic or an aqueous continuous phase solution (Rao and

Geckeler, 2011). Supercritical fluid technology has become necessary in the process of synthesis of polymeric NPs in order to eliminate the organic solvents used in classical methods, or toxic solvents that cannot be completely removed from the finished product (Jung and Perrut, 2001).

Dendrimers are polymeric, three-dimensional, nanoscale (10–100 nm), highly branched structures (Pal *et al.*, 2011). These NMs are composed of a central core from which a series of branches branch off, to which several functional groups are terminally attached. As a result of this type of arrangement, spaces called internal cavities are formed in the dendrimer structure, in which functional molecules can be encapsulated, thus conferring the property of an ideal carrier (Santos *et al.*, 2019). According to Sherje *et al.* (2018), the functional terminal elements of these nanoscale structures play a role in the interaction with external/target molecules. The method by which a nanometric dendrimer structure can be obtained must offer the possibility of the best possible control in each construction phase, given that these globular structures are built in cascade. Thus, two main synthesis techniques have been developed: divergent and convergent (Parat and Felder-Flesch, 2016; Santos *et al.*, 2019).

The process of divergent synthesis of a dendrimer starts from the nucleus (core), which, upon activation, exhibits the ability to bind a monomer to it. This is the first step in the synthesis process, followed by activation of the terminal group of the monomer in the generation thus formed in order to couple new monomers. These two steps can be repeated to obtain a dendrimer with a number of generations (Tomalia and Fréchet, 2002; Parat and Felder-Flesch, 2016). Although, the method is widely used, it has a number of disadvantages; for example: (i) many reactions take place at the level of the single functional backbone molecule, which can lead to deficient branching and thus partial functionalization of the dendrimer (Mendes *et al.*, 2017); and (ii) it is a method that requires many reagents, leading to difficulty in purification of the final product (Klajnert and Bryszewska, 2001). In order to overcome these drawbacks, another technique has been developed for the synthesis of dendrimers, namely convergent synthesis. In contrast to the divergent process, the dendrimer is formed from the functional groups, responsible for the interaction with the target molecule or group, toward the nucleus (Nanjwade *et al.*, 2009). As this synthesis method involves fewer coupling reactions, related to each growth phase, dendritic structures with a much higher purity can be obtained due to better control over the target structure. Moreover, the technique allows the construction of a heterogeneous dendrimer composed of different monomers, which can lead to the appearance of a multifunctional character (Šebestík *et al.*, 2012; Walter and Malkoch, 2012; Lyu *et al.*, 2019). However, as with the divergent process, there are a number of disadvantages that characterize this technique, such as the impossibility of constructing a high-generation dendrimer, difficulty in modifying peripheral clusters, and low yield.

As well as the disadvantages mentioned for each method, a notable difficulty encountered in these classical synthesis processes is the uncertainty of the existence of a perfect, error-free dendritic structure. In order to overcome these drawbacks and to minimize reaction time and production costs, the idea of developing accelerated dendrimeric synthesis approaches (e.g. double exponential growth, the hypercore or two-step convergent method, and the hyper-monomer method) has been pursued (Maraval *et al.*, 2000; Balaji and Lewis, 2009; Walter and Malkoch, 2012).

Three types of colloidal systems are known, micelle, liposome, and nano-emulsion, which, due to their properties such as larger surface area, encapsulation, protection and delivery of bioactive substances (Espitia *et al.*, 2019), and their longer shelf life, higher kinetic stability (Hazra, 2017), and good wettability (Qin *et al.*, 2019), can perfectly fit the target applications in agriculture.

Micelles are structures formed by the assembly of amphiphilic molecules and consist of a hydrophilic polar region called the head and a hydrophobic non-polar region called the tail (Patel *et al.*, 2014). Various polymers such as polyethylene glycol (PEG), poly(*N*-isopropylacrylamide) (pNIPAM) and poly(*N*-vinyl pyrrolidone) (PVP), which are neutral, water-soluble, and non-toxic polymers, are used to form micelles with average particle sizes ranging from 5 to 100 nm (Alexander-Bryant *et al.*, 2013). Micelles are used mainly for the transport and delivery of poorly water-soluble active compounds. They are formed in aqueous solutions, with the polar head outwards and the non-polar tail inwards, forming the core (Xu et al., 2013). However, the best-known techniques for obtaining micelles are solvent evaporation, oil-in-water emulsion, dialysis techniques, and solid dispersion (Joseph et al., 2017). Micelles also ensure the release of macromolecules by conferring both chemical and physical stability to these encapsulated molecules, improving their bioavailability (Xu et al., 2013).

Liposomes are lipid-like, double-layered vesicles of nanometer size. As their structure involves both aqueous and lipid phases, these nanosystems exhibit the ability to either simultaneously or separately incorporate both hydrophilic and hydrophobic substances (Khorasani et al., 2018). For their synthesis, thin-film hydration-sonication methods, reverse phase evaporation methods, and detergent removal methods have been developed (Akbarzadeh et al., 2013; Zhang, 2017).

Nano-emulsions are formulations with remarkable rheological properties and optical stability (Espitia et al., 2019). Typically, nano-emulsions are optically clear systems that scatter less visible light due to their different refractive index, thus also lending themselves to optical studies (Shah et al., 2010). Nano-emulsions are used for the synthesis of nanocapsules and NPs with a wide range of applications (Muñoz-Espí and Álvarez-Bermúdez, 2018). Nano-emulsions can be found as oil droplets dispersed in the aqueous phase (oil–water nano-emulsions), as water droplets dispersed in the oil phase (water–oil nano-emulsions), and as oil and water droplets dispersed in bicontinuous nano-emulsions (Sadurní et al., 2005; Kumar and Singh, 2012). Synthesis of nano-emulsions is initiated by dissolving the compound of interest in the lipophilic part of the nano-emulsion, while the aqueous part of the nano-emulsion is mixed with a surfactant and then with a co-surfactant at a slow rate under continuous stirring until the transparent system is formed. A pseudoternary phase diagram is used to determine the required amount of surfactant and co-surfactant, as well as the percentage of oily phase. After the formation of the transparent system, ultrasound can be used to adjust the particle size dispersed in

the system. Carbomers can be added to these transparent systems as gelling agents to form specific gels of stable nano-emulsions (Kumar and Singh, 2012).

The two main techniques for synthesizing nano-emulsions are high-energy and low-energy synthesis techniques. The main high-energy synthesis techniques include high-pressure homogenization and ultrasound, which uses excess energy to break up droplets of approximately 100 nm in size (Delmas et al., 2011). Low-energy nano-emulsion synthesis techniques are achieved without excessive shear forces and rely on the low interfacial tension property of the system, with energy input provided by a magnetic stirrer (Tadros et al., 2004; Solans et al., 2005). Thus, the main application of nano-emulsions is related to the formation of NPs, in which case nano-emulsions can easily be considered as true nano-sized reactors with various applications, including nanopesticide formulation. For example, a stable and effective antifungal D-limonene nano-emulsion was obtained through emulsion polymerization by Feng et al. (2020). Nanometric liposomes loaded with limonene were also obtained using the thin-film dehydration method in order to enhance the shelf life and postharvest quality of strawberries (Dhital et al., 2018).

2.2.2 Synthesis of quantum dots

Quantum dots (QDs) are semiconductor nanocrystals with special optical and electronic properties (Cotta, 2020). For the synthesis of QDs, two types of techniques are applied: use top-down and bottom-up approaches. In the top-down approach, electron beam lithography or X-ray lithography, molecular beam epitaxy or ion implantation are the processing methods applied to obtain QDs of up to 30 nm diameter or arrays of zero-dimension dots. Bottom-up approaches are self-assembly techniques through which QDs are prepared via wet chemical methods (precipitation methods) and vapor-phase methods in an atom-by-atom growth process (Valizadeh et al., 2012).

The interaction of QDs with plants or with agri-environments has been assessed by numerous researchers and it has been shown that these NMs could have extensive application in agriculture. Based on their fluorescence, QDs are suitable for biolabeling applications (Cotta, 2020). Cadmium telluride/cadmium sulfide (CdTe/CdS) core/shell QDs in aqueous phase, and cadmium selenide/zinc sulfide (CdSe/ZnS) core/shell and CdSe/ZnSe/ZnS core/double shell QDs in high-boiling organic solvents were synthesized by Ung *et al.* (2012) and successfully used to trace residual parathion methyl pesticide. Niobium-modified stannic oxide (SnO_2) QDs have been green synthesized by a sequence of procedures involving hydrolysis, oxidation, and hydrothermal treatment in aqueous solution. Niobium-modified SnO_2 QDs worked as a photocatalyst in the *in situ* degradation of the polyethylene agricultural wastes (Shao *et al.*, 2021). In another green-synthesis approach, copper QDs were obtained using aqueous extracts of *Rubia cordifolia* roots and showed good antibacterial properties (Mariselvam *et al.*, 2014). Zinc oxide QDs obtained by the sol–gel method promoted the uptake and translocation of nutrients in hydroponic culture of lettuce, improving biomass and nutritional quality (Liang *et al.*, 2021).

2.2.3   Synthesis of metal and metal oxide NPs

Metal NP synthesis is performed using various methods that are classified as top-down and bottom-up methods. In top-down method approaches, bulk material is reduced to nanometric dimensions by physical, chemical, and/or mechanical treatments, while in bottom-up approaches, atoms and molecules are converted into NPs. Mechanical milling, such as ball milling and mechano-chemical synthesis are top-down methods for the preparation of metallic NPs. Solid-state methods (physical vapor deposition, chemical vapor deposition), liquid-state methods (sol–gel method, chemical reduction, hydrothermal and solvothermal methods), gas-phase methods (spray pyrolysis, laser pyrolysis, flame pyrolysis), green-synthesis methods and another few bottom-up methods have recently been considered by Jamkhande *et al.* (2019) in a very comprehensive work.

Among the methods listed, the most used for the production of NPs with applications in agriculture are bottom-up liquid-state and green-synthesis methods. The sol–gel method was used to extract mesoporous silica NPs from amorphous rice husk ash, promoting sustainable waste management and the recovery of secondary products from agrosystems (Dorairaj *et al.*, 2022). Mesoporous zinc aluminosilicate ($ZnO.Al_2O_3.Si_{10}O_{20}$) was synthesized from *Oryza sativa* L. husks using a co-precipitation method and the obtained nanocomposite was used as a urea-loaded nanofertilizer with high nitrogen-recovery efficiency (Naseem *et al.*, 2020). Zinc oxide–zinc ferrate ($ZnO–ZnFe_2O_4$) NPs that intensified the immune response and plant growth, and enhanced waste-water remediation were synthesized by a biohydrothermal method using *Psidium guajava* leaf extract (Sahoo *et al.*, 2021).

2.2.4   Synthesis of carbon nanotubes

Depending on the number of graphite atomic layers that make up their constitution, two types of carbon nanotubes (CNTs) are distinguished: single-walled CNTs and multi-walled CNTs. Various methods have been developed for the production of CNTs, such as arc discharge, laser ablation, vapor-phase growth, and chemical vapor deposition. In the arc discharge method, two electrodes are placed in a vacuum chamber supplied with inert gas. By optimizing the synthesis conditions by controlling the arc current between the electrodes and the gas pressure, respectively, the CNTs condense on the cathode electrode. The laser-ablation method uses pulsed laser ablation of graphite, in the presence of a catalyst, at a controlled temperature and pressure in the reaction chamber. For chemical vapor deposition, synthesis of CNTs is completed by the decomposition of the gas, via heat and plasma. These three methods require a carbon precursor, while the method of vapor-phase growth for the synthesis of CNTs is accomplished in its absence. All methods require a carbon source (Patel *et al.*, 2020; Shoukat and Khan, 2021).

2.3 Applications of Nanoformulations

The world faces numerous and unforeseen challenges caused by climate change, a global increasing population, land unavailability, industrialization and pollution, pest management at different stages of agricultural production and postharvest management (Ndlovu et al., 2020; Neme et al., 2021). Nanotechnologies could play a key role in overcoming food security challenges, increasing agricultural production and minimizing losses through food processing (Abobatta, 2018; Sivarethinamohan and Sujatha, 2021). Although only recently emerging, nanopesticides, nanofertilizers, nanosensors, and smart target delivery systems based on biological and environmental features have been developed through nanotechnologies. Nano-agromaterials have a notable influence on the germination percentage and physiological processes of plants. They can increase crop productivity by stimulating plant metabolic processes (e.g. photosynthesis, pigment production, energy transfer), and can give targeted and controlled release of agrochemicals so that there is efficiency within agroecosystems, without overdoses and leaching of NPs (Srivastava et al., 2021). Nanotechnologies have also been successfully used for genome sequencing, gene delivery and molecular diagnostics (Elingarami et al., 2013; Ikawati et al., 2021).

2.3.1 Nanofertilizers

Although the mechanisms underlying the improvement of nutrient uptake by plants are well known, the development of fertilizers that ensure optimal nutrient uptake has not yet been achieved in current agriculture. It is well known that the essential nutrients for the development of agricultural production and hence yield are nitrogen, phosphorus and potassium. Unfortunately, there are data reporting that, after soil application of these macronutrients, as much as 50% or even 90% is lost, leading to a notable resource deficit (Wang et al., 2016; Avila-Quezada et al., 2022). The overuse of chemical fertilizers is also a major concern, with attempts to reduce their use in favor of environmentally friendly and equally efficient alternatives (Zulfiqar et al., 2019). In the most recent reviews, there is an appreciation of nanofertilizers as the most effective alternative to overcome the negative impact of conventional fertilizers and as an intelligent solution for sustainable agriculture (Babu et al., 2022; Nongbet et al., 2022). NMs ensure a slow and controlled release of fertilizers, stimulate seed germination, improve water absorption, promote biomass enhancement and diminish the negative effect of various toxicants on plant growth (Al-Mamun et al., 2021). NPs used as nanofertilizers can be obtained following physical, chemical, or biosynthetic processes. They can be used in different NFs: by encapsulating nutrients in NMs (e.g. chitosan, polyvinyl alcohol, starch, alginate, PEG, carboxymethyl cellulose and nanoporous materials), by applying NMs as a thin polymer coating over active compounds, or by using nano-emulsions (Arora et al., 2022).

Nanofertilizers are classified into three groups: (i) nanofertilizers with macronutrients (hydroxyapatite, nano-urea, nano-calcium, nano-sulfur) in their structure; (ii) nanofertilizers with micronutrients (e.g. zinc oxide NPs (ZnONPs), copper oxide NPs (CuONPs), boron NPs (BNPs), manganese NPs (MnNPs)); and (iii) nanobiofertilizers (Kalia and Sharma, 2019; Zulfiqar et al., 2019). The first category includes nanofertilizers in which nutrients such as phosphorus, nitrogen, potassium, magnesium, and calcium are encapsulated. Micronutrient-based nanofertilizers are used to maintain optimal metabolic processes. Thus, it was demonstrated that foliar administration of two types of zinc and boron-based nanofertilizers on pomegranate (Punica granatum cv. Ardestani) increased fruit yield by more than 30% (Khot et al., 2012). Also, using zinc NPs (ZnNPs) fertilizers on rice, wheat and maize, the yield of production experienced a notable increase (Monreal et al., 2016). Another experimental study revealed that irrigation application of a stabilized maghemite solution led to improved chlorophyll content in rapeseed (Brassica napus), thus contributing to good and rapid plant growth, unlike the matrix used, which in this case was chelated iron (Palmqvist et al., 2017).

Macronutrients and micronutrients can react with clay colloids in the soil matrix and form precipitates that cannot be absorbed by plants. Due to their large surface area and sorption capacity, NPs could become functional ion reservoirs through these controlled-release mechanisms and thus lead to increased efficiency of the use of micro-/macronutrients and to a reduction of waste through leaching, degradation, and volatilization (Achari and Kowshik, 2018). In short, nanobiofertilizers are defined as biofertilizers encapsulated in NPs, which are intended to improve soil productivity and stimulate plant growth stages (Akhtar *et al.*, 2022). By encapsulating microorganisms, NMs provide beneficial conditions for root functioning and increase tolerance to abiotic stress (Das and Beegum, 2022). Elements of a bioorganic nature in nanobiofertilizer complexes have been shown to be synergistically effective in a diverse range of mechanisms, such as synthesis of phytohormones and siderophores, and nitrogen fixation in roots (Mala *et al.*, 2017). In addition, NFs increase the stability of biofertilizers by ensuring their protection against drying, heat, and ultraviolet (UV) inactivation (Zulfiqar *et al.*, 2019).

The essential nanoencapsulated nutrient controlled released is a process that is highly dependent on the activation of physical factors such as heat, sound waves, or magnetic fields (Ribeiro and Carmo, 2019), biological factors such as microorganisms, which have the ability to biodegrade the polymer coating leading to the release of the encapsulated nutrient as well as fixation in the soil (Ramzan *et al.*, 2020), or chemical factors like pH or moisture content (Weeks and Hettiarachchi, 2019). In addition to controlled-release, nanostructured fertilizers use mechanisms for targeted delivery to improve nutrient-use efficiency in plants (Arora *et al.*, 2022). Nanofertilizers enhance the rate of photosynthesis by expanding the leaf surface area and improving light absorption potentials, leading to increased crop plant biomass and improved yield (Kalia and Kaur, 2019). Although there is little information about the fate of nanofertilizers at the level of the food chain, including in humans (Morales-Díaz *et al.*, 2017), they do not seem to pose a threat to human health or ecosystems (Tolaymat *et al.*, 2017).

2.3.2 Nanopesticides

Globally, about 40% of agricultural losses occur due to insects, weeds, fungi, bacteria, and parasites, which is about US$220 billion annually (Zhang and Goss, 2022). Various materials are used for the synthesis of nanopesticides, including metals, (non)metallic oxides, carbon, silicates, dendrimers, proteins, polymers, and lipids, in the form of nano-emulsions, solid lipid NPs, mesoporous NPs, and nanoclays (Zhao *et al.*, 2017). Nanopesticides are formulations specifically designed for the controlled and targeted delivery of antimicrobials, insecticides, and fungicides. Nanopesticides are of interest because a small amount of active substance is required to achieve the desired effect and they imply little or no phytotoxicity. The most commonly used formulations of nanopesticides include nanospheres, nanocapsules, micelles, and nanogels (Yadav *et al.*, 2022). Various nanotechnology-based approaches have been used to obtain nanopesticides. For example, tebuconazole nanofungicide was formulated using low-energy nano-emulsion technology (Díaz-Blancas *et al.*, 2016), a nano-emulsion of *Mentha piperita* essential oil was obtained via the thin-film residue method, using high-energy emulsification technology (Massoud *et al.*, 2018), and a thymol nano-emulsion was obtained by the sonication method (Kumari *et al.*, 2018); all of these nano-emulsions showed increased encapsulation efficiency, high release performance, and promising pesticidal properties.

Dendrimers have been studied for double-stranded RNA (dsRNA) delivery into insect cells to obtain a selective insect pest control. Poly(amidoamine) (PAMAM) dendrimers were found to be an effective polymeric nanocarrier for dsRNA delivery and mediated RNA interference (Lu *et al.*, 2019; Edwards *et al.*, 2020).

Chitosan is a polysaccharide obtained by deacetylation of chitin from crustacean shells. Due to its characteristics (biocompatibility, lack of toxicity, and the presence of cationic groups), chitosan can be conjugated with dsRNA and thus used in pest control (Gurusamy *et al.*, 2020). Some studies have shown that PEG polymer-based NFs were more effective in pest control compared with commercial formulations due to slower release of active components (Zhu et al., 2020; Ikawati et al., 2021). NFs used as pesticides must present a series of characteristics: increased efficiency, biodegradability in soil and environment, good dispersion, wettability, and a photogenerative nature. Many studies have been conducted to obtain nanopesticides that are less harmful to the environment; these nanopesticides have a reduced number of active compounds, which also implies a reduced volume of pesticide used, and the specificity of the target ensures efficacy (Kumar et al., 2019).

NFs have been developed to provide an effective and sustainable way to control pests and increase crop production (Zhang and Goss, 2022). The use of nanopesticides instead of conventional ones reduces evaporation and leaching losses. Furthermore, nanoencapsulated pesticides have increased efficiency in terms of permeability, stability, and solubility, ensuring a controlled release of active compounds and improving pest control (Kumar et al., 2019). Although the use of nanotechnology in agriculture, and especially nanopesticides, is at an early stage, according to Wang et al. (2022), nanopesticides are 30% more effective against target organisms than conventional formulations and 40% less toxic to non-target species.

2.4 Interactions of Nanoformulations with Biological Systems

In agriculture, NMs are developed to combat the ageing and reduction of seed productivity that may occur because of their improper storage and implicit biochemical damage at the cellular level, and to increase yield, germination rate, and plant physiology. For seed priming, metal NPs from iron (FeNPs), silver (AgNPs), titanium dioxide (TiO$_2$NPs), gold (AuNPs), copper (CuNPs), iron disulfide (FeS$_2$NPs), zinc (ZnNPs) and zinc oxide (ZnONPs) or CNTs stimulate seed growth, germination, the antioxidant system, and plant tolerance to stress (Shukla et al., 2019). Metallic and non-metallic oxide NPs (AuNPs, AgNPs, ZnONPs, iron oxide NPs (Fe$_2$O$_3$NPs), TiO$_2$NPs, silicon dioxide NPs (SiO$_2$NPs), aluminum oxide NPs (Al$_2$O$_3$NPs), selenium NPs (SeNPs), CNTs, and QDs) play an important role in the growth

and development of plants, but also in alleviating the stress exerted by various factors (Aqeel et al., 2022). Some NPs play a significant role in improving stress resistance, such as resistance to drought, temperature, heavy metals, UVB radiation, and flooding (Hassanisaadi et al., 2022).

2.4.1 Interaction with soil microbiota

Soil is the main source of macro- and micromineral nutrients for plants; both soil characteristics and other factors influence the mineral bioavailability. Microorganisms (bacteria, fungi, actinomycetes, viruses, and protozoa) represent less than 1% of the total mass of the soil but play an essential role in the various ecosystems, as they form the basis of the food chain (Kumar et al., 2018). Additionally, the microbiota in the soil influence its quality because microorganisms play an important role in the processes of decomposition of organic matter and in the cycle of nutrients. The roles of soil microorganisms are defined by their characteristics. Algae provide soil fertility by adding organic matter, increasing water-holding capacity, helping break down rocks and, with lichens, establishing the soil structure. Fungi play a role in the decomposition of matter and the cycle of nutrients (nitrogen and phosphorus), bacteria mobilize nutrients from dead plants/animals, and protozoa maintain the balance of soil microbiota (Singh and Gurjar, 2022). Therefore, any factor that affects the activity, populations, and biomass of microbiota also influences soil quality and sustainability (Dinesh et al., 2012). At the same time, some microorganisms live in symbiosis with plants, providing them with the necessary nutrients (Mahawar and Prasanna, 2018). The soil microbiota is considered the second genome of plants because microorganisms play a role in controlling plant diseases by reducing the activity of pathogens (Santaella and Plancot, 2020).

There are few studies on the potential impact of NPs on symbiotic associations, but Tian et al. (2019) considered that the influence of NPs on mycorrhizae or formation of nodules varies depending on the type of NPs, speciation, size, and concentration of NPs, fungal species, and soil physico-chemical properties. Thus,

NPs can have negative or positive effects on the development of mycorrhizae and rhizobial symbioses. NPs of essential minerals improve the colonization of plants by beneficial microbiota by improving the adhesion of microbial cells to the plant surface, and by stimulating enzyme activity and synthesis of secondary metabolites, exopolysaccharides and antibiotics essential for the optimal functioning of the rhizosphere (Achari and Kowshik, 2018).

NPs concentration is a critical factor that determines NPs toxicity on mycorrhizae, as less toxic NPs (e.g. ZnONPs) are inhibitory at high concentrations and highly toxic NPs (e.g. AgNPs) can be stimulatory at low concentrations (Tian *et al.*, 2019). Irregular soil exposure to NPs could have adverse effects on the microbiota and, in some cases, a negative impact on edaphic factors, leading to infertility and toxicity (Javed *et al.*, 2019). Metal NPs have a negative impact on the population of beneficial soil bacteria and fungi due to their sustained deposition, low biodegradability, and persistence (Ameen *et al.*, 2021). Other critical factors that influence NPs toxicity to microorganisms are the application characteristics, especially the time but also the type of NPs. Metal and metal oxides NPs have more negative effects on soil microorganisms than organic NPs (Çiçek, 2021). Recent studies (Mishra *et al.*, 2020; Ottoni *et al.*, 2020) mentioned that the use of biogenic NMs reduces the negative impact on soil microbiota compared with their analogs derived from chemical synthesis. Basay *et al.* (2021) showed that microorganisms are very sensitive to environmental stress. NPs can enter microbial cells by endocytosis or cell penetration and many of them (e.g. AgNPs, ZnNPs, CuNPs) have antimicrobial properties (Kumar *et al.*, 2018). Research shows that the main mechanism by which NFs cause toxicity is oxidative stress, which occurs due to increased reactive oxygen species (ROS) production, but also through direct physical damage or via toxic ions released after dissolution of NPs (Abbas *et al.*, 2020). NPs can also affect bacterial cell respiration and cause cell disintegration (Ameen *et al.*, 2021). Most metal NPs (e.g. from Cu, Zn, Ag, magnesium (Mg), Au, Ti) cause dents on the cell-wall surface and eventually lead to cell death (Mahawar and Prasanna, 2018).

Santaella and Plancot (2020) considered that the chronic toxicity of NPs is kinetic and is influenced by the transformations that occur at the soil level and lead to the transient adjustment and adaptation of the microbiota. The impact of NFs on the soil microbiota may also be influenced by other organisms living in the soil (e.g. nematodes). Within the soil microbiota, bacterial populations represent approximately 15% of the microorganism populations, and they have a direct or indirect beneficial influence on plant growth (Ameen *et al.*, 2021). NMs are located at the level of soil microaggregates and are in direct contact with most native microbial communities, as 40–70% of the total soil bacteria are found at the level of these aggregates (Antisari *et al.*, 2013). At the cellular level, the toxicity of NPs on beneficial bacterial strains depends on the shape, size, chemical composition, and concentration of NPs but also on the exposure time. Some bacterial cells can also become partially or totally resistant to some NPs due to structural changes in the cell envelope, physiological changes, or even point mutations (Ameen *et al.*, 2021). Denitrification is the activity of microorganisms most affected by NPs, but NPs originating from green synthesis have less influence on the nitrogen cycle (Guilger *et al.*, 2017; Santaella and Plancot, 2020).

The importance of fungi in the soil is well known because they play a role in the decomposition of plant matter in the soil and the production of nutrients necessary for plant growth (Ameen *et al.*, 2021). The biomass of fungi and bacteria in the soil is 102–104 times higher than the biomass of other microorganisms (archaea, protists), including viruses (Khan *et al.*, 2022). The negative effects of metal NPs on symbiotic associations (plant–fungi, plant–bacteria) have been highlighted in various studies. McKee and Filser (2016) pointed out that metallic NPs disrupt biogeochemical circuits and can disrupt the ratio between bacteria and fungi, with some of these effects being evident even at low concentrations.

2.4.2   Interaction with plants

NPs have different morphological and physiological impacts on plants by improving plant growth and yield mainly by increasing mineral nutrition, chlorophyll content, and antioxidant enzymes, by stimulating the soil/rhizosphere, and by suppressing or inducing resistance to agricultural pests (Achari and Kowshik, 2018). The effect of NPs varies depending on the concentration and method of application, the physico-chemical properties (e.g. composition, size), and the plant species tested. Tombuloglu *et al.* (2022) found that the size of iron oxide NPs affects translocation and mineral nutrition of *Cucurbita maxima* L. Translocation from roots to leaves through vascular bundles was significant for NPs smaller than 40 nm, an important aspect for establishing the right sizes for the use of iron oxide NPs as nanofertilizer in the remediation of iron deficiency. Foliar application of calcium NPs (CaNPs) at 100 mg l^{-1} improved drought stress in hydroponic *B. napus* L. The favourable response of plants to drought stress correlated with the increase in the efficiency of photosystem II and the content of chlorophyll, enzymes, and non-enzymatic components involved in antioxidant mechanisms, but also with the decrease in ROS (Ayyaz *et al.*, 2022). In another study, edaphic fertilization and foliar application of SeNPs and ZnNPs led to an increased content of phenols and flavonoids, through which the antioxidant activity was improved (Montaño-Herrera *et al.*, 2022). However, Salama *et al.* (2022) reported that in *Phaseolus vulgaris* L. plants grown in the field, the foliar application of a concentration of only 40 mg l^{-1} of a combination of ZnONPs, manganese oxide NPs (MnO$_2$NPs) and molybdenum oxide NPs (MoO$_3$NPs) improved vegetative characters and yield. In addition, the polymorphisms evidenced by Start codon-targeted polymerase chain reaction (SCoT-PCR) analysis showed that higher concentrations of nanofertilizer are genotoxic. These results agree with those of Şuţan *et al.* (2021) who reported an increased frequency of chromosomal aberrations, nuclear abnormalities, and cytological changes in *Allium cepa* L. meristematic root tips after treatment with biosynthesized silver oxide NPs (Ag$_2$ONPs). In addition, the polymorphisms evidenced by SCoT-PCR analysis showed that higher concentrations of nanofertilizer are genotoxic. Other studies have also shown the genotoxic potential of metal NPs (Şuţan *et al.*, 2018, Şuţan *et al.*, 2019; Khan *et al.*, 2019; Heikal *et al.*, 2020). Besides genomic changes, damage induced by NPs can also occur at the proteome level. A study on soybean found that

Al$_2$O$_3$NPs, ZnONPs, and AgNPs caused protein degradation (Hossain *et al.*, 2016).

Seed germination is the first and most important stage in the life cycle of the plant, and is significantly influenced by the soil and its microbiota and by water availability but also by genetic factors (Ali *et al.*, 2021). Uniform and rapid seed germination is essential in commercial agriculture, as germination is the basis of growth and development processes that ensure plant productivity (Shukla *et al.*, 2019). Nanotechnology and NMs have improved the germination process, resulting in enhanced stress tolerance, plant growth, and crop fruiting rate (Ali *et al.*, 2021). Seed priming with citrate-coated ZnONPs and citrate-coated Fe$_3$O$_4$NPs alleviated lead-induced stress in *Basella alba* L. by reducing uptake of lead and ROS generation, and by activating the antioxidant system (Gupta *et al.*, 2022). However, according to Khot *et al.* (2012), in order to increase the seed germination rate, NMs must penetrate the seeds. Gour *et al.* (2019) reported that CTNs have several ecotoxicological effects in plants, including the release of toxic ions and the impairment of physiological processes, and by their accumulation in the organs associated with photosynthesis can lead to stomatal obstruction and foliar heating. Mittal *et al.* (2019) claimed that carbon-based NMs can stimulate seed germination and increase crop yield when used as fertilizers, plant growth stimulators, and soil conditioners. However, high concentrations of NPs at the root level lead to morphological and physiological alterations of crop plants, and inhibit seed germination, root growth and development, and the absorption of raw sap, which leads to inhibition of leaf development and ultimately to a decrease in biomass (Usman *et al.*, 2020). NP toxicity occurs because of the production of ROS, which leads to oxidative stress, negatively influencing membrane permeability and nutrient accumulation, as well as DNA damage (Murali *et al.*, 2022).

2.5 Conclusions

The numerous and varied synthesis methods presented in this chapter have not yet been fully tested for obtaining nano-based agri-inputs that increase the crops' yield, help to use resources in a more efficient way, accurately signal the plants' needs in real time, and alleviate the impact of agricultural residues. Moreover, the potential benefits of nanotechnology have yet to be tested on a large scale in field crops. This limitation is due to concerns about the absorption, translocation, bioavailability, and toxicity of NPs, and the legal framework required to regulate their use, as well as the inability of the agricultural sector to implement nanotechnologies. NPs uptake by plant roots could be considered the first level toward bioaccumulation, and their translocation and accumulation in edible plant tissues could have negative effects on human health. Thus it is likely that, for the next few years or decades, agriculture and nanotechnology will evolve together to achieve sustainable goals.

Acknowledgement

The authors thank the Romanian Ministry of Education and Research, CNCS – UEFISCDI, for the financial support through the Project number PN-III-P4-ID-PCE-2020–0620, within PNCDI III.

References

Abbas, Q., Yousaf, B., Ullah, H., Ali, M.U., Ok, Y.S. *et al.* (2020) Environmental transformation and nano-toxicity of engineered nano-particles (ENPs) in aquatic and terrestrial organisms. *Critical Reviews in Environmental Science and Technology* 50(23), 2523–2581. DOI: 10.1080/10643389.2019.1705721.

Abobatta, W.F. (2018) Nanotechnology application in agriculture. *Acta Scientific Agriculture* 2, 99–102

Achari, G.A. and Kowshik, M. (2018) Recent developments on nanotechnology in agriculture: plant mineral nutrition, health, and interactions with soil microflora. *Journal of Agricultural and Food Chemistry* 66(33), 8647–8661. DOI: 10.1021/acs.jafc.8b00691.

Akbarzadeh, A., Rezaei-Sadabady, R., Davaran, S., Joo, S.W., Zarghami, N. *et al.* (2013) Liposome: classification, preparation, and applications. *Nanoscale Research Letters* 8(1), 102. DOI: 10.1186/1556-276X-8-102.

Akhtar, N., Ilyas, N., Meraj, T.A., Pour-Aboughadareh, A., Sayyed, R.Z. *et al.* (2022) Improvement of plant responses by nanobiofertilizer: a step towards sustainable agriculture. *Nanomaterials* 12(6), 965. DOI: 10.3390/nano12060965.

Alexander-Bryant, A.A., Vanden Berg-Foels, W.S. and Wen, X. (2013) Bioengineering strategies for designing targeted cancer therapies. *Advances in Cancer Research* 118, 1–59. DOI: 10.1016/B978-0-12-407173-5.00002-9.

Ali, S.S., Al-Tohamy, R., Koutra, E., Moawad, M.S., Kornaros, M. *et al.* (2021) Nanobiotechnological advancements in agriculture and food industry: applications, nanotoxicity, and future perspectives. *Science of The Total Environment* 792, 148359. DOI: 10.1016/j.scitotenv.2021.148359.

Al-Mamun, Md.R., Hasan, Md.R., Ahommed, Md.S., Bacchu, Md.S., Ali, Md.R. *et al.* (2021) Nanofertilizers towards sustainable agriculture and environment. *Environmental Technology & Innovation* 23, 101658. DOI: 10.1016/j.eti.2021.101658.

Ameen, F., Alsamhary, K., Alabdullatif, J.A. and ALNadhari, S. (2021) A review on metal-based nanoparticles and their toxicity to beneficial soil bacteria and fungi. *Ecotoxicology and Environmental Safety* 213, 112027. DOI: 10.1016/j.ecoenv.2021.112027.

Amna, R.Y., Ahmad, J., Qamar, S. and Qureshi, M.I. (2023) Engineered nanomaterials for sustainable agricultural production, soil improvement and stress management: an overview. In: Husen, A. (ed.) *Engineered Nanomaterials for Sustainable Agricultural Production, Soil Improvement and Stress Management*. Academic Press, London, pp. 1–23.

Antisari, L.V., Carbone, S., Gatti, A., Vianello, G. and Nannipieri, P. (2013) Toxicity of metal oxide (CeO$_2$, Fe$_3$O$_4$, SnO$_2$) engineered nanoparticles on soil microbial biomass and their distribution in soil. *Soil Biology and Biochemistry* 60, 87–94. DOI: 10.1016/j.soilbio.2013.01.016.

Anton, N., Benoit, J.P. and Saulnier, P. (2008) Design and production of nanoparticles formulated from nano-emulsion templates—a review. *Journal of Controlled Release* 128(3), 185–199. DOI: 10.1016/j. jconrel.2008.02.007.

Aqeel, U., Aftab, T., Khan, M.M.A., Naeem, M. and Khan, M.N. (2022) A comprehensive review of impacts of diverse nanoparticles on growth, development and physiological adjustments in plants under changing environment. *Chemosphere* 291(Pt 1), 132672. DOI: 10.1016/j.chemosphere.2021.132672.

Arora, S., Murmu, G., Mukherjee, K., Saha, S. and Maity, D. (2022) A comprehensive overview of nanotechnology in sustainable agriculture. *Journal of Biotechnology* 355, 21–41. DOI: 10.1016/j. jbiotec.2022.06.007.

Avila-Quezada, G.D., Ingle, A.P., Golińska, P. and Rai, M. (2022) Strategic applications of nano-fertilizers for sustainable agriculture: benefits and bottlenecks. *Nanotechnology Reviews* 11(1), 2123–2140. DOI: 10.1515/ntrev-2022-0126.

Ayyaz, A., Fang, R., Ma, J., Hannan, F., Huang, Q. *et al.* (2022) Calcium nanoparticles (Ca-NPs) improve drought stress tolerance in *Brassica napus* by modulating the photosystem II, nutrient acquisition and antioxidant performance. *NanoImpact* 28, 100423. DOI: 10.1016/j.impact.2022.100423.

Babu, S., Singh, R., Yadav, D., Rathore, S.S., Raj, R. *et al.* (2022) Nanofertilizers for agricultural and environmental sustainability. *Chemosphere* 292, 133451. DOI: 10.1016/j.chemosphere.2021.133451.

Baker, S., Satish, S., Prasad, N. and Chouhan, R.S. (2019) Nano-agromaterials: influence on plant growth and crop protection. *Industrial Applications of Nanomaterials* 2009, 341–363. DOI: 10.1016/ b978-0-12-815749-7.00012-8.

Balaji, B.S. and Lewis, M.R. (2009) Double exponential growth of aliphatic polyamide dendrimers *via* AB$_2$ hypermonomer strategy. *Chemical Communications* 2009(30), 4593–4595. DOI: 10.1039/b903948a.

Basay, C.P., Paterno, E.S., Delmo-Organo, N., Villegas, L.C. and Fernando, L.M. (2021) Effects of nano-formulated plant growth regulator on culturable bacterial population, microbial biomass and enzyme activities in two soil types. *Mindanao Journal of Science and Technology* 19, 145–163.

Beig, B., Niazi, M.B.K., Sher, F., Jahan, Z., Malik, U.S. *et al.* (2022) Nanotechnology-based controlled release of sustainable fertilizers. A review. *Environmental Chemistry Letters* 20(4), 2709–2726. DOI: 10.1007/s10311-022-01409-w.

Çiçek, S. (2021) Effects of nanoparticles on soil microorganisms. *World Journal of Agriculture and Soil Science* 6(5), WJASS.MS.ID,000649. DOI: 10.33552/WJASS.2021.06.000649.

Cotta, M.A. (2020) Quantum dots and their applications: what lies ahead? *ACS Applied Nano Materials* 3(6), 4920–4924. DOI: 10.1021/acsanm.0c01386.

Das, S. and Beegum, S. (2022) Nanofertilizers for sustainable agriculture. In: Ghosh, S., Thongmee, S., and Kumar, A. (eds.) *Agricultural Nanobiotechnology*. Woodhead Publishing, Sawston, UK, pp. 355–370.

Delmas, T., Piraux, H., Couffin, A.-C., Texier, I., Vinet, F. *et al.* (2011) How to prepare and stabilize very small nanoemulsions. *Langmuir* 27(5), 1683–1692. DOI: 10.1021/la104221q.

Dhital, R., Mora, N.B., Watson, D.G., Kohli, P. and Choudhary, R. (2018) Efficacy of limonene nano coatings on post-harvest shelf life of strawberries. *LWT—Food Science and Technology* 97, 124–134. DOI: 10.1016/j.lwt.2018.06.038.

Díaz-Blancas, V., Medina, D.I., Padilla-Ortega, E., Bortolini-Zavala, R., Olvera-Romero, M. *et al.* (2016) Nanoemulsion formulations of fungicide tebuconazole for agricultural applications. *Molecules* 21(10), 1271. DOI: 10.3390/molecules21101271.

Dinesh, R., Anandaraj, M., Srinivasan, V. and Hamza, S. (2012) Engineered nanoparticles in the soil and their potential implications to microbial activity. *Geoderma* 173–174, 19–27. DOI: 10.1016/j. geoderma.2011.12.018.

Dorairaj, D., Govender, N., Zakaria, S. and Wickneswari, R. (2022) Green synthesis and characterization of UKMRC-8 rice husk-derived mesoporous silica nanoparticle for agricultural application. *Scientific Reports* 12(1), 20162. DOI: 10.1038/s41598-022-24484-z.

Edwards, C.H., Christie, C.R., Masotti, A., Celluzzi, A., Caporali, A. *et al.* (2020) Dendrimer-coated carbon nanotubes deliver dsRNA and increase the efficacy of gene knockdown in the red flour beetle *Tribolium castaneum. Scientific Reports* 10(1), 12422. DOI: 10.1038/s41598-020-69068-x.

Elingarami, S., Li, X. and He, N. (2013) Applications of nanotechnology, next generation sequencing and microarrays in biomedical research. *Journal of Nanoscience and Nanotechnology* 13(7), 4539–4551. DOI: 10.1166/jnn.2013.7522.

Elmer, W. and White, J.C. (2018) The future of nanotechnology in plant pathology. *Annual Review of Phytopathology* 56, 111–133. DOI: 10.1146/annurev-phyto-080417-050108.

Espitia, P.J.P., Fuenmayor, C.A. and Otoni, C.G. (2019) Nanoemulsions: synthesis, characterization, and application in bio-based active food packaging. *Comprehensive Reviews in Food Science and Food Safety* 18(1), 264–285. DOI: 10.1111/1541-4337.12405.

Feng, J., Wang, R., Chen, Z., Zhang, S., Yuan, S. *et al.* (2020) Formulation optimization of D-limonene-loaded nanoemulsions as a natural and efficient biopesticide. *Colloids and Surfaces A: Physicochemical and Engineering Aspects* 596, 124746. DOI: 10.1016/j.colsurfa.2020.124746.

Gour, N., Upadhyaya, P. and Patel, J. (2019) Nanomaterials as therapeutic and diagnostic tool for controlling plant diseases. *Comprehensive Analytical Chemistry* 84, 225–261. DOI: 10.1016/bs.coac.2019.04.003.

Guilger, M., Pasquoto-Stigliani, T., Bilesky-Jose, N., Grillo, R., Abhilash, P.C. *et al.* (2017) Biogenic silver nanoparticles based on trichoderma harzianum: synthesis, characterization, toxicity evaluation and biological activity. *Scientific Reports* 7, 44421. DOI: 10.1038/srep44421.

Guo, D., Muhammad, N., Lou, C., Shou, D. and Zhu, Y. (2019) Synthesis of dendrimer functionalized adsorbents for rapid removal of glyphosate from aqueous solution. *New Journal of Chemistry* 43(1), 121–129. DOI: 10.1039/C8NJ04433C.

Gupta, N., Singh, P.M., Sagar, V., Pandya, A., Chinnappa, M. *et al.* (2022) Seed priming with ZnO and Fe_3O_4 nanoparticles alleviate the lead toxicity in *Basella alba* L. through reduced lead uptake and regulation of ROS. *Plants* 11(17), 2227. DOI: 10.3390/plants11172227.

Gurusamy, D., Mogilicherla, K. and Palli, S.R. (2020) Chitosan nanoparticles help double-stranded RNA escape from endosomes and improve RNA interference in the fall armyworm, *Spodoptera frugiperda*. *Archives of Insect Biochemistry and Physiology* 104(4), e21677. DOI: 10.1002/arch.21677.

Hassanisaadi, M., Barani, M., Rahdar, A., Heidary, M., Thysiadou, A. *et al.* (2022) Role of agrochemical-based nanomaterials in plants: biotic and abiotic stress with germination improvement of seeds. *Plant Growth Regulation* 97(2), 375–418. DOI: 10.1007/s10725-021-00782-w.

Hazra, D.K. (2017) Nano-formulations: high definition liquid engineering of pesticides for advanced crop protection in agriculture. *Advances in Plants & Agriculture Research* 6(3), 211. DOI: 10.15406/apar.2017.06.00211.

Heikal, Y.M., Şuţan, N.A., Rizwan, M. and Elsayed, A. (2020) Green synthesized silver nanoparticles induced cytogenotoxic and genotoxic changes in *Allium cepa* L. varies with nanoparticles doses and duration of exposure. *Chemosphere* 243, 125430. DOI: 10.1016/j.chemosphere.2019.125430.

Hossain, Z., Mustafa, G., Sakata, K. and Komatsu, S. (2016) Insights into the proteomic response of soybean towards Al_2O_3, ZnO, and AG nanoparticles stress. *Journal of Hazardous Materials* 304, 291–305. DOI: 10.1016/j.jhazmat.2015.10.071.

Ikawati, S., Himawan, T., Abadi, A.L. and Tarno, H. (2021) Toxicity nanoinsecticide based on clove essential oil against *Tribolium castaneum* (Herbst). *Journal of Pesticide Science* 46, 22–228. DOI: 10.1584/jpestics.D20-059.

Jamkhande, P.G., Ghule, N.W., Bamer, A.H. and Kalaskar, M.G. (2019) Metal nanoparticles synthesis: an overview on methods of preparation, advantages and disadvantages, and applications. *Journal of Drug Delivery Science and Technology* 53, 101174. DOI: 10.1016/j.jddst.2019.101174.

Javed, Z., Dashora, K., Mishra, M., Fasake, V.D. and Srivastava, A. (2019) Effect of accumulation of nanoparticles in soil health—a concern on future. *Frontiers in Nanoscience and Nanotechnology* 5(2), 1000182. DOI: 10.15761/FNN.1000182.

Jhanzab, H.M., Qayyum, A., Bibi, Y., Sher, A., Hayat, M.T. *et al.* (2022) Chemo-blended Ag & Fe nanoparticles effect on growth, physiochemical and yield traits of wheat (*Triticum aestivum*). *Agronomy* 12(4), 757. DOI: 10.3390/agronomy12040757.

Joseph, M., Trinh, H.M., and Mitra, A.K. (2017) Peptide and protein-based therapeutic agents. In: Mitra, A.K., Cholkar, K., and Mandal, A. (eds.) *Emerging Nanotechnologies for Diagnostics, Drug Delivery and Medical Devices*. Elsevier, Amsterdam, pp. 145–167.

Jung, J. and Perrut, M. (2001) Particle design using supercritical fluids: literature and patent survey. *The Journal of Supercritical Fluids* 20(3), 179–219. DOI: 10.1016/S0896-8446(01)00064-X.

Kalia, A. and Kaur, H. (2019) Nano-biofertilizers: harnessing dual benefits of nano-nutrient and bio-fertilizers for enhanced nutrient use efficiency and sustainable productivity. In: Pudake, R.N., Chauhan, N. and Kole, C. (eds) *Nanoscience for Sustainable Agriculture*. Springer, pp. 51–75. DOI: 10.1007/978-3-319-97852-9.

Kalia, A., Sharma, S.P., and Kaur, H. (2019) Nanoscale fertilizers: harnessing boons for enhanced nutrient use efficiency and crop productivity. In: Abd-Elsalam, K. and Prasad, R. (eds.) *Nanobiotechnology Applications in Plant Protection*. Springer, Cham, pp. 191–208.

Kalia, A. and Sharma, S.P. (2019) Nanomaterials and vegetable crops: realizing the concept of sustainable production. In: Pudake, R.N., Chauhan, N., and Kole, C. (eds.) *Nanoscience for Sustainable Agriculture.* Springer, Cham, pp. 323–355.

Khan, S.T., Adil, S.F., Shaik, M.R., Alkhathlan, H.Z., Khan, M. *et al.* (2022) Engineered nanomaterials in soil: their impact on soil microbiome and plant health. *Plants* 11(1), 109. DOI: 10.3390/plants11010109.

Khan, Z., Shahwar, D., Ansari, M.K. and Chandel, R. (2019) Toxicity assessment of anatase (TiO$_2$) nanoparticles: a pilot study on stress response alterations and DNA damage studies in *Lens culinaris* medik. *Heliyon* 5, e02059. DOI: 10.1016/j.heliyon.2019.e02069.

Khorasani, S., Danaei, M. and Mozafari, M.R. (2018) Nanoliposome technology for the food and nutraceutical industries. *Trends in Food Science & Technology* 79, 106–115. DOI: 10.1016/j.tifs.2018.07.009.

Khot, L.R., Sankaran, S., Maja, J.M., Ehsani, R. and Schuster, E.W. (2012) Applications of nanomaterials in agricultural production and crop protection: a review. *Crop Protection* 35, 64–70. DOI: 10.1016/j.cropro.2012.01.007.

Klajnert, B. and Bryszewska, M. (2001) Dendrimers: properties and applications. *Acta Biochimica Polonica* 48(1), 199–208.

Kumar, P., Burman, U. and Kaul, R.K. (2018) Ecological risks of nanoparticles: effect on soil microorganisms. In: Tripathi, D.K., Ahamd, P., Sharma, S., Chauhan, D.K. and Dubey, N.K. (eds) *Nanomaterials in Plants, Algae, and Microorganisms: Concepts and Controversies*, Vol. 1. Academic Press, London, pp. 429–452.

Kumar, S., Nehra, M., Dilbaghi, N., Marrazza, G., Hassan, A.A. *et al.* (2019) Nano-based smart pesticide formulations: emerging opportunities for agriculture. *Journal of Controlled Release* 294, 131–153. DOI: 10.1016/j.jconrel.2018.12.012.

Kumari, S., Kumaraswamy, R.V., Choudhary, R.C., Sharma, S.S., Pal, A. *et al.* (2018) Thymol nanoemulsion exhibits potential antibacterial activity against bacterial pustule disease and growth promotory effect on soybean. *Scientific Reports* 8(1), 6650. DOI: 10.1038/s41598-018-24871-5.

Kumar, S.L.H. and Singh, V. (2012) Nanoemulsification – a novel targeted drug delivery tool. *Journal of Drug Delivery and Therapeutics* 2(4), 40–45. DOI: 10.22270/jddt.v2i4.185.

Liang, Z., Pan, X., Li, W., Kou, E., Kang, Y. *et al.* (2021) Dose-dependent effect of ZnO quantum dots for lettuce growth. *ACS Omega* 6(15), 10141–10149. DOI: 10.1021/acsomega.1c00205.

Lu, C., Li, Z., Chang, L., Dong, Z., Guo, P. *et al.* (2019) Efficient delivery of dsRNA and DNA in cultured silkworm cells for gene function analysis using PAMAM dendrimers system. *Insects* 11(1), 12. DOI: 10.3390/insects11010012.

Lyu, Z., Ding, L., Huang, A.Y.-T., Kao, C.-L. and Peng, L. (2019) Poly(amidoamine) dendrimers: covalent and supramolecular synthesis. *Materials Today Chemistry* 13, 34–48. DOI: 10.1016/j.mtchem.2019.04.004.

Mahawar, H. and Prasanna, R. (2018) Prospecting the interactions of nanoparticles with beneficial microorganisms for developing green technologies for agriculture. *Environmental Nanotechnology, Monitoring & Management* 10, 477–485. DOI: 10.1016/j.enmm.2018.09.004.

Mala, R., Celsia, A.V., Bharathi, S.V., Blessina, S.R. and Maheswari, U. (2017) Evaluation of nano structured slow release fertilizer on the soil fertility, yield and nutritional profile of *Vigna radiata. Recent Patents on Nanotechnology* 11, 50–62. DOI: 10.2174/1872210510666160727093554.

Mallakpour, S. and Behranvand, V. (2016) Polymeric nanoparticles: recent development in synthesis and application. *eXPRESS Polymer Letters* 10, 895–913. DOI: 10.3144/expresspolymlett.2016.84.

Maraval, V., Laurent, R., Donnadieu, B., Mauzac, M., Caminade, A.-M. *et al.* (2000) Rapid synthesis of phosphorus-containing dendrimers with controlled molecular architectures: first example of surface-block, layer-block, and segment-block dendrimers issued from the same Dendron. *Journal of the American Chemical Society* 122(11), 2499–2511. DOI: 10.1021/ja992099j.

Mariselvam, R., Ranjitsingh, A.J.A., Padmalatha, C. and Selvakumar, P.M. (2014) Green synthesis of copper quantum dots using *Rubia cardifolia* plant root extracts and its antibacterial properties. *Journal of Academia and Industrial Research* 3, 191–194.

Massoud, M.A., Adel, M.M., Zaghloul, O.A., Mohamed, M.I.E. and Abdel-Rheim, K.H. (2018) Eco-friendly nanoemulsion formulation of *Mentha piperita* against stored product pest *Sitophilus oryzae. Advances in Crop Science and Technology* 06(06), 404. DOI: 10.4172/2329-8863.1000404.

McKee, M.S. and Filser, J. (2016) Impacts of metal-based engineered nanomaterials on soil communities. *Environmental Science* 3(3), 506–533. DOI: 10.1039/C6EN00007J.

Mejias, J.H., Salazar, F., Pérez Amaro, L., Hube, S., Rodriguez, M. *et al.* (2021) Nanofertilizers: a cutting-edge approach to increase nitrogen use efficiency in grasslands. *Frontiers in Environmental Science* 9, 635114. DOI: 10.3389/fenvs.2021.635114.

Mendes, L.P., Pan, J. and Torchilin, V.P. (2017) Dendrimers as nanocarriers for nucleic acid and drug delivery in cancer therapy. *Molecules* 22(9), 1401. DOI: 10.3390/molecules22091401.

Mishra, S., Yang, X. and Singh, H.B. (2020) Evidence for positive response of soil bacterial community structure and functions to biosynthesized silver nanoparticles: an approach to conquer nanotoxicity? *Journal of Environmental Management* 253, 109584. DOI: 10.1016/j.jenvman.2019.109584.

Mittal, J., Osheen, S., Gupta, A. and Kumar, R. (2019) Carbon nanomaterials in agriculture. In: Pudake, R.N., Chauhan, N. and Kole, C. (eds) *Nanoscience for Sustainable Agriculture*. Springer, Cham, pp. 153–171. DOI: 10.1007/978-3-319-97852-9.

Monreal, C.M., DeRosa, M., Mallubhotla, S.C., Bindraban, P.S. and Dimkpa, C. (2016) Nanotechnologies for increasing the crop use efficiency of fertilizer-micronutrients. *Biology and Fertility of Soils* 52(3), 423–437. DOI: 10.1007/s00374-015-1073-5.

Montaño-Herrera, A., Santiago-Saenz, Y.O., López-Palestina, C.U., Cadenas-Pliego, G., Pinedo-Guerrero, Z.H. *et al.* (2022) Effects of edaphic fertilization and foliar application of Se and Zn nanoparticles on yield and bioactive compounds in *Malus domestica* L. *Horticulturae* 8(6), 542. DOI: 10.3390/horticulturae8060542.

Morales-Díaz, A.B., Ortega-Ortíz, H., Juárez-Maldonado, A., Cadenas-Pliego, G., González-Morales, S. *et al.* (2017) Application of nanoelements in plant nutrition and its impact in ecosystems. *Advances in Natural Sciences: Nanoscience and Nanotechnology* 8(1), 013001. DOI: 10.1088/2043-6254/8/1/013001.

Muñoz-Espí, R., and Álvarez-Bermúdez, O. (2018) Application of nanoemulsions in the synthesis of nanoparticles. In: Jafari, S.M. and McClements, D.J. (eds.) *Nanoemulsions: Formulation, Applications, and Characterization*. Academic Press, London, pp. 477–515.

Murali, M., Gowtham, H.G., Singh, S.B., Shilpa, N., Aiyaz, M. *et al.* (2022) Fate, bioaccumulation and toxicity of engineered nanomaterials in plants: current challenges and future prospects. *The Science of the Total Environment* 811, 152249. DOI: 10.1016/j.scitotenv.2021.152249.

Nanjwade, B.K., Bechra, H.M., Derkar, G.K., Manvi, F.V. and Nanjwade, V.K. (2009) Dendrimers: emerging polymers for drug-delivery systems. *European Journal of Pharmaceutical Sciences* 38(3), 185–196. DOI: 10.1016/j.ejps.2009.07.008.

Naseem, F., Zhi, Y., Farrukh, M.A., Hussain, F. and Yin, Z. (2020) Mesoporous $ZnAl_2Si_{10}O_{24}$ nanofertilizers enable high yield of *Oryza sativa* L. *Scientific Reports* 10(1), 10841. DOI: 10.1038/s41598-020-67611-4.

Ndlovu, N., Mayaya, T., Muitire, C., and Munyengwa, N. (2020) Nanotechnology applications in crop production and food systems. *International Journal of Plant Breeding and Crop Science Research* 7, 624–634.

Neme, K., Nafady, A., Uddin, S. and Tola, Y.B. (2021) Application of nanotechnology in agriculture, post-harvest loss reduction and food processing: food security implication and challenges. *Heliyon* 7(12), e08539. DOI: 10.1016/j.heliyon.2021.e08539.

Nile, S.H., Baskar, V., Selvaraj, D., Nile, A., Xiao, J. *et al.* (2020) Nanotechnologies in food science: applications, recent trends, and future perspectives. *Nano-Micro Letters* 12(1), 45. DOI: 10.1007/s40820-020-0383-9.

Nongbet, A., Mishra, A.K., Mohanta, Y.K., Mahanta, S., Ray, M.K. *et al.* (2022) Nanofertilizers: a smart and sustainable attribute to modern agriculture. *Plants* 11, 2587. DOI: 10.3390/plants11192587.

Ottoni, C.A., Lima Neto, M.C., Léo, P., Ortolan, B.D., Barbieri, E. *et al.* (2020) Environmental impact of biogenic silver nanoparticles in soil and aquatic organisms. *Chemosphere* 239, 124698. DOI: 10.1016/j.chemosphere.2019.124698.

Pal, S.L., Jana, U., Manna, P.K., Mohanta, G.P., and Manavalan, R. (2011) Nanoparticle: an overview of preparation and characterization. *Journal of Applied Pharmaceutical Science* 1, 228–234.

Palmqvist, N.G.M., Seisenbaeva, G.A., Svedlindh, P. and Kessler, V.G. (2017) Maghemite nanoparticles acts as nanozymes, improving growth and abiotic stress tolerance in *Brassica napus*. *Nanoscale Research Letters* 12, 631. DOI: 10.1186/s11671-017-2404-2.

Parat, A. and Felder-Flesch, D. (2016) General introduction on dendrimers, classical versus accelerated syntheses and characterizations. In: *Dendrimers in Nanomedicine*. Jenny Stanford Publishing, New York, pp. 1–22.

Patel, A., Cholkar, K. and Mitra, A.K. (2014) Recent developments in protein and peptide parenteral delivery approaches. *Therapeutic Delivery* 5(3), 337–365. DOI: 10.4155/tde.14.5.

Patel, D.K., Kim, H.-B., Dutta, S.D., Ganguly, K. and Lim, K.-T. (2020) Carbon nanotubes-based nanomaterials and their agricultural and biotechnological applications. *Materials* 13(7), 1679. DOI: 10.3390/ma13071679.

Qin, H., Zhou, X., Gu, D., Li, L. and Kan, C. (2019) Preparation and characterization of a novel waterborne lambda-cyhalothrin/alkyd nanoemulsion. *Journal of Agricultural and Food Chemistry* 67(38), 10587–10594. DOI: 10.1021/acs.jafc.9b03681.

Ramzan, S., Rasool, T., Bhat, R.A., Ahmad, P., Ashraf, I. et al. (2020) Agricultural soils a trigger to nitrous oxide: a persuasive greenhouse gas and its management. *Environmental Monitoring and Assessment* 192(7), 436. DOI: 10.1007/s10661-020-08410-2.

Rao, J.P. and Geckeler, K.E. (2011) Polymer nanoparticles: preparation techniques and size-control parameters. *Progress in Polymer Science* 36(7), 887–913. DOI: 10.1016/j.progpolymsci.2011.01.001.

Ribeiro, C. and Carmo, M. (2019) Why nonconventional materials are answers for sustainable agriculture. *MRS Energy & Sustainability* 6(1), 7. DOI: 10.1557/mre.2019.7.

Sadurní, N., Solans, C., Azemar, N. and García-Celma, M.J. (2005) Studies on the formation of O/W nano-emulsions, by low-energy emulsification methods, suitable for pharmaceutical applications. *European Journal of Pharmaceutical Sciences* 26(5), 438–445. DOI: 10.1016/j.ejps.2005.08.001.

Sahoo, S.K., Panigrahi, G.K., Sahoo, A., Pradhan, A.K. and Dalbehera, A. (2021) Bio-hydrothermal synthesis of ZnO–ZnFe$_2$O$_4$ nanoparticles using *Psidium guajava* leaf extract: role in waste water remediation and plant immunity. *Journal of Cleaner Production* 318, 128522. DOI: 10.1016/j.jclepro.2021.128522.

Salama, D.M., Abd El-Aziz, M.E., Shaaban, E.A., Osman, S.A. and Abd El-Wahed, M.S. (2022) The impact of nanofertilizer on agro-morphological criteria, yield, and genomic stability of common bean (*Phaseolus vulgaris* L.). *Scientific Reports* 12(1), 18552. DOI: 10.1038/s41598-022-21834-9.

Santaella, C. and Plancot, B. (2020) Interactions of nanoenabled agrochemicals with soil microbiome. In: Fraceto, L.F., S.S. de Castro, V.L., Grillo, R., Avila, D. and Lima, R. (eds) *Nanopesticides*. Springer, Cham, pp. 137–163. DOI: 10.1007/978-3-030-44873-8.

Santos, A., Veiga, F. and Figueiras, A. (2019) Dendrimers as pharmaceutical excipients: synthesis, properties, toxicity and biomedical applications. *Materials* 13(1), 65. DOI: 10.3390/ma13010065.

Šebestík, J., Reiniš, M. and Ježek, J. (2012) Synthesis of dendrimers: convergent and divergent approaches. In: *Biomedical Applications of Peptide-, Glyco- and Glycopeptide Dendrimers, and Analogous Dendrimeric Structures*. Springer, Vienna, pp. 55–81.

Shah, P., Bhalodia, D. and Shelat, P. (2010) Nanoemulsion: a pharmaceutical review. *Systematic Reviews in Pharmacy* 1, 24–32. DOI: 10.4103/0975-8453.59509.

Shao, J., Deng, K., Chen, L., Guo, C., Zhao, C. et al. (2021) Aqueous synthesis of Nb-modified SnO$_2$ quantum dots for efficient photocatalytic degradation of polyethylene for *in situ* agricultural waste treatment. *Green Processing and Synthesis* 10(1), 499–506. DOI: 10.1515/gps-2021-0046.

Sherje, A.P., Jadhav, M., Dravyakar, B.R. and Kadam, D. (2018) Dendrimers: a versatile nanocarrier for drug delivery and targeting. *International Journal of Pharmaceutics* 548, 707–720. DOI: 10.1016/j.ijpharm.2018.07.030.

Shoukat, R. and Khan, M.I. (2021) Carbon nanotubes: a review on properties, synthesis methods and applications in micro and nanotechnology. *Microsystem Technologies* 27(12), 4183–4192. DOI: 10.1007/s00542-021-05211-6.

Shukla, P., Chaurasia, P., Younis, K., Qadri, O.S., Faridi, S.A. et al. (2019) Nanotechnology in sustainable agriculture: studies from seed priming to post-harvest management. *Nanotechnology for Environmental Engineering* 4(1), 11. DOI: 10.1007/s41204-019-0058-2.

Singh, D. and Gurjar, B.R. (2022) Nanotechnology for agricultural applications: facts, issues, knowledge gaps, and challenges in environmental risk assessment. *Journal of Environmental Management* 322, 116033. DOI: 10.1016/j.jenvman.2022.116033.

Sivarethinamohan, R. and Sujatha, S. (2021) Unlocking the potentials of using nanotechnology to stabilize agriculture and food production. In: *AIP Conference Preceedings*, AIP Publishing LLC, NewYork, p. 20022 (Vol. 2327). DOI: 10.1063/5.0039418.

Solans, C., Izquierdo, P., Nolla, J., Azemar, N. and Garcia-Celma, M.J. (2005) Nano-emulsions. *Current Opinion in Colloid & Interface Science* 10(3–4), 102–110. DOI: 10.1016/j.cocis.2005.06.004.

Srivastava, P., Singh, R., Bhadouria, R., Pal, D.B., Singh, P. et al. (2021) Engineered nanoparticles in smart agricultural revolution: an enticing domain to move carefully. In: Singh, P., Singh, R., Verma, P., Bhadouria, R. and Kumar, A., et al (eds) *Plant-Microbes-Engineered Nano-Particles (PM-ENPs) Nexus in Agro-Ecosystems*. Springer, Cham, pp. 3–18. DOI: 10.1007/978-3-030-66956-0.

Șuțan, N.A., Fierăscu, I., Șuțan, C., Soare, L.C., Neblea, A.M. *et al.* (2021) *In vitro* mitodepressive activity of phytofabricated silver oxide nanoparticles (Ag$_2$O-NPs) by leaves extract of *Helleborus odorus* Waldst. & Kit. ex Willd. *Materials Letters* 286, 129194. DOI: 10.1016/j.matlet.2020.129194.

Șuțan, N.A., Manolescu, D.S., Fierascu, I., Neblea, A.M., Sutan, C. *et al.* (2018) Phytosynthesis of gold and silver nanoparticles enhance *in vitro* antioxidant and mitostimulatory activity of *Aconitum toxicum* Reichenb. rhizomes alcoholic extracts. *Materials Science and Engineering C* 93, 746–758. DOI: 10.1016/j.msec.2018.08.042.

Șuțan, N.A., Vilcoci, D.S., Fierascu, I., Neblea, A.M., Șuțan, C. *et al.* (2019) Influence of the phytosynthesis of noble metal nanoparticles on the cytotoxic and genotoxic effects of *Aconitum toxicum* Reichenb. leaves alcoholic extract. *Journal of Cluster Science* 30(3), 647–660. DOI: 10.1007/s10876-019-01524-9.

Tadros, T., Izquierdo, P., Esquena, J. and Solans, C. (2004) Formation and stability of nano-emulsions. *Advances in Colloid and Interface Science* 108–109, 303–318. DOI: 10.1016/j.cis.2003.10.023.

Tian, H., Kah, M. and Kariman, K. (2019) Are nanoparticles a threat to mycorrhizal and rhizobial symbioses? A critical review. *Frontiers in Microbiology* 10, 1660. DOI: 10.3389/fmicb.2019.01660.

Tolaymat, T., Genaidy, A., Abdelraheem, W., Dionysiou, D. and Andersen, C. (2017) The effects of metallic engineered nanoparticles upon plant systems: an analytic examination of scientific evidence. *Science of the Total Environment* 579, 93–106. Available at: https://doi.org/10.1016/j.scitotenv.2016.10.229

Tomalia, D.A. and Fréchet, J.M.J. (2002) Discovery of dendrimers and dendritic polymers: a brief historical perspective. *Journal of Polymer Science Part A: Polymer Chemistry* 40(16), 2719–2728. DOI: 10.1002/pola.10301.

Tombuloglu, H., Slimani, Y., Akhtar, S., Alsaeed, M., Tombuloglu, G. *et al.* (2022) The size of iron oxide nanoparticles determines their translocation and effects on iron and mineral nutrition of pumpkin (*Cucurbita maxima* L.). *Journal of Magnetism and Magnetic Materials* 564, 170058. DOI: 10.1016/j.jmmm.2022.170058.

Tripathi, A. and Prakash, S. (2022) Nanobiotechnology: emerging trends, prospects, and challenges. In: Ghosh, S., Thongmee, S. and Kumar, A. (eds) *Agricultural Nanobiotechnology*. Woodhead Publishing, Sawston, UK, pp. 1–21.

Ung, T.D.T., Tran, T.K.C., Pham, T.N., Nguyen, D.N., Dinh, D.K. *et al.* (2012) CdTe and CdSe quantum dots: synthesis, characterizations and applications in agriculture. *Advances in Natural Sciences* 3(4), 043001. DOI: 10.1088/2043-6262/3/4/043001.

Usman, M., Farooq, M., Wakeel, A., Nawaz, A., Cheema, S.A. *et al.* (2020) Nanotechnology in agriculture: current status, challenges and future opportunities. *Science of The Total Environment* 721, 137778. DOI: 10.1016/j.scitotenv.2020.137778.

Valizadeh, A., Mikaeili, H., Samiei, M., Farkhani, S.M., Zarghami, N. *et al.* (2012) Quantum dots: synthesis, bioapplications, and toxicity. *Nanoscale Research Letters* 7(1), 480. DOI: 10.1186/1556-276X-7-480.

Walter, M.V. and Malkoch, M. (2012) Simplifying the synthesis of dendrimers: accelerated approaches. *Chemical Society Reviews* 41(13), 4593–4609. DOI: 10.1039/c2cs35062a.

Wang, P., Lombi, E., Zhao, F.J. and Kopittke, P.M. (2016) Nanotechnology: a new opportunity in plant sciences. *Trends in Plant Science* 21(8), 699–712. DOI: 10.1016/j.tplants.2016.04.005.

Wang, D., Saleh, N.B., Byro, A., Zepp, R., Sahle-Demessie, E. *et al.* (2022) Nano-enabled pesticides for sustainable agriculture and global food security. *Nature Nanotechnology* 17(4), 347–360. DOI: 10.1038/s41565-022-01082-8.

Weeks, J.J. and Hettiarachchi, G.M. (2019) A review of the latest in phosphorus fertilizer technology: possibilities and pragmatism. *Journal of Environmental Quality* 48(5), 1300–1313. DOI: 10.2134/jeq2019.02.0067.

Xu, W., Ling, P. and Zhang, T. (2013) Polymeric micelles, a promising drug delivery system to enhance bioavailability of poorly water-soluble drugs. *Journal of Drug Delivery* 2013, 340315. DOI: 10.1155/2013/340315.

Yadav, J., Jasrotia, P., Kashyap, P.L., Bhardwaj, A.K., Kumar, S. *et al.* (2022) Nanopesticides: current status and scope for their application in agriculture. *Plant Protection Science* 58(1), 1–17. DOI: 10.17221/102/2020-PPS.

Zhang, H. (2017) Thin-film hydration followed by extrusion method for liposome preparation. *Methods in Molecular Biology* 1522, 17–22. DOI: 10.1007/978-1-4939-6591-5_2.

Zhang, Y. and Goss, G.G. (2022) Nanotechnology in agriculture: comparison of the toxicity between conventional and nano-based agrochemicals on non-target aquatic species. *Journal of Hazardous Materials* 439, 129559. DOI: 10.1016/j.jhazmat.2022.129559.

Zhao, X., Cui, H., Wang, Y., Sun, C., Cui, B. *et al.* (2017) Development strategies and prospects of nano-based smart pesticide formulation. *Journal of Agricultural and Food Chemistry* 66(26), 6504–6512. DOI: 10.1021/acs.jafc.7b02004.

Zhu, H., Shen, Y., Cui, J., Wang, A., Li, N. *et al.* (2020) Avermectin loaded carboxymethyl cellulose nano-particles with stimuli-responsive and controlled release properties. *Industrial Crops and Products* 152, 112497. DOI: 10.1016/j.indcrop.2020.112497.

Zielińska, A., Carreiró, F., Oliveira, A.M., Neves, A., Pires, B. *et al.* (2020) Polymeric nanoparticles: pro-duction, characterization, toxicology and ecotoxicology. *Molecules* 25(16), 3731. DOI: 10.3390/molecules25163731.

Zulfiqar, F., Navarro, M., Ashraf, M., Akram, N.A. and Munné-Bosch, S. (2019) Nanofertilizer use for sustainable agriculture: advantages and limitations. *Plant Science* 289, 110270. DOI: 10.1016/j.plantsci.2019.110270.

3 Nanoemulsion Formulations for Food Processing and Enhancing the Nutritional Quality and Shelf Life of Food

Daniela Nicuță[1], Ana-Maria Roșu[2]* and Roxana-Elena Voicu[1]

[1]*Department of Biology, Ecology and Environmental Protection, Faculty of Sciences, "Vasile Alecsandri" University of Bacău, Bacău, Romania;* [2]*Department of Chemical and Food Engineering, Faculty of Engineering, "Vasile Alecsandri" University of Bacău, Bacău, Romania*

Abstract

Nano-emulsions are used in different industries, especially in the food industry to enhance the nutritional quality and shelf life of foods. Nanoparticles are characterized as ultrafine particles with a larger surface area, and less sensitivity to physical and chemical changes. These new nanoproducts must be characterized and optimized for use in the food industry. Nano-emulsions are composed of two phases: an oil phase and an aqueous phase. They are treated as one of the most encouraging systems to permit solubility, bioavailability, and functionality of bioactive compounds. For this reason, they are used in essential oils extracted from plants. In food processing, nano-emulsions are used as coating materials, for encapsulation and formulation of new food products, and to improve food quality and taste. They also help to preserve fresh as well as processed foods.

Keywords: nano-emulsions, nutritional quality, encapsulation, films, safety

3.1 Nano-Emulsions for Food Technology: Basic Knowledge

Nanotechnology is a revolutionary method that can be used in agriculture and the food industry, because it is able to intervene positively in the food sector by affecting the sensory properties of food (e.g. taste, colour, texture), increasing the storage period, and improving solubility in water of encapsulated bioactive components, and their thermal stability and bioavailability (McClements *et al.*, 2007, 2009; Huang *et al.*, 2010; Silva *et al.*, 2012). According to recent research, nano-emulsions are potentially the most effective systems to protect the functional compound. Nanoemulsification is an ideal approach to prepare bioactive compounds in nanoform (Jin *et al.*, 2016; Karthik *et al.*, 2017). Emulsions are relatively stable mixtures containing immiscible liquids, usually water and oil. Depending on the dispersed or continuous phase of oil or water, this mixture can be water in oil or oil in water. Emulsions can be divided into macroemulsions, nano-emulsions, and microemulsions according to their thermodynamic stability and physical characteristics. Macroemulsions (conventional emulsions) are characterized by small particles with average

---

**Corresponding author: ana.rosu@ub.ro

© CAB International 2023. *Nanoformulations for Sustainable Agriculture and Environmental Risk Mitigation* (eds Z. Khan and N.A. Șuțan)
DOI: 10.1079/9781800623095.0003

sizes between 100 nm and 100 μm, thermodynamic and kinetic instability, high dispersion (>40%), and an opaque or cloudy appearance. The last aspect is determined by the fact that the particle sizes are similar to the wavelength of light ($r \approx \lambda$) (McClements and Rao, 2011; Gupta et al., 2016). Conventional emulsions have the tendency to break down over time, because the free energy of the separated phases of emulsion, oil and water phases is less than the energy of the native emulsion itself (McClements and Rao, 2011).

Nano-emulsions (mini-emulsions) are characterized by relatively small droplet sizes with average radii in the range of 10–100 nm, thermodynamic instability, usually low dispersion (<10–20%), and a transparent, translucent, or slightly creamy appearance ($r \ll \lambda$) (McClements and Rao, 2011; Solans and Solé, 2012; Gupta et al., 2016; Jin et al., 2016). Microemulsions are thermodynamically stable systems under certain environmental conditions (temperature, composition) that contain spherical or lamellar particles with sizes in the range of 10–100 nm, with a dispersion of less than 10% (Gupta A. et al., 2016). Because the size of the particles is lower than the wavelength of light ($r \ll \lambda$), they are transparent.

3.1.1 The composition of nano-emulsions for the food industry

Nano-emulsions consist of an aqueous stage, an oily phase, and stabilizers, which are intended to modulate the formation of nano-emulsions. In the food sector, the most commonly used are oil-in-water nano-emulsions, whose particles are of the core-shell type. The body is made of a lipophilic material and the outer layer is composed of a surface-active amphiphilic substance (Gupta A. et al., 2016; Jin et al., 2016; Salem and Ezzat, 2018).

The oil phase

In the formulation of nano-emulsions used in the food industry, the oil phase can be represented by various nonpolar molecules (e.g. monoacylglycerols, diacylglycerols, triacylglycols, free fatty acids, different types of oils (essential, aromatic),

fat substitutes, waxes, liposoluble or fat-soluble vitamins), and various lipophilic nutraceuticals (e.g. curcumin, phytosterols, carotenoids, coenzyme Q, phytostanols) (McClements and Rao, 2011). Due to their functional and nutritional properties, as well as their low cost, the oils used the most in the food industry for the formulation of nano-emulsions are triacylglycol oils derived from sunflower, soybean, maize, olive, flaxseed, seaweed, and fish (McClements and Rao, 2011; Salem and Ezzat, 2018). It should be noted that these oils have long-chain, medium-chain, and short-chain triacylglycerols. Due to their relatively low polarity, interfacial tension, and high viscosity, low- and medium-chain triacylglycerol oils are difficult to use for the formulation of nano-emulsions using phase-inversion temperature or high-pressure homogenization processes.

Although essential oils and aromatic oils are characterized by interfacial tension, high polarity, and decreased viscosity—properties that support the development of small droplets using phase-inversion temperature or high-pressure homogenization methods—the nano-emulsions formed in this way have a very stable low value of interfacial tension, either because of the low surface oil tension or due to Ostwald ripening (Witthayapanyanon et al., 2006).

The aqueous phase

In the nano-emulsion formulation, the aqueous phase contains water, and the water:oil ratio is an important factor in its formation and stability. The aqueous phase may have various polar molecules, acids, carbohydrates, alcoholic co-solvents, minerals, proteins, and bases (McClements and Rao, 2011), which affect the viscosity, pH, and polarity of the water, and thus the physico-chemical properties and behavioral phase of the formed nanosystem (Jadhav et al., 2020).

Stabilizers

As described earlier, nano-emulsions are thermodynamically unstable systems—both the oil and the aqueous phase can break down because of Ostwald ripening, coalescence, flocculation, and gravitational separation (Kabalnov, 2001; Aswathanarayan and Vittal, 2019). However, the stability of the nanosystem can be improved by the addition of a stabilizing agent. Stabilizers can disperse the particles and form a single-layer, multilayer, or solid particles nano-emulsion. The most commonly used stabilizers are emulsifiers, weighting agents, baking retarders, and texture improvers. The correct choice of emulsifier type in the preparation of a nano-emulsion is very important to increase the strength of the new nanosystem to stressful environments (ionic strength, pH, heating, long-term or cooling storage). A hydrophilic–lipophilic equilibrium value above 10 is considered one criterion for choice of emulsifier in the oil-in-water emulsion formulations. Emulsifiers lower the interfacial tension, prevent collisions and droplet coalescence, and thus increase the kinetic stability of the nano-emulsions (Mason *et al.*, 2006; Aswathanarayan and Vittal, 2019). Emulsifiers are surface-active molecules capable of absorbing droplet surfaces and protecting them from aggregation (McClements and Rao, 2011).

Commonly used emulsifiers are surfactants that contain hydrophilic as well as hydrophobic moieties. Due to their amphiphilic structure, surfactants adsorb to the oil–water interface, which results in the two immiscible liquids having the interfacial stretching reduced (Waglewska and Bazylińska, 2021). Small-molecule surfactants are extremely mobile at the interface and are suitable for nano-emulsion preparation using both low- and high-energy methods. This category principally includes glycolipids, lecithins, monoglycerides, fatty acids, and fatty alcohols. High-mass surfactants can include proteins such as milk proteins (whey, lactoferrin, α-lactalbumin, β-lactoglobulin, bovine serum albumin, casein), proteins from fish and meat (myosin, gelatin), egg proteins (ovalbumin), vegetable proteins (proteins from soy, pea, rice), and polysaccharides (e.g. arabic gum, modified starch) (Singhal *et al.*, 2016; Mustafa *et al.*, 2018; Ozogul *et al.*, 2022). The protein molecule can penetrate the lipid stage to different degrees. The specific connection is mostly electrostatic (surfactant head groups bind to the oppositely charged groups of the protein). Saturation binding of anionic surfactants is not influenced by pH, and it appears to be controlled by cooperative hydrophobic interactions.

Two of the most important groups of amphiphilic materials are polysaccharides and low-molecular-weight surfactants, which can be used successfully to stabilize emulsions. Their ability to control rheology and stability among a large range of compositions is among the most important aspects of polymer–surfactant systems (Kralova and Sjöblom, 2009). Depending on their electrical properties, surfactants can be ionic (negatively or positively charged), nonionic, or zwitterionic. Negatively charged ionic surfactants can be used in low-energy and high-energy approaches (e.g. diacetyl tartaric acid esters of mono- and diglycerides (DATEM), citric acid esters of mono- and diglycerides (CITREM)). The most commonly used positively charged emulsifier is sodium lauryl sulfate, but it can cause irritation. Nonionic surfactants have low toxicity and do not cause irritation (e.g. sucrose monopalmitate, monooleate, sorbitan, polyoxyethylene ether (Brij-97) polysorbates (e.g. Tween-20, Tween-80, Lens solution)). Zwitterionic surfactants have two or more ionizable groups, oppositely charged on the same molecule. Thus, they can have different charges such as negative, neutral or positive depending on the pH of the solution (e.g. lecithin). The aggregation of droplets is prevented by ionic surfactants through electrostatic repulsion, while nonionic surfactants reduce aggregation by steric hindrance, thermal fluctuation interactions, and hydration. In the nutrition and food industry, different types of emulsifiers are polysaccharides, low-molecular-weight phospholipids, surfactants, and proteins (McClements, 2004; McClements and Rao, 2011; Silva *et al.*, 2015; Salem and Ezzat, 2018; Waglewska and Bazylińska, 2021).

In the preparation of nano-emulsions, weighting agents are used to prevent sedimentation or gravity separation. The following weighting agents are commonly used in the nutritional and food industry: ester gum (e.g.), sucrose acetate isobutyrate (SAIB), damar gum (DG), and brominated vegetable oil (BVO) (McClements, 2004). In a recent study, sedimentation or creaming took place when the density of the droplets was higher or lower than the density of the aqueous phase (Chanamai and McClements, 2000). The weight concentrations needed to equalize the densities of the oil and aqueous phases were 55% by weight (wt%) EG, 55 wt% DG, 25 wt% BVO, and 45 wt% SAIB. The efficiency of droplet diminution during homogenization also depends on the type of balancing agent due to the difference in viscosity in the oil phase (BVO >SAIB > DG/EG). The effect of sucrose (0–13 wt%) on the stability of oil-in-water emulsions with 1 wt% soybean has also been studied. The viscosity of the aqueous phase (cream retardation) and density contrast between the droplets (cream acceleration) was increased by sucrose. The results showed that the two effects cancelled each other out and the cream stability was not affected by sucrose concentration.

Nano-emulsions have limited practical applications due to their degradation through the process known as Ostwald ripening. The role of surfactants is key to the mass transfer between droplets in the dispersed phase. It is known that if the solubility of the dispersed phase substance in dispersion medium is not extremely low, then the major contribution at Ostwald ripening is molecular diffusion. If the solubility of the substances is very low in the dispersed phase, then micelles and nanodroplets transport the substances through the dispersion medium. Due to the instability of the interfaces, nanodroplets and slightly larger droplets are formed on the surface of the original emulsion droplets (Marangoni effect) (Koroleva and Yurtov, 2021). A retarder of maturation is a hydrophobic component that is soluble in the oil stage but insoluble in water. Commonly used ripening retarders include mineral oils, medium-chain triglycerides, and EGs (Sonneville-Aubrun et al., 2004; Aswathanarayan and Vittal, 2019). The type and concentration of maturation inhibitors have an important effect on the properties of the newly formed nanostructure. Studies on physical properties and antimicrobial activity of thyme oil nano-emulsions showed that increasing concentrations of maturation inhibitors in the lipid stage decreased the antibiotic activity of the nano-emulsion (Chang et al., 2012). For instance, for nano-emulsions with 60 wt% maize oil in the lipid phase, the minimum inhibitory concentration (MIC) of thyme oil needed to inhibit the growth of *Zygosaccharomyces bailii* was 375 µg ml^{-1}, while for nano-emulsions with 90 wt% maize oil in the lipid phase, even 6000 µg ml^{-1} of thyme oil could

not inhibit the growth of this yeast. It was also observed that medium-chain triacylglycerols reduced the antimicrobial activity of thyme oil more than maize oil at the same concentration in the lipid phase. For example, when the proportion of maturation inhibitor in the lipid phase was 70% by weight, the MICs of thyme oil for nano-emulsions containing maize oil and medium-chain triacylglycerols were 750 and 3000 µg ml^{-1}, respectively.

Texture modifiers are materials that are added to the continuous phase of nano-emulsions to thicken and gel them in order to limit the movement of oil droplets, increasing the creaminess of the aqueous phase. Texture modifiers used in the food industry are proteins (e.g. milk, eggs, vegetable proteins), and polysaccharides (e.g. alginate, starch, xanthan gum, pectin, carrageenan, guar gum) (McClements and Rao, 2011; Ozogul et al., 2022).

3.2   Production of Nano-Emulsions to Enhance the Nutritional Quality and Shelf Life of Food

3.2.1   High-energy methods

The formation of nano-emulsions using high energy is achieved through high-shear agitation operation, high-pressure homogenizers and sometimes ultrasonic generators. An apparatus that supplies the available energy and has a homogeneous flow rate can produce the smallest emulsion sizes (Solans et al., 2005). Nano-emulsions obtained with high energy are based on a different composition compared with classic emulsions—a surfactant and a functional component, as well as an increased amount of energy. The apparatus is divided into three major zones used for the classification of nano-emulsions.

A high-pressure homogenizer is used in the production of fine emulsions. Using this method, the mixture is highly pressurized and pumped through a resistive valve. An extremely high-shear stress results in the formation of fine emulsion drops (Gadhave, 2014). The nano-emulsions produced using this device do not exhibit good dispersion compared with high-energy devices, in terms of the size of the drops (Anton and Vandamme, 2009). High-energy methods involve the use of a mechanical device, such as a homogenizer with high-pressure valves, a microfluidizer and an ultrasonicator (Aswathanarayan and Vittal, 2019). All of these devices are used to create a disruptive force acting on the dispersed phase, turning it into small droplets of nano-emulsion (Maali and Hamed Mosavian, 2013). The apparatus used in the high-pressure valve homogenization method (HPVH) is composed of a pressure valve, a volumetric pump and several homogenization chambers.

Application of HPVH cannot be used for glutinous lipids (Windhab *et al.*, 2005). Thus, the process of HPVH involves the efficient decomposition of the drops and increases their stability (Stang *et al.*, 2001). This technique is suitable for the production of nano-emulsions in the nutritional industry, due to the ease of application, the reproducibility and the high flow involved in the use of HPVH (Tabilo-Munizaga *et al.*, 2019). The use of high pressure provides better-quality nano-emulsion and is much more effective (Liu *et al.*, 2019). Some examples of the use of HPVH are for the preparation of bergamot, mandarin, carvacrol and lemon nano-emulsions (Liu *et al.*, 2019). Thus, carvacrol nano-emulsion demonstrates an inhibitory activity against microorganisms such as *Escherichia coli* O157:H7 and *Salmonella Typhimurium*. In a working protocol, carvacrol nano-emulsion was encapuslated into a chemically modified substrate polysaccharide, chitosan, which was used as a stabilizer to form a bioactive coating. The carvacrol nano-emulsion gave protection to green beans against *E. coli* and *S. Typhimurium* during storage at 4°C for up to 13 days (Severino *et al.*, 2015).

Carotenoid-rich nano-emulsion solutions can also be obtained with the help of HPVH (400–800 bar). This nano-emulsion is encapsulated in an emulsifier of sucrose monostearate. This procedure demonstrated a long stability against anti-oxidant activity of plants during storage at 4°C (McClements *et al.*, 2009).

Viscosity is a characteristic that is affected by high-pressure homogenization (Tabilo-Munizaga *et al.*, 2019). High-pressure homogenizers were first used in the milk business to eliminate fat globules. Now, these homogenizers are widely used in the food processing industry. Thus, nano-emulsions are intended for the supplement nutrient industry and canning industry. However, utilizing this procedure can potentially lead to the degradation of nutrients that are temperature sensitive. Researchers have shown that this procedure can only be reclaimed within the short residence time (3–40 Ms) of the emulsion (Jafari *et al.*, 2008).

3.2.2 Microfluidization method

Another important process is microfluidization. The basic principle of microfluidization is comparable to HPVH, except that a microchannel is used with a width of 50–300 μm. This size is necessary to form droplets (Donsì, 2018). Speed is one of the most specific parameters in this method.

The coarse emulsion is pushed under high pressure (up to 270 MPa) into a microchannel at a speed of 400 m s^{-1}. To augment the viscosity of continuous phases, it is necessary to add a fast adsorption emulsifier, as it has the ability to reduce the proportion of recoalescence (Jafari *et al.*, 2008; Donsì *et al.*, 2009; Severino *et al.*, 2015). Some examples of microfluidization are nano-emulsions obtained from thyme and sage oil dispersed in sodium alginate solution.

Nano-emulsions obtained with sage essential oil show biofilm properties. This gives the products better resistance to water vapor (Severino *et al.*, 2015). Another example is microfilms made from thyme essential oil. These have an antibacterial impact on pathogens such as *E. coli* and have a use in the food sector. Thus, microfluidizing devices can be used to prepare nano-emulsions with active ingredients. These nano-emulsions lead to the formation of biofilms that exhibit different functional and physical properties (Anton and Vandamme, 2011). The nano-emulsion of ginger essential oil is another excellent illustration that can be created using microfluidization. The nano-emulsion from ginger essential oil is activated when it is enclosed in montmorillonite. The finished product has an impact on the food's antioxidant activity (Gadhave, 2014).

3.2.3 Ultrasonication method

A new and certified method in the study is ultrasonic emulsification. If only a small amount of the product is used, this technique can successfully reduce the size of the drops. A recent study showed that the polymerizable nano-emulsions obtained show the effectiveness on the dispersion process. The procedure is dependent on the ultrasound duration at various metrics. These parameters depend on whether the monomer is hydrophobic, which prolongs the required sonication time (Solans et al., 2005). There are certain factors that impact the formation of nano-emulsion. These include: the viscosity ratio of the dispersed and continuous phase, the concentration of a parameter such as emulsifier, and the amplitude of the applied waves (Shariffa et al., 2017). Most of the nano-emulsions produced by ultrasonication have wide-ranging distributions. The ultrasound method involves research and experiments at the laboratory level. To date, all attempts have led to small quantities of the product, and ultrasonic devices are not yet marketed at an industrial level (Gadhave, 2014). As a result of variations in parameters such as the ultrasonic time and the concentration of surfactant, nano-emulsions with a droplet diameter of 29.3 nm have been produced. These nano-emulsions are formatted from the oily phase represented by basil oil and nonionic emulsifier (Tween 80 and water). When exposed to ultrasound for 15 min, they exhibited remarkable stability, (Tabilo-Munizaga et al., 2019).

The bioactivity of the essential oil is improved by the nano-emulsion. Increased antibacterial activity can be observed by reducing the droplet size of the nano-emulsion. Thus, small encapsulation nano-emulsions can be more simply introduced into basil seed biofilms. A high concentration of nano-emulsion can affect the microstructure of the biofilm by improving its mechanical properties (Tadros et al., 2004). Studies have shown that the productivity of HPVH can be compared to that of emulsification to develop capsaicin nano-emulsions (*Capsicum oleoresin*). These nano-emulsions obtained by HPVH are 79% effective and induce significant antimicrobial activity. In contrast, nano-emulsions created using ultrasonic technology have characteristics that are both physical and chemical, with a droplet size of 65 nm.

3.2.4 Phase-inversion temperature method

Another way of obtaining nano-emulsions is by means of the phase-inversion temperature method. Most industries employ this technique. This method is based on modifications in the solubility of nonionic polyoxyethylene surfactants with temperature (McClements et al., 2009).

At a low temperature, the substrate monolayer leads to the formation of a micellar solution phase. This solution can coexist as an excess oily phase. At an intermediate temperature, a bicontinuous microemulsion is obtained, which may contain comparable amounts of water and oil phases. An example of this is the nano-emulsion of cinnamon oil, which utilizes the phase-inversion temperature method, when a nonionic surfactant and water are heated above the system phase-reversal temperature. Through continuous cooling and stirring, small oil droplets with an average diameter of 101 nm are formed (Ruiz-Montañez et al., 2017).

3.2.5 Phase-inversion composition method

The phase-inversion composition method is used for the preparation of food-grade nano-emulsions fortified with vitamin E. They have a particle diameter of 40 nm. This technique is one of the most effective in manufacturing high-surfactant nano-emulsions compared with the microfluidization method. This technique is advantageous for the formulation of nano-emulsions with surfactants (Tesch et al., 2003).

Phase-inversion composition is a technique that optimizes the change in composition at a certain temperature (usually ambient temperature). The technique involves progressive dilution with the water or oily phase. This phase change is determined by a constant known as Gibbs free energy of emulsions. This results in the spontaneous reversal of the surfactant between positive and negative. Phase-inversion composition is a classical method, used for the production of water-in-water emulsion by dilution of the water-in-water emulsion (Donsì *et al.*, 2009).

3.2.6 Spontaneous emulsification method

Studies have shown that spontaneous emulsification occurs when water is added to a solution. In the absence of surfactant, there is a small concentration of oil in a water-miscible solvent. Thus, oil drops are produced according to the ratio of excess oil to the water-soluble solvent. Studies have shown that this method can be used as an alternative to other methods such as ultrasonic and high-shear methods (Gadhave, 2014). However, this method has some limitations, such as the low concentration of diffusible oil and the solvent that can be dissolved in water in all proportions. This method makes it difficult to remove the solvent (Aswathanarayan and Vittal, 2019). The oil:emulsion ratio or the surfactant:emulsion ratio in the nano-emulsion system, as well as the preparation conditions such as the stirring rate, can affect the characteristics of the emulsion (Sugumar *et al.*, 2015).

3.2.7 Novel nano-emulsion preparation techniques

Standard nanoemulsification is the breaking down of larger droplets or reversal of solvents. Novel emulsification techniques are increasingly being developed to enlarge the variety of ways of obtaining new materials. In addition to these new materials, new operating conditions are also involved in order to reduce production costs (Tabilo-Munizaga *et al.*, 2019). Pickering nano-emulsions are obtained by condensation of vapors. The high-energy method prevents particle adsorption on droplets, compared with the low-energy method, which cannot be used to obtain Pickering nano-emulsions. The new approach of obtaining Pickering nano-emulsions by condensation of vapors through a single-step process has multiple benefits over conventional techniques. One of the main advantages is the use of a minimum concentration of nanoparticles (Sugumar *et al.*, 2015).

3.3 Applications, Benefits and Limitations of Nano-Emulsion Formulas Used for Maintaining Food Quality

After discovering methods to obtain nano-emulsions, their beneficial uses have since been revealed, and have sparked great interest in essential studies and applications in various fields, such as healthcare, cosmetics, the food industry, and agrochemical and pharmaceutical products, as well as biotechnology (Borthakur *et al.*, 2016). There are many interesting possibilities for new scientific areas and applications in the field of nano-emulsions (Mason *et al.*, 2006). A study conducted by Maali and Hamed Mosavian (2013) regarding the use of nano-emulsions from 2000 to 2010 established that the production of nano-emulsions was of particular interest to the cosmetics and pharmaceutical industry, while, over the same time frame, their utilization in the food industry was meager. This was due to the problems of stability of nano-emulsions in food (Gutiérrez *et al.*, 2008).

In recent years, interest in innovative nanoscience-based technology has increased in terms of improving the health, safety, and quality of food. Because of this, creating and utilizing nano-emulsions has become a top priority in the food sector and can be formulated with existing food and technology (Das *et al.*, 2020). In particular, oil-in-water nano-emulsions, which hold small drops of oil (<200 nm) dispersed in water, can be used as distribution systems for various hydrophobic substances in food, including nutraceuticals, antioxidants, nutrients, antimicrobials, colors, and flavors (Das *et al.*, 2020). Improved processes for obtaining nano-emulsions have shown that the smaller size of oil droplets ($r < 100$ nm) determine a number of potential advantages over conventional emulsions. These advantages correlate to the high stability of droplet conglomeration and obtained disruption, the ability to modulate the texture of the product, increased optical clarity, and augmented bioavailability of lipophilic components (McClements and Rao, 2011). The size of the drops present in nano-emulsions depends mainly on the pH of the environment (Borthakur *et al.*, 2016).

Nano-emulsions have been demonstrated to be a significant delivery system for protecting, encapsulating, and improving the bioavailability of bioactive lipophilic drugs, nutraceuticals, etc. (Huang *et al.*, 2010). The application of emulsion technology in the food industry can lead to new product structures and new physico-chemical properties such as flavor, texture, stability, taste, and extension of shelf life. Furthermore, the decrease in the average diameter of nanospikes has been shown to have an important role in the speed of the digestion process of lipids exposed to pancreatic lipase due to the increase in the lipid surface via a reduction in droplet size (Salvia-Trujillo *et al.*, 2017). Nano-emulsions have a much higher surface-to-volume ratio than regular emulsions. This is due to the presence of a smaller number of molecules in the dispersed phase compared with other common emulsions, and this difference is based on the molecular weight of the particles (Mason *et al.*, 2006). The physico-chemical properties such as rheology, stability, and optical behaviour of nano-emulsions are connected to particle size, composition, physical state, concentration, encapsulation, and interfacial properties. Jin *et al.* (2016) highlighted that nano-emulsions have sparked great interest among researchers, which had led to more than 5000 articles on this subject being published at that time. However, there is little work on the use of nano-emulsions in food (Jin *et al.*, 2016). In the food industry, stable kinetic and optically transparent isotropic nano-emulsions have been used as possible candidates as a result of their excellent properties compared with conventional emulsions, such as superior stability, high voltages between oil interfaces in water, bioavailability, nontoxicity, and nonirritating behaviour (McClements and Rao, 2011).

3.3.1 Nano-emulsions for the dairy industry

In many countries, dairy products are considered an important component of basic nutrition and are therefore consumed at a relatively high level (Diaconescu *et al.*, 2011). Milk contains almost all the nutrients needed to maintain life (Belitz *et al.*, 2009) and is consumed on its own or used as a raw material to produce various products, such as yogurt, cheese, cream, ice cream, and butter (Widyastuti and Febrisiantosa, 2013). Along with food control systems that are essential to preserve the health and security of consumers, food industry researchers are looking for new technologies to produce high-quality products, in terms of both nutrition and organoleptics, and to create an extended shelf life. In the food industry, nano-emulsions are one of the most commonly used supply systems (Dasgupta *et al.*, 2019). The fat molecules (generally in the form of globules) in milk, which is an oil-in-water emulsion, represent the dispersed phase, while water serves as the dispersal medium. Fat droplets are emulsified in milk whey. Among dairy products, there is also the water-in-oil emulsion type—also called an oily emulsion—which is present in cold butter and cream (Panghal *et al.*, 2019). Emulsions in dairy products have lower stability from variations in ionic power and pH, as well as during mechanical and thermal processing stages, such as mixing, heating, or cooling. Recently, there has been an increased interest in technologies producing nano-emulsions. They offer certain advantages to the milk industry because the lipophilic components can be soluble and take different forms such as sprays, foams, creams, and liquids. Nano-emulsions are also being used to improve the bioavailability of valuable nutrients in dairy products. Currently, the use of nano-emulsions in the food processing industries, including dairy, is under research to improve the properties of ingredients in the food matrix (Panghal *et al.*, 2019).

Nano-emulsions allow sustainable food processing, as they prevent the alteration of their components due to oxidation, temperature, pH variations and enzymatic reactions (Borthakur *et al.*, 2016). The properties of emulsion droplets depend on different factors including total soluble solids, ionic strength, pH, temperature, and the type of emulsifier present at the interface (mostly protein: casein or whey protein). The fat globules are separated after prolonged storage or centrifugation in the form of cream/sour cream (Belitz *et al.*, 2009). The stability against gravitational separation, flocculation, coalescence, and property modification is due to the fine droplet size in the nano-emulsion. In the emulsion, there should be a homogeneous distribution of small droplets from the dispersion phase. Panghal *et al.* (2019) stated that the stability of homogenized milk is accomplished by reducing the uniform dispersion and size of the fat cells, which helps in preventing or slowing the rate of cream formation, and therefore prolongs the shelf life of dairy products.

The proteins present in milk behave as an active surface component due to their amphiphilic structure, stabilizing the emulsion droplets, and thus increasing the coalescence. There are different methods for preventing coalescence by using proteins and emulsifiers. The basic nano- and micro-sized structures of dairy products are represented by casein mica, fat globules, and whey proteins. They occur in different composition of products such as complex liquids, foams (cream and ice cream), cheese, gel networks (yogurt), and emulsions (butter) (Bimlesh *et al.*, 2018; Dasgupta *et al.*, 2019). In addition, the casein and whey proteins found in milk act as good emulsifying agents. At the air–water/oil–water interface adsorption of these proteins occurs, avoiding coalescence by maintaining the interfacial membranes. The organization of casein micelles is an example of self-assembly structures that create stable nano-emulsions. To avoid partial displacement and coalescence of the fat globules during subsequent operations, such as freezing or agitation, surfactants can also be added to ice cream mixtures (Dasgupta *et al.*, 2019).

In milk and milk-based products, various organoleptic characteristics such as mouth sensation, taste, aroma, consistency, and rheological characteristics, are considered as determinants of quality (Jin *et al.*, 2016). Obtaining the desired quality parameters in a product can be done by sizing the drops in emulsion as little as possible and distributing them on the surface. Using foreign stabilizing agents is not necessary as the stabilization of the emulsion in milk is accomplished by the inherent capacity of emulsification of milk proteins. The demand for nano-emulsions in the dairy industry is increasing by the day, with research emphasizing their particular properties. This is because consumers' needs have changed, and they prefer more functional foods that give them health benefits, rather than those that provide them with energy. Thus, we are trying to obtain functional drinks and foods from milk

fortified with vitamin-rich nano-emulsions that have nutrient bioavailability and an improved physical stability, suggesting health and nutritional benefits (Panghal *et al.*, 2019). Currently, functional foods have bioactive compounds and antioxidants that give nutritional enrichment and an increased shelf life, stability of products, etc. (Panghal *et al.*, 2019). An example of the development of a stable nutraceutical product is kenaf seed oil—an important medicinal plant used in India, China, Thailand, and Malaysia (Monti and Alexopoulou, 2013). Kenaf seed oil can be added as a supplement to beverages, such as dairy products, to enhance their nutritional value.

Ice cream

The main ingredients of ice cream are cow's milk and sugar. Recent advancements in nano-technologies have allowed the production of ice cream with improved nutritional, organoleptic, antioxidant, and low-fat properties (Azari-Anpar *et al.*, 2017; Romulo et al., 2021). Making ice cream and choosing the right ingredients are the key factors in obtaining the desired product (Panghal *et al.*, 2019). Consequently, the giant company Unilever developed a low-fat (1%) ice cream while preserving the taste and quality of the product (Dasgupta *et al.*, 2019). van der Hee *et al.* (2009) designed two formulas for calcium-fortified and low-fat ice cream (in comparison with regular ice cream). From their study, they established that young adults who consumed the two types of ice cream obtained the same calcium absorption as that attained from milk consumption. Therefore, when comparing the results of the bioavailability of calcium in ice cream vs milk, they showed that calcium-fortified ice cream can contribute to daily calcium needs (van der Hee *et al.*, 2009). Sharifi *et al.* (2013) produced ice cream with alginate nanoparticles loaded with iron (Fe) and zinc (Zn) salts. Their research reported obtaining alginate nanoparticles with an average size of 90–135 nm, of smooth regular shape, and without aggregation. They used Fourier-transform infrared spectroscopy analysis to show that the loading efficiency of Zn/Fe was 70–85%, and the release profile of nanoparticles showed a constant balance (Sharifi *et al.*, 2013).

Previous research on the benefits of plant oils has encouraged the enrichment of ice cream with such biocompounds. The inclusion of bioactive ingredients, such as phenolic compounds from pomegranate peel (Çam *et al.*, 2014), purple rice bran oil (Alfaro *et al.*, 2015), whey protein (Danesh *et al.*, 2017), and vitamin D$_3$ (Tipchuwong *et al.*, 2017) has been reported. In addition, ice cream was improved by supplementing it with the olein fraction obtained from chia oil (*Salvia hispanica* L.), which increased the concentration of XXIV−3 fatty acids, while also providing antioxidant qualities (Ullah *et al.*, 2017). Unsaturated fats, such as hazelnut oil and olive oil, can be used successfully to produce ice cream with the aim of producing healthier products based on vegetable oils (Güven *et al.*, 2018). Microencapsulated linseed oil has been used to fortify ice cream with α-linolenic acid (XXIV−3 fatty acid), which has proved to be stable over the oxidation process during storage. The free fatty acid content increased: a 100g serving of freshly prepared ice cream provided approximately 45% of recommended daily allowance (1.4g α-linolenic acid day^{-1}). However, this amount decreased with increasing storage period (Gowda *et al.*, 2018). In a study conducted by Mohammed *et al.* (2020), an ice cream was developed using a *Nigella sativa* oil nano-emulsion. This nano-emulsion was stabilized by combinations of sodium caseinate, acacia gum, and Tween-20, at different proportions. The results showed that the *Nigella sativa* oil nano-emulsion improved the physical properties of the ice cream, and the final product was approved by consumers. This product has multiple benefits and can be used as a functional ice cream (Mohammed *et al.*, 2020).

Yogurt

Yogurt is a traditional dish originating from Central Asia that has now spread throughout the world. It is made from sheep, cow, or buffalo milk heated to 85°C for 30 min (or to 90–95°C for 5–10 min), cooled to 40–45°C, and inoculated with *Lactobacillus delbrueckii* subsp. *bulgaricus* and *Streptococcus thermophilus* culture (about 1% by weight of milk). About 20–30% of the lactose is converted into galactose and glucose by β-galactosidase produced by the bacteria. Glucose follows the glycolytic pathway

by turning into pyruvic acid and then into lactic acid. Simultaneously, *Streptococcus thermophilus* metabolizes the lactose into lactic acid. The proteins of some species of *Lactobacillus* convert milk proteins into peptides and amino acids. As a result of the fermentative processes, other organic compounds (e.g. acetaldehyde, acetone, ethanol) are produced, which subsequently establish organoleptic properties in the yogurt. Through fermentation, the milk is first coagulated as casein micelles and then disaggregated by releasing calcium salts due to lactic acid accumulation. The fermentation process stops once it reaches 2–4°C, after which the yogurt is preserved at temperatures below 8°C until it is opened (Ghiorghiță and Nicuță-Petrescu, 2005).

Yogurt is recommended to be consumed daily because of the health benefits. As well as the basic ingredients mentioned above, the composition of yogurt can be improved with various supplements to increase its nutritional value and improve its health benefits without disrupting its organoleptic properties. Fish oil, for example, can be used in yogurt preparations, but because of its low water solubility, it is difficult to add to fortified and functional foods, and this could lead to poor bioavailability. However, using nano-emulsion technology, the water solubility of bioactive lipophilic compounds and oils can be improved (Panghal *et al.*, 2019). Fortified yogurt with fish oil/γ-oryzanol-enriched nano-emulsions can be developed by incorporating 13g of fish oil- or γ-oryzanol-enriched nano-emulsions into a 100g yogurt sample. The mixture is then refrigerated at a low temperature of 4°C for 21 days in a hermetically sealed glass container (Zhong *et al.*, 2018).

McCowen *et al.* (2010) demonstrated that yogurt enhanced with ω−3 polyunsaturated fatty acids is a beneficial solution to increase the ω−3 fatty acid content in plasma lipids and lower arachidonic acid concentrations (McCowen *et al.*, 2010). Research on the fortification of yogurt with nano-emulsions containing linseed oil (Almasi *et al.*, 2021) or seaweed oil (Lane *et al.*, 2014) have shown it to be a source for increasing solubility and bioavailability, and ensuring a daily intake of ω−3. Fortification of foods with the essential vitamins A, D, and E is beneficial to human health. Among the lipophilic vitamins, vitamin D can be incorporated into drinking yogurt, as well as into drinking milk or milk used to prepare cheese (Öztürk, 2017). The introduction of various plant extracts, especially from plants with phytotherapeutic properties, into food using nano-emulsions is a new development. El-Naggar *et al.* (2020) studied in mice the effects of yogurt fortified with the nano-emulsions of *Malva parviflora* leaf extract regarding the attenuation of ulcerative colitis induced by acetic acid. The results showed that the fortified yogurt had the ability to inhibit the creation of reactive oxygen species, and to capture free radicals, thereby improving antioxidant defense enzymes against acetic acid-induced ulcerative colitis. Recently, with regard to the development of functional dairy products with natural flavor, various nano-emulsions using essential oils (lemon, lemongrass, cinnamon, mint, and cloves) have been used for a more appealing taste in yogurt (Salama *et al.*, 2021). Notably, the use of peppermint essential oil products by Bakry *et al.* (2019) has been reported to hide the unpleasant fishy taste of tuna oil incorporated into yogurt, as well as to increase its functionality. As a preventative measure against the growth of fungi, garlic essential oils have also been used (Milanović *et al.*, 2021). According to El-Sayed *et al.* (2022), yogurt rich with probiotic jelly candy and with grape seed extract nano-emulsions is promising for human health.

Cheese

Cheese, an age-old food product prepared from different types of milk, has a constantly increasing market demand. Cheese is an essential part of our diet due to its composition, which is rich in protein, calcium, minerals, and vitamins, and its wide diversity of textures, and flavors (Costa *et al.*, 2018). During the cheese-making process, the milk must contain at least 50% lactic acid bacteria in the total microflora. The milk undergoes successive operations such as filtration, homogenization, pasteurization, and concentration. It is then coagulated and pressed to remove whey and form a cheese paste. Finally, the cheese undergoes fermentation and maturation (Ghiorghiță and Nicuță-Petrescu, 2005). Cheese consumption has increased significantly over the years around the world, which led the dairy industry to evolve into a

global business. Continuing research in this field is important to increase the product shelf life and promote product quality and safety (Costa *et al.*, 2018). Simultaneously, the demand for consumers to eat foods without preservatives or containing only natural preservatives is increasing. This obliges the food industry to use preservatives made from plant or microbial origins in the production process instead of artificial preservatives. Essential oils are natural, volatile substances obtained from plants that can be used as food preservatives while also helping to improve the organoleptic properties of cheese products. Along with their beneficial properties for human health, plant extracts and essential oils possess antioxidant, antibacterial, and antifungal properties due to the bioactive compounds they contain. In-depth studies are currently being done to demonstrate the potential uses of essential oils as natural antibacterial agents with the goal of reducing pathogenic bacteria, molds, and yeasts that degrade cheese (Pan *et al.*, 2014).

Thymol nano-emulsions maintained by a mixture of lecithin and gelatin have shown strong antibacterial properties against *Listeria monocytogenes* and *E. coli* O157:H7 in cantaloupe juice and milk (Xue *et al.*, 2017). Moreover, it was found that the antibacterial properties of nano-emulsions are influenced by the fat levels in dairy products. Thus, increased inhibition of *L. monocytogenes* in skimmed milk compared with regular milk with a higher amount of fat was reported after 48 hours of storage (Xue *et al.*, 2017). The specific location of the active compounds (in the aqueous phase, oily phase, or interface) seems to play a major role in determining the antimicrobial properties of the emulsion (Terjung *et al.*, 2011). Christaki *et al.* (2022) obtained oil-in-water nano-emulsions with oregano essential oil as the dispersed phase and the corresponding aqueous extracts as the continuous phase. These nano-emulsions were used as delivery systems for bioactive compounds from oregano, a natural antifungal agent. The nano-emulsions were also tested on unprocessed cheese (made from whey). The results showed that the shelf life of the cheese was prolonged due to the antifungal activity of the compounds in oregano. Similar conclusions have also been drawn by Bedoya-Serna *et al.* (2018) on the use of oregano oil nano-emulsions as a potential alternative for preserving Minas Padrão cheese against fungal contamination.

The results described above can be utilized by the food industry in the development of novel systems in which both lipophilic bioactive compounds and hydrophiles from medicinal and aromatic plants can be used at the same time (Christaki *et al.*, 2022). However, the organoleptic properties of cheese are modified by the oil-in-water nano-emulsions of oregano. The authors stated that bioactive compounds from medicinal and aromatic plants used in delivery systems may prove effective in extending the shelf life of products. However, cheese preserved with essential oils is still very limited on the commercial market. Feta cheese, in particular, is preserved with essential oils (oregano, black pepper, sage, and mandarin), which are incorporated into the brine (Christaki *et al.*, 2022). Antimicrobial edible shells of nano-emulsion-based ingredients made from oregano oil were shown to extend the shelf life of low-fat cheese by discouraging the growth of molds, bacteria, and yeasts (Artiga-Artigas *et al.*, 2017). Another study showed that encapsulation of mint (*Mentha piperita*), sea buckthorn (*Hippophae rhamnoides*), basil (*Ocimum basilicum*), and fennel (*Foeniculum vulgare*) oils in sodium alginate hydrogels improved the quality and shelf-life properties of yogurt due to their antioxidant and antimicrobial effects (Constantinescu *et al.*, 2019). One or more antimicrobials can be introduced in different locations within nano-emulsions: polar antimicrobials in the aqueous phase, antimicrobial amphiphiles in the interfacial region, and nonpolar antimicrobials in oil droplets. An improvement in nutritional value and sensory properties of some cheeses was achieved by incorporating natural components such as apricot pulp (Kasapoglu *et al.*, 2020), tomato juice (Mehanna *et al.*, 2017), or extracts of medicinal plants (Li *et al.*, 2022). Adding bioactive compounds to processed cheese could affect not only taste but also consistency (Li *et al.*, 2022). Studies have reported the use of nano-emulsions for the production of processed cheese products. For example, El-Sayed and Shalaby (2021) prepared a curcumin-based nano-emulsion for cheese production. It exhibited high antimicrobial properties and retained the organoleptic properties of the cheese during storage (60 days) at 4°C. The product

was accepted by consumers, and therefore the authors suggested placing the product on the market (El-Sayed and Shalaby, 2021). Another study by El-Sayed and El-Sayed (2021) showed that cumin oil-based nano-emulsions are a solution for preserving the qualities of soft white cheese during storage.

In fresh cheese, the incorporation of bioactive compounds from *Opuntia oligacantha* through microcapsules or nano-emulsions was tested, with the aim of developing a functional food (Pérez-Soto *et al.*, 2021). The tests covering the two methods were an effective alternative for maintaining and/or improving the characteristics of this food during storage. The physicochemical characteristics were not influenced by the addition of the compounds. Microbiological analyses demonstrated the antibacterial and antifungal activity of the bioactive compounds incorporated by both methods. The antioxidant capacity was also higher compared with the control sample. Texture profile analysis revealed that the rigidity of the cheese was affected by the addition of bioactive compounds, but there was no effect on firmness, elasticity, or chewability. Both methods of incorporation of bioactive compounds proved to be effective in maintaining the qualities of the fresh cheese. The benefits of using nano-emulsion technologies within the dairy industry and related research on their potential applications for preserving and improving food qualities continue to expand.

3.3.2 Nano-emulsions for confectionery: encapsulation of flavor and coloring agents

Flavorings and dyes can be considered very valuable items in any food formula, as they help to improve organoleptic qualities. They are usually unstable, and maintaining them in food is often a major concern for food manufacturers. The flavor is due to highly scented organic molecules, found mostly in gaseous or liquid form. Some of these substances can also be solid, having a distinct odor such as vanillin and menthol. These molecules are lipophilic and have low molecular weight (between 100 and 250 Da). They belong to the group of compounds including alcohols, hydrocarbons, aldehydes, esters, ketones, sulfides, and acids (Gupta *et al.*, 2016). The flavor and taste of emulsified foods are important factors that determine the quality of food. The perception of flavors depends on the location where the molecules are distributed in the food emulsion: in the surrounding liquid (continuous phase), inside the droplets (dispersed phase), at the oil–water interface, or in the vapor phase above the emulsion. Thus, the key factors that should be considered when selecting flavors used in food emulsions are: their distribution in the oily/aqueous/interface regions and their transmission speed from the product to the taste and smell receptors of the body during consumption. For example, nonvolatile molecules responsible for taste are experienced more intensely when in the aqueous phase than in the oil phase (McClements, 2016).

Development of functional foods has a number of drawbacks due to problems with the bioavailability, solubility, and low stability of the bioactive compounds. Most of the bioactive ingredients are predisposed to degradation during food processing or to oxidative damage during storage (Shahidi and Zhong, 2010). Some bioactive substances have reduced solubility and fast metabolism, which decreases their bioavailability, while other compounds are unstable and sensitive to processing (McClements and Yan, 2010). These challenges can be eliminated by using nano-emulsions on the basis of which bioactive molecules—micronutrients, dyes, and flavorings—are encapsulated and introduced into the food matrix (protein or polysaccharide) in order to give it protection against evaporation, reaction, or migration in a food. Environmentally biodegradable polymers as carrier matrices have gained more attention due to their biocompatibility and biodegradability. They can be either synthetic (polyanhydrides, polyesters, and polyphosphazenes), or natural polysaccharides, such as hyaluronic acid, plant or animal proteins, chitosan, and alginate (Gupta S. *et al.*, 2016). Encapsulation is used to preserve flavor in food during storage and to protect flavor from unwanted interactions in food. Flavor encapsulation is also useful against light-induced reactions and/or oxidation in order to increase the shelf life of flavors, and/or to allow their controlled release during food consumption.

Several methods of encapsulation of flavors can be used: spray drying, spray cooling, spray chilling, extrusion, freeze drying, coacervation, and molecular inclusion (Yadav *et al.*, 2015; Choi and McClements, 2020; Dammak *et al.*, 2020). The appropriate choice of microencapsulation technique depends on the mechanisms for releasing flavors from the food matrices, the product's final use, and the processing conditions involved in the manufacturing of the product (Madene *et al.*, 2006). The protective coating of the nanocapsulation must not react with the encapsulated substance. It is present in a form that is easy to handle (i.e. allows the complete disposal of the solvent), has low viscosity at high concentrations, and provides maximum protection of the active ingredient against external factors (Madene *et al.*, 2006). It is recommended nanoencapsulation is performed in an oily or emulsifying phase, as it ensures the stability, bioavailability, and controlled release of bioactive molecules to tissues and cells in the body (Fakhar *et al.*, 2022). This process can also improve the nutritional content of foods, without affecting their taste and texture. It also improves the slow release of flavors and enhances the stability and solubility of the ingredients in functional foods (Chen *et al.*, 2006; McClements and Rao, 2011; Fakhar *et al.*, 2022). Most often, lipophilic components such as phytosterols, $\omega-3$ oils, oil-soluble vitamins, and carotenoids are encapsulated (McClements *et al.*, 2007; McCowen *et al.*, 2010; Öztürk, 2017). Therefore, it is important to understand how the characteristics of nano-emulsions influence the release and absorption of the encapsulated components. In the food industry, it is recommended to identify as many natural emulsifiers as possible to replace synthetic emulsifiers that may have some toxic effects on the human body. Chemical stability studies should be conducted with regard to the active agents encapsulated in natural emulsifiers (Dammak *et al.*, 2020). Many studies have shown that encapsulating lipophilic active ingredients in nano-emulsions increases their bioavailability (Salvia-Trujillo *et al.*, 2017). The ability of nano-emulsions to make gels can be utilized in functional food design, where nano-emulsion gels become a constituent part of the food (Donsì, 2018).

The flavor delivery in food processing is a very important aspect, as it must consider the preservation of highly volatile molecules throughout the shelf life of the product. It is also important to analyse their controlled release during food preparation and consumption, as this influences how they are perceived by consumers. The application of essential oils in food as natural antimicrobial agents has increased rapidly in recent years. However, if incorporated, they can affect the organoleptic properties of food due to their volatility and prominent taste (Donsì, 2018). Gels and gelling agents are used to obtain the desired textural and sensory properties of confectionery products used in the preparation of jams, jellies, cake creams, and ice cream (Yadav et al., 2015). Synthetic dyes used in the preparation of sweets can be replaced by natural dyes, obtained from plant pigments, such as chlorophyll, carotenoids, and anthocyanins. The efficient extraction of these natural compounds can be achieved by different methods, such as negative-pressure cavitation, ultra-high pressure, ohmic heating, high-voltage electrical discharges, mechano-chemical methods, pulsed electric fields, and high-pressure homogenization (Ghosh et al., 2022). Lycopene microcapsules were prepared by an atomized drying technique, and modified starch was used as an encapsulation agent, and these microcapsules applied to a cake were able to release the pigment during preparation of the dessert (Rocha et al., 2012). Apple puree fortified with shell/grape pellet phenolics, encapsulated in maltodextrin, is an ingredient that could meet the demand for durable and beneficial coloring agents (Lavelli et al., 2016).

Peel extract from pomegranate (*Punica granatum* L.) is important for human health due to its phenolic compounds, which confer antimutation, antiviral, antimicrobial, and antioxidant properties. The encapsulation possibility of this extract was tested and used as a coating material for maltodextrin/whey protein isolate (95% whey protein) (50:50), maltodextrin/skimmed milk powder (50:50), and maltodextrin/gum arabic (50:50). Pomegranate bark extract was added to the hydrated coating material and the process of extraction and encapsulation was assisted by ultrasound. This encapsulated phenolic extract was shown to be effective in increasing the shelf life of groundnut butter, despite the decrease in solubility of the raw extract in a matrix with high lipid content (Kaderides et al., 2015). As well as preserving flavor, releasing it from the encapsulated material is also

essential. Preferably, all encapsulated aromas should provide controlled release. However, a slow or delayed release is sometimes desirable (Baranauskienė et al., 2007). β-Carotene nano-emulsions stabilized with modified starch and spray dried have demonstrated good stability during storage. Nano-emulsions prepared from carotenoids extracted from melon had increased colour stability and water solubility. The oil-in-water carotenoid nano-emulsions encapsulated in porcine gelatin, isolated with whey protein coating, had an average particle size of 70–160 nm. Gelatin was able to increase the water solubility of the carotenoids. Yogurt that contains these nano-emulsions as a natural coloring agent has been shown to be stable for 60 days (Aswathanarayan and Vittal, 2019).

3.3.3 Nano-emulsions as natural preservatives

In the food industry, a primary problem is the preservation of food quality for long periods of time. Food spoils due to chemical degradation and microbial contamination. For example, lipid oxidation is a chemical reaction that occurs in food emulsions and causes the quality of the product to be altered because of the production of an unpleasant flavor, loss of beneficial polyunsaturated lipids, or formation of potentially toxic products (McClements and Rao, 2011). Spoiled food leads to increased economic losses, waste, environmental contamination, and food safety concerns (Ozogul et al., 2022). In order to preserve the quality of products, the food industry usually uses synthetic preservatives. The efficacy of these antioxidants depends on their chemical structure, solution conditions (pH, ionic strength, and temperature), and the physico-chemical environment (oil, water, or interface region). Many synthetic food additives are common antioxidants used in food processing. However, as consumer demand for natural foods has increased, the industry is looking for more 'label-friendly' alternatives (McClements, 2016). For this reason, much research is based on evaluating the effectiveness of natural antioxidants, including tocopherols and fruit and plant extracts. Certain additives are used for the prevention of lipid oxidation in food, such as phosphoric acid, ethylenediaminetetraacetic acid (EDTA), proteins, citric acid, polyphosphates, organic acids, and polysaccharides. The antioxidant agent choice depends on the specific type of food (McClements, 2016; Hassoun and Çoban, 2017).

The use of natural substances as food preservatives is a challenge as they generally have low water solubility, chemical instability, or low efficiency. To avoid these inconveniences, natural substances must be transformed into nano-emulsions to be used as food preservatives, (Ozogul *et al.*, 2022). Nano-emulsion delivery systems must be suitable for the food matrix in which they are introduced and cannot adversely affect the organoleptic characteristics of the final product. Nano-emulsions are considered to be natural antimicrobials due to the fact that water is unavailable to microorganisms, as it is bound up in the structure of the nano-emulsion (Ozogul *et al.*, 2017).

It has been observed that nano-emulsions of essential oils can control microbial multiplication and preserve food quality, as well as extend the shelf life of food. Essential oils have antimicrobial and antioxidant properties as a result of their bioactive compounds. For example, common essential oils such as sage, oregano, thyme, orange and lemon contain bioactive compounds including α-pinene, β-pinene, carvacrol, thymol, D-limonene, and eugenol. Nano-emulsions containing these antimicrobials are efficient against both Gram-positive and Gram-negative bacteria (Burt, 2004). Numerous studies have reported on use of antimicrobial nano-emulsions based on essential oils and plant extracts for preserving different types of food, such as fresh fruits, vegetables, meat, fish, and dairy products (Table 3.1). Plant-based nano-emulsions can also be used as antimicrobial agents to improve the quality and shelf life of these food types.

3.3.4 Nano-emulsions for food packaging

The purpose of packaging is to maintain the quality of food during storage and transport, to increase the shelf life for consumption, and to prevent food spoilage due to the action of microorganisms, chemical contaminants, or physical factors (e.g. oxygen, humidity, light) (Caon *et al.*, 2017). The demand for fresh food with a longer shelf life and a lower content of synthetic additives and preservatives is increasing rapidly. Keeping food in good-quality conditions and increasing its shelf life can be achieved by using bioactive agents with antioxidant or antimicrobial properties. As the packaging is protective, solutions are being sought to incorporate bioactive agents into it rather than adding them directly to the food. Essential oils produced from plants are used as a source of bioactive compounds. The introduction of natural plant oils into film-forming solutions for food packaging results in some inconveniences, such as reduced miscibility and phase separation in the film-forming process, as well as changes in film transparency. Additionally, some bioactive compounds are sensitive to certain physical factors such as temperature, ultraviolet radiation, pH, oxygen, and humidity (Norcino *et al.*, 2020).

The significant properties of nano-emulsions such as stability, high physico-chemical activity, and improved biological properties indicate the advantages of their use in the production of new food-packaging materials. Obtaining coating films and various packaging membranes based on nano-emulsions involves an additional step, namely the 'solidification' of nano-emulsions (Espitia *et al.*, 2018). This process is achieved by adding a support (bio)polymer (called a matrix) to the continuous phase of the nano-emulsion, which favors dispersion and stability of the droplets before the solidification stage. During the formation of nano-emulsions, viscosity increases in the continuous phase, and the rate of coalescence between drops is reduced (Espitia *et al.*, 2018). Most food-packaging systems combine biopolymers and natural antimicrobial compounds. These systems are designed so that the oil droplets of an oil-in-water nano-emulsion remain encapsulated (Espino-Manzano *et al.*, 2020). Packaging materials containing nano-emulsions may exhibit antimicrobial activity if one of the phases constituting the nano-emulsion contains an active biological compound. Usually, the biological activity is ensured by the hydrophobic phase, which in general includes essential oils. In the process of forming such packaging systems, the incorporation of nano-emulsions containing essential oils

Table 3.1. Types of essential oils with antimicrobial role in nano-emulsions for food preservation. Adapted from Ozogul *et al.* (2022).

Type of essential oil used in nano-emulsion	Type of food product	Reported results on antimicrobial efficiency	Reference
Thyme, rosemary, laurel, sage	Fish fillets (rainbow trout)	Lower the values of biochemical properties and effective in slowing bacterial growth. Greatly improved the organoleptic quality of rainbow trout fillets, apart from rosemary, which gave a bitter taste	Ozogul *et al.* (2017)
Geraniol or linalool	Meat	Significantly reduced *Escherichia coli* K12 and *Listeria innocua* *Pseudomonas lundensis* was demonstrated to be more resistant to both nano-emulsions	Balta *et al.* (2017)
Cinnamon	Chicken breast	Suppressed microbial growth in chicken breasts during storage at 4°C	Wang *et al.* (2021)
Rosemary, cinnamon, oregano	Fresh celery inoculated with *Escherichia coli* and *Listeria monocytogenes*	Documented for the potential use of essential oils encapsulated in oil-in-water nano-emulsions for decreasing bacterial counts on celery. Oregano essential oil nano-emulsion was the most effective against *E. coli* and *Listeria monocytogenes*	Dávila-Rodríguez *et al.* (2019)
Citrus essential oil (CEO)[a]	Silvery pomfret	CEO reduced the stability and rheological characteristics of the nano-emulsion systems. Influenced the physico-chemical properties of coatings. Nano-emulsion coating containing CEO has great potential in seafood preservation	Wu *et al.* (2016)

[a]Nano-emulsion based on chitosan nanoparticles loaded with citrus essential oil.

can cause changes in their chemical, morphological, thermal, and mechanical properties. In a study by Norcino *et al.* (2020), the production of pectin-based films containing nano-emulsions with copaiba essential oil modified the matrices of the nano-emulsions, which in turn, provided active functionality. It preserved and even improved their physico-mechanical properties (increasing their extensibility and decreasing their rigidity), as well as antimicrobial properties (Norcino *et al.*, 2020). Using nano-emulsions with active ingredients to create edible films with different physical and functional properties has been demonstrated using nano-emulsions with thyme, lemongrass, or sage oil as the dispersed phase and sodium alginate solution as the continuous phase (Acevedo-Fani *et al.*, 2015).

Covering food with edible packaging is an efficient process for keeping product fresh as well as maintaining the quality and extending the shelf life. Among edible coatings, pectin appears to be the preferred candidate for its nontoxic, odorless, and biodegradable properties. Pectin coatings also hinder gas exchange between the food and the outside environment. Pectin can also be a good carrier for bioactive ingredients (e.g. oregano essential oil and resveratrol), thereby also further increasing the health benefits (Xiong *et al.*, 2020).

In a study by Filho *et al.* (2021), starch-based coating films containing carnauba wax nano-emulsion, cellulose nanocrystals, and essential oils from *Mentha spicata* and *Cymbopogon martinii* were considered as potential active packaging materials. Such films are new materials used in covering fresh fruits and vegetables, as well as for other food products (e.g. bread or cheese), which quickly become damaged due to the growth of fungi on their surface. A number of studies have reported good physical, chemical, antioxidant, and antimicrobial properties of nano-emulsion films (Tongnuanchan *et al.*, 2013; Alexandre *et al.*, 2016; Gharibzahedi and Mohammadnabi, 2017; Noori *et al.*, 2018; Dammak and Sobral, 2019; Khan *et al.*, 2020).

A chitosan-based film embedded with natural oil nano-emulsions would be an ideal solution for a clever food covering (Elshamy *et al.*, 2021). According to Elshamy *et al.* (2021), the incorporation of thyme oil nano-emulsion into the chitosan-based film did not significantly alter its appearance and transparency; instead, it improved the light barrier properties as well as the physical and mechanical properties of the film and increased its thermal stability without causing a structural alteration in its matrix. The introduction of thyme nano-emulsion into the chitosan-based film also remarkably improved its antimicrobial activity against food-borne pathogens. Active banana starch film, loaded with orange oil and curcumin nano-emulsions, also showed improved properties in terms of water vapor permeability and tear resistance (Sanchez *et al.*, 2022).

Gelatin is a component that can also be used to produce films with antioxidant properties, for example if it is based on nano-emulsions containing *Brassica rapa* oil and ginger essential oil.

In conclusion, films and coatings based on nano-emulsions used for food preservation have demonstrated their applicability through various means. They have been shown to be promising both for preserving the quality of fresh or processed foods, and for transporting natural bioactive compounds from essential oils, which are antimicrobials and natural antioxidants (de Oliveira Filho *et al.*, 2021). To preserve the quality of food, organic or inorganic nanoparticles can be used, but rather than introducing them directly, they can be used in packaging materials to obtain the antimicrobial effect without modifying the properties of the packaged food (Jaiswal *et al.*, 2019).

3.3.5 Nanoemulsifed nutraceuticals

Nutraceutical or bioceutical is a term that was introduced by de Felice and the Foundation for Innovation in Medicine in 1989 (Espín *et al.*, 2007). Nutraceutical refers to a dietary supplement able to provide a concentrated form of a bioactive compound and used to improve health. In recent years, nutraceuticals have come to the attention of scientists, food manufacturers, and consumers because they have been shown to improve human health and performance (Das *et al.*, 2020). The list of nutraceutical compounds (e.g. bioactive peptides, vitamins, probiotics, antioxidants) is very long, and they are represented mainly by phytochemicals: glucosinolates and terpenoids (carotenoids, monoterpenes, and phytosterols), sulfur-containing compounds, and various groups of polyphenols. Plant polyphenols (e.g. curcumin, epigallocatechin gallate, resveratrol) and carotenoids (e.g. β-carotene, zeaxanthin, lycopene, lutein) are the focus of interest in the food industry and among consumers because of their important health benefits, such as reducing risk factors for the development of cancer, lowering blood pressure, regulating blood sugar concentration, regulating the digestive system, strengthening the immune system, lowering cholesterol levels, and playing an important role as antioxidants (Huang *et al.*, 2010; Puttasiddaiah *et al.*, 2022).

The beneficial effects of phytochemicals depend on how they enter the body and the degree of assimilation at the cellular level (Zheng *et al.*, 2018). These processes are significantly influenced by the physico-chemical properties of the bioactive compounds, such as water solubility, lipophilicity, partition coefficient, and crystallinity (McClements and Rao, 2011). Active components that are poorly soluble in oil or water pose a problem for the route of administration, transport, and in terms of reaching their targets. For these reasons, many nutraceuticals have poor oral bioavailability. For example, Manach *et al.* (2005) analysed the oral bioavailability of 97 polyphenolic compounds and showed that proanthocyanidins, catechins, curcumin, and anthocyanins were the least absorbed polyphenols. The use of β-carotene in functional foods is limited due to its chemical instability, lower water solubility and higher melting point (Mehmood *et al.*, 2021). Thus, the issue of delivery of bioactive components into the human body is an important consideration. To ensure the effective amount of intact drug at the desired site in the body, it is important that the preparation is not unstable (Huang *et al.*, 2010). To overcome the instability and poor water solubility and improve the bioavailability of nutraceuticals, it is possible to incorporate the substance of interest into a food matrix using a nano-emulsion. For nutraceutical-based delivery systems, nano-emulsions offer the following advantages: (i) high kinetic or thermodynamic stability compared with conventional emulsions and suspensions; (ii) both hydrophilic and lipophilic compounds can be incorporated; and (iii) due to the small droplet sizes of nano-emulsions (<100 μm), bioactive compounds can be transported much more easily across cell membranes, resulting in increased plasma concentrations and high bioavailability (Choi and McClements, 2020; Zheng and McClements, 2020). Several delivery systems have been investigated to increase the bioavailability of curcumin, which is highly unstable and hydrophobic and difficult to incorporate into liquid food systems. For this reason, it is recommended to encapsulate curcumin

in carrier oil in the form of a nano-emulsion, which ensures the slow release of curcumin after ingestion (Sari *et al.*, 2015). Other research has involved the use of organogel-based nano-emulsions for the oral delivery of curcumin and to enhance its bioavailability (Yu and Huang, 2012). The organogel was used as an oil phase in the curcumin nano-emulsion formulation. Tween-20 was used as an emulsifier to ensure maximum bioavailability of curcumin. Tests on the oral bioavailability of curcumin in the nano-emulsion showed that it was nine times higher compared with unformulated curcumin. The authors of the study suggested that this new formulation approach could also be used for the oral delivery of other poorly soluble, high-loading nutraceuticals, which has significant implications for the functional food, dietary supplement, and pharmaceutical industries.

The degree of absorption of bioactive compounds in the gastrointestinal tract is another important aspect. One such study was conducted by Qazi *et al.* (2021), who investigated the effects of two isocaloric milk gels (containing curcumin nano-emulsions) on fat digestion, as well as on the bioavailability of curcumin during gastrointestinal digestion. Although both gels had the same composition and rheological properties, their different disintegration behaviour during gastric digestion directly affected gastric emptying of the oil droplets and associated lipophilic curcumin embedded in the emulsion. These observations indicate that the release of bioactive compounds from functional foods depends on the complex processes in the gastrointestinal tract and should be considered when developing various foods containing nutraceuticals (Qazi *et al.*, 2021).

Chuacharoen *et al.* (2019) investigated the effects of surfactant concentration on the physico-chemical properties of entrapped curcumin nano-emulsions relevant to commercial applications. The functionality of nanoemulsified curcumin in inhibiting lipid oxidation was also determined. Curcumin-loaded nano-emulsions were prepared with three different concentrations of the surfactants lecithin and Tween-80. Increasing the amount of surfactant decreased the particle size and increased the encapsulation efficiency, and also affected the stability of the droplets and the functionality of the encapsulated curcumin during processing and storage time. These results showed that the use of an appropriate amount of surfactant can provide beneficial protection of encapsulated curcumin under various conditions of use by maintaining the physico-chemical stability of the nano-emulsion and the functionality of the bioactive compound. These results are important for the use of curcumin nano-emulsions in commercial foods (Chuacharoen *et al.*, 2019). Recently, research has been conducted to evaluate the effects of different emulsifiers on the behaviour of curcumin-loaded nano-emulsions during digestion, safety, and absorption, with the goal of providing important information for the development of nano-emulsions that provide improved curcumin safety and functionality. The research has shown that the emulsifier used has a strong influence on the behaviour of the nano-emulsion, the bioaccessibility and stability of curcumin during digestion, and the viability of cells during the digestion process, as well as on their cytotoxicity (de Oliveira Filho *et al.*, 2021). The field of nanodimensional delivery of bioactive compounds has developed rapidly. Most commonly, lipid-based nanoformulations (e.g. nano-/Pickering/double emulsions, nanostructured lipid carriers, solid lipid nanoparticles, surfactant-based nanovectors) and biopolymeric nanostructures are used, as well as biopolymeric nanostructures (e.g. nanofibers, nano-hydrogels/organogels/oleogels, nanotubes, cyclodextrins, and protein–polysaccharide nanocomplexes/conjugates) (Falsafi *et al.*, 2022).

Resveratrol is an important polyphenolic compound with a wide range of applications in functional foods, but the main obstacle to its use is its instability and low solubility. Nano-emulsions prepared with fish oil and various concentrations of resveratrol were found to be stable at room temperature for 1 month. The authors recommend that these nano-emulsions can be used in dairy products and beverages (Ron *et al.*, 2010; Shehzad *et al.*, 2020). *In vivo* and *in vitro* studies have shown that the bioavailability and bioaccessibility of vitamins is increased by the use of nano-emulsion-based delivery systems. Although nano-emulsification of lipophilic vitamins offers many advantages, improvements are still needed to increase the

efficacy of these food encapsulation systems (Öztürk, 2017).

3.4 Benefits and Limitations of Nano-Emulsions

The extraction and application of nano-emulsions in the food industry is a topic of interest among researchers in this field, as well as manufacturers and consumers. Humankind has recognized the benefits of high-quality foods and safe products with a long shelf life and, above all, food with specific nutrients that contribute to maintaining health. Studies conducted in recent years to meet these consumer requirements highlight the advantages of nano-emulsions in all areas of the food and beverage industry. These benefits include the facts that: (i) nano-emulsions are kinetically stable colloidal systems due to the small droplets; (ii) nano-emulsions have improved functional, physical, and chemical properties; (iii) nano-emulsion composition and structure can be controlled for bioactive lipophilic compound encapsulation and delivery; and (iv) nano-emulsions have high bioavailability—they represent important systems for the delivery of dietary supplements, colorants, flavorings, and compounds with antimicrobial and antifungal activity. In addition, some nano-emulsion formulations containing active ingredients can be used to develop coating and packaging films with biodegradable materials to improve the quality, functional properties, nutritional value, and shelf life of foods.

Although nano-emulsions have potential advantages over conventional emulsions, there are a number of important issues that need to be overcome for them to be readily used in the food industry. For example, both the manufacturing process and the characterization of the resulting nano-emulsions need to be considered, as well as acceptability of the sensory properties, toxicity/safety during consumption and, last but not least, the cost of the products. Formulations of nano-emulsions often use emulsifiers to keep them stable, with the emulsifier serving as both an additive and a preservative for the food and/or active ingredients. However, most of the ingredients used in the formulation of nano-emulsions, regardless of the manufacturing method, are synthetic surfactants, synthetic polymers, synthetic oils, or organic solvents, which are not beneficial for health. For this reason, solutions are being sought to use natural ingredients in the formulation of food nano-emulsions that are both legally acceptable and label friendly.

Although nano-emulsions have proved to be quite effective in terms of product safety, bioavailability, and bioactivity of phytochemical compounds and various bioactive substances, their digestibility and stability are still a matter of concern regarding food safety. The use of appropriate, verified processing procedures is recommended to bridge the gap between laboratory experiments and industrial processes. Therefore, further studies are needed to provide realistic and reliable information on the production and consumption of nano-emulsion-based foods.

References

Acevedo-Fani, A., Salvia-Trujillo, L., Rojas-Graü, M.A. and Martín-Belloso, O. (2015) Edible films from essential-oil-loaded nanoemulsions: physicochemical characterization and antimicrobial properties. *Food Hydrocolloids* 47, 168–177. DOI: 10.1016/j.foodhyd.2015.01.032.

Alexandre, E.M.C., Lourenço, R.V., Bittante, A.M.Q.B.B., Moraes, I.C.F. and Sobral, P.J. do A. (2016) Gelatin-based films reinforced with montmorillonite and activated with nanoemulsion of ginger essential oil for food packaging applications. *Food Packaging and Shelf Life* 10, 87–96. DOI: 10.1016/j.fpsl.2016.10.004.

Alfaro, L., Hayes, D., Boeneke, C., Xu, Z., Bankston, D. *et al.* (2015) Physical properties of a frozen yogurt fortified with a nano-emulsion containing purple rice bran oil. *LWT - Food Science and Technology* 62(2), 1184–1191. DOI: 10.1016/j.lwt.2015.01.055.

Almasi, K., Esnaashari, S.S., Khosravani, M. and Adabi, M. (2021) Yogurt fortified with omega-3 using nanoemulsion containing flaxseed oil: investigation of physicochemical properties. *Food Science & Nutrition* 9(11), 6186–6193. DOI: 10.1002/fsn3.2571.

Anton, N. and Vandamme, T.F. (2009) The universality of low-energy nano-emulsification. *International Journal of Pharmaceutics* 377, 142–147. DOI: 10.1016/j.ijpharm.2009.05.014.

Anton, N. and Vandamme, T.F. (2011) Nano-emulsions and micro-emulsions: clarifications of the critical differences. *Pharmaceutical Research* 28(5), 978–985. DOI: 10.1007/s11095-010-0309-1.

Artiga-Artigas, M., Acevedo-Fani, A. and Martín-Belloso, O. (2017) Improving the shelf life of low-fat cut cheese using nanoemulsion-based edible coatings containing oregano essential oil and mandarin fiber. *Food Control* 76, 1–12. DOI: 10.1016/j.foodcont.2017.01.001.

Aswathanarayan, J.B. and Vittal, R.R. (2019) Nanoemulsions and their potential applications in food industry. *Frontiers in Sustainable Food Systems* 3, 95. DOI: 10.3389/fsufs.2019.00095.

Azari-Anpar, M., Khomeiri, M., Ghafouri-Oskuei, H. and Aghajani, N. (2017) Response surface optimization of low-fat ice cream production by using resistant starch and maltodextrin as a fat replacing agent. *Journal of Food Science and Technology* 54(5), 1175–1183. DOI: 10.1007/s13197-017-2492-0.

Bakry, A.M., Chen, Y.Q. and Liang, L. (2019) Developing a mint yogurt enriched with omega-3 oil: physiochemical, microbiological, rheological, and sensorial characteristics. *Journal of Food Processing and Preservation* 43(12), e14287. DOI: 10.1111/jfpp.14287.

Balta, I., Brinzan, L., Stratakos, A.Ch., Linton, M., Kelly, C. *et al.* (2017) Geraniol and linalool loaded nanoemulsions and their antimicrobial activity. *Bulletin of University of Agricultural Sciences and Veterinary Medicine Cluj-Napoca. Animal Science and Biotechnologies* 74(2), 157. DOI: 10.15835/buasvmcn-asb:0025.

Baranauskienė, R., Bylaitė, E., Žukauskaitė, J. and Venskutonis, R.P. (2007) Flavor retention of peppermint (*Mentha piperita* L.) essential oil spray-dried in modified starches during encapsulation and storage. *Journal of Agricultural and Food Chemistry* 55, 3027–3036. DOI: 10.1021/jf062508c.

Bedoya-Serna, C.M., Dacanal, G.C., Fernandes, A.M. and Pinho, S.C. (2018) Antifungal activity of nanoemulsions encapsulating oregano (*Origanum vulgare*) essential oil: *in vitro* study and application in Minas Padrão cheese. *Brazilian Journal of Microbiology* 49(4), 929–935. DOI: 10.1016/j.bjm.2018.05.004.

Belitz, H.D., Grosch, W. and Schieberle, P. (2009) Milk and dairy products. In: *Food Chemistry*. Springer, Berlin/ Heidelberg, pp. 498–545.

Bimlesh, K.S.S., Rajan, M., Rajesh, S. and Richa, K. (2018) Physico-chemical and antimicrobial properties of d-limonene oil nanoemulsion stabilized by whey protein-maltodextrin conjugates. *Journal of Food Science and Technology* 55(7), 2749–2757. DOI: 10.1007/s13197-018-3198-7.

Borthakur, P., Boruah, P.K., Bhagyasmeeta Sharma, B., and Das, M.R. (2016) Nanoemulsion: preparation and its application in food industry. In: Grumezescu, M.A. (ed.) *Emulsions Nanotechnology in the Agri-Food Industry*, Vol. 3, pp. 153–191.

Burt, S. (2004) Essential oils: their antibacterial properties and potential applications in foods—a review. *International Journal of Food Microbiology* 94, 223–253. DOI: 10.1016/j.ijfoodmicro.2004.03.022.

Çam, M., İçyer, N.C. and Erdoğan, F. (2014) Pomegranate peel phenolics: microencapsulation, storage stability and potential ingredient for functional food development. *LWT - Food Science and Technology* 55(1), 117–123. DOI: 10.1016/j.lwt.2013.09.011.

Caon, T., Martelli, S.M., and Fakhouri, F.M. (2017) New trends in the food industry: application of nanosensors in food packaging. In: Grumezescu, A.M. (ed) *Nanobiosensors*. Academic Press, London, pp. 773–804.

Chanamai, R. and McClements, D.J. (2000) Impact of weighting agents and sucrose on gravitational separation of beverage emulsions. *Journal of Agricultural and Food Chemistry* 48, 5561–5565. DOI: 10.1021/jf0002903.

Chang, Y., McLandsborough, L. and McClements, D.J. (2012) Physical properties and antimicrobial efficacy of thyme oil nanoemulsions: influence of ripening inhibitors. *Journal of Agricultural and Food Chemistry* 60(48), 12056–12063. DOI: 10.1021/jf304045a.

Chen, L., Remondetto, G.E. and Subirade, M. (2006) Food protein-based materials as nutraceutical delivery systems. *Trends in Food Science & Technology* 17(5), 272–283. DOI: 10.1016/j.tifs.2005.12.011.

Choi, S.J. and McClements, D.J. (2020) Nanoemulsions as delivery systems for lipophilic nutraceuticals: strategies for improving their formulation, stability, functionality and bioavailability. *Food Science and Biotechnology* 29(2), 149–168. DOI: 10.1007/s10068-019-00731-4.

Christaki, S., Moschakis, T., Hatzikamari, M. and Mourtzinos, I. (2022) Nanoemulsions of oregano essential oil and green extracts: characterization and application in whey cheese. *Food Control* 141, 109190. DOI: 10.1016/j.foodcont.2022.109190.

Chuacharoen, T., Prasongsuk, S. and Sabliov, C.M. (2019) Effect of surfactant concentrations on physicochemical properties and functionality of curcumin nanoemulsions under conditions relevant to commercial utilization. *Molecules* 24, 2744. DOI: 10.3390/molecules24152744.

Constantinescu, A.M., Tiţa, M.A. and Georgescu, C. (2019) Comparative analysis of yoghurts obtained with bioactive compounds. *Bulletin of the Transilvania University of Braşov* 12(61), 73–84. DOI: 10.31926/but.fwiafe.2019.12.61.2.6.

Costa, J.M., Maciel, C.L., Teixeira, A.J., Vicente, A.A. and Cerqueira, A.M. (2018) Use of edible films and coatings in cheese preservation: opportunities and challenges. *Food Research International* 107, 84–92. DOI: 10.1016/j.foodres.2018.02.013.

Dammak, I. and Sobral, P.J. do A. (2019) Active gelatin films incorporated with eugenol nanoemulsions: effect of emulsifier type on films properties. *International Journal of Food Science & Technology* 54(9), 2725–2735. DOI: 10.1111/ijfs.14183.

Dammak, I., Sobral, P.J. do A., Aquino, A., Neves, M.A. and Conte-Junior, C.A. (2020) Nanoemulsions: using emulsifiers from natural sources replacing synthetic ones—a review. *Food Science and Food Safety* 19(5), 2721–2746. DOI: 10.1111/1541-4337.12606.

Danesh, E., Goudarzi, M. and Jooyandeh, H. (2017) Short communication: Effect of whey protein addition and transglutaminase treatment on the physical and sensory properties of reduced-fat ice cream. *Journal of Dairy Science* 100(7), 5206–5211. DOI: 10.3168/jds.2016-12537.

Dasgupta, N., Ranjan, S. and Gandhi, M. (2019) Nanoemulsions in food: market demand. *Environmental Chemistry Letters* 17(2), 1003–1009. DOI: 10.1007/s10311-019-00856-2.

Das, A.K., Nanda, P.K., Bandyopadhyay, S., Banerjee, R., Biswas, S., *et al.* (2020) Application of nanoemulsion-based approaches for improving the quality and safety of muscle foods: a comprehensive review. *Comprehensive Reviews in Food Science and Food Safety* 19(5), 2677–2700. DOI: 10.1111/1541-4337.12604.

Dávila-Rodríguez, M., López-Malo, A., Palou, E., Ramírez-Corona, N. and Jiménez-Munguía, M.T. (2019) Antimicrobial activity of nanoemulsions of cinnamon, rosemary, and oregano essential oils on fresh celery. *LWT—Food Science and Technology* 112, 108247. DOI: 10.1016/j.lwt.2019.06.014.

de Oliveira Filho, J.G., Albiero, B.R., Cipriano, L., de Oliveira Nobre Bezerra, C.C., Oldoni, F.C.A. *et al.* (2021) Arrowroot starch-based films incorporated with a carnauba wax nanoemulsion, cellulose nanocrystals, and essential oils: a new functional material for food packaging applications. *Cellulose* 28(10), 6499–6511. DOI: 10.1007/s10570-021-03945-0.

Diaconescu, C., Vidu, L., Urdes, L. and Dragomir, N. (2011) *Tehnici avansate de apreciere a calităţii laptelui şi produselor lactate*. Valahia University Press, Romania.

Donsì, F. (2018) Applications of nanoemulsions in foods. In: Jafari, S.M. and McClements, D.J. (eds) *Nanoemulsions: Formulation, Applications, and Characterization*. Academic Press, London, pp. 349–377.

Donsì, F., Ferrari, G. and Maresca, P. (2009) High-pressure homogenization for food sanitization. In: Barbosa-Cánovas, G., Mortimer, A., Lineback, D., Spiess, W., Buckle, K. *et al.* (eds) *Global Issues in Food Science and Technology*. Academic Press, Burlington, MA, pp. 309–352.

El-Naggar, M.E., Hussein, J., El-sayed, S.M., Youssef, A.M., El Bana, M. *et al.* (2020) Protective effect of the functional yogurt based on *Malva parviflora* leaves extract nanoemulsion on acetic acid-induced ulcerative colitis in rats. *Journal of Materials Research and Technology* 9(6), 14500–14508. DOI: 10.1016/j.jmrt.2020.10.047.

El-Sayed, H.S. and El-Sayed, S.M. (2021) A modern trend to preserve white soft cheese using nano-emulsified solutions containing cumin essential oil. *Environmental Nanotechnology, Monitoring & Management* 16, 100499. DOI: 10.1016/j.enmm.2021.100499.

El-Sayed, M.I. and Shalaby, T.I. (2021) Production of processed cheese supplemented with curcumin nanoemulsion. *American Journal of Food and Nutrition* 9(3), 96–105. DOI: 10.12691/ajfn-9-3-1.

El-Sayed, S.M., El-Sayed, H.S., Elgamily, H.M. and Youssef, A.M. (2022) Preparation and evaluation of yogurt fortified with probiotics jelly candy enriched with grape seeds extract nanoemulsion. *Journal of Food Processing and Preservation* 46(7), e16713. DOI: 10.1111/jfpp.16713.

Elshamy, S., Khadizatul, K., Uemura, K., Nakajima, M. and Neves, M.A. (2021) Chitosan-based film incorporated with essential oil nanoemulsion foreseeing enhanced antimicrobial effect. *Journal of Food Science and Technology* 58(9), 3314–3327. DOI: 10.1007/s13197-020-04888-3.

Espín, J.C., García-Conesa, M.T. and Tomás-Barberán, F.A. (2007) Nutraceuticals: facts and fiction. *Phytochemistry* 68(22–24), 2986–3008. DOI: 10.1016/j.phytochem.2007.09.014.

Espino-Manzano, S.O., León-López, A., Aguirre-Álvarez, G., González-Lemus, U., Prince, L. *et al.* (2020) Application of nanoemulsions (W/O) of extract of *Opuntia oligacantha* C.F. Först and orange oil in gelatine films. *Molecules* 25(15), 3487. DOI: 10.3390/molecules25153487.

Espitia, P.J.P., Fuenmayor, C.A. and Otoni, C.G. (2018) Nanoemulsions: synthesis, characterization, and application in bio-based active food packaging. *Comprehensive Reviews in Food Science and Food Safety* 18(1), 264–285. DOI: 10.1111/1541-4337.12405.

Fakhar, I., Farhan, S., Muhammad, A., Ali, I. and Muhammad, A.K. (2022) Food grade nanoemulsions: promising delivery systems for functional ingredients. *Journal of Food Science and Technology* (in Press) 60(5), 1461–1471. DOI: 10.1007/s13197-022-05387-3.

Falsafi, S.R., Rostamabadi, H., Babazadeh, A., Tarhan, Ö., Rashidinejad, A. *et al.* (2022) Lycopene nanodelivery systems; recent advances. *Trends in Food Science & Technology* 119, 378–399. DOI: 10.1016/j.tifs.2021.12.016.

Gadhave, A.D. (2014) Nanoemulsions: formation, stability and applications. *International Journal for Research in Science & Advanced Technologies* 2, 38–43.

Gharibzahedi, S.M.T. and Mohammadnabi, S. (2017) Effect of novel bioactive edible coatings based on jujube gum and nettle oil-loaded nanoemulsions on the shelf-life of Beluga sturgeon fillets. *International Journal of Biological Macromolecules* 95, 769–777. DOI: 10.1016/j.ijbiomac.2016.11.119.

Ghiorghiţă, G. and Nicuţă-Petrescu, D. (2005) Biotehnologiile azi. In: *Biotechnologies Used in Food Industry*. Junimea, Iaşi, Romania, pp. 253–258.

Ghosh, S., Sarkar, T., Das, A. and Chakraborty, R. (2022) Natural colorants from plant pigments and their encapsulation: an emerging window for the food industry. *LWT—Food Science and Technology* 153, 112527. DOI: 10.1016/j.lwt.2021.112527.

Gowda, A., Sharma, V., Goyal, A., Singh, A.K. and Arora, S. (2018) Process optimization and oxidative stability of omega-3 ice cream fortified with flaxseed oil microcapsules. *Journal of Food Science and Technology* 55(5), 1705–1715. DOI: 10.1007/s13197-018-3083-4.

Gupta, A., Eral, H.B., Hatton, T.A. and Doyle, P.S. (2016) Nanoemulsions: formation, properties and applications. *Soft Matter* 12(11), 2826–2841. DOI: 10.1039/C5SM02958A.

Gupta, S., Khan, S., Muzafar, M., Kushwaha, M., Yadav, A.K., and Gupt, A.P. (2016) Encapsulation: entrapping essential oil/flavors/aromas in food. In: Grumezescu, A.M. (ed.) *Encapsulations: Nanotechnology in the Agri-Food Industry*, Vol. 2. Academic Press, London, pp. 229–268.

Gutiérrez, J.M., González, C., Maestro, A., Solè, I., Pey, C.M. *et al.* (2008) Nano-emulsions: new applications and optimization of their preparation. *Current Opinion in Colloid & Interface Science* 13(4), 245–251. DOI: 10.1016/j.cocis.2008.01.005.

Güven, M., Kalender, M. and Taşpinar, T. (2018) Effect of using different kinds and ratios of vegetable oils on ice cream quality characteristics. *Foods* 7(7), 104. DOI: 10.3390/foods7070104.

Hassoun, A. and Çoban, Ö.E. (2017) Essential oils for antimicrobial and antioxidant applications in fish and other seafood products. *Trends in Food Science & Technology* 68, 26–36. DOI: 10.1016/j.tifs.2017.07.016.

Huang, Q., Yu, H. and Ru, Q. (2010) Bioavailability and delivery of nutraceuticals using nanotechnology. *Journal of Food Science* 75(1), 50–57. DOI: 10.1111/j.1750-3841.2009.01457.x.

Jadhav, R.P., Koli, V.W., Kamble, A.B. and Bhutkar, M.A. (2020) A review on nanoemulsion. *Asian Journal of Research in Pharmaceutical Science* 10(2), 103–108. DOI: 10.5958/2231-5659.2020.00020.X.

Jafari, S.M., Assadpoor, E., He, Y.H. and Bhandari, B. (2008) Re-coalescence of emulsion droplets during high-energy emulsification. *Food Hydrocolloids* 22(7), 1191–1202. DOI: 10.1016/j.foodhyd.2007.09.006.

Jaiswal, L., Shankar, S. and Rhim, J.W. (2019) Applications of nanotechnology in food microbiology. *Methods in Microbiology* 46, 43–60. DOI: 10.1016/bs.mim.2019.03.002.

Jin, W., Xu, W., Liang, H., Li, Y., Liu, S. *et al.* (2016) Nanoemulsions for food: properties, production, characterization, and applications. In: Grumezescu, M.A. (ed.) *Emulsions Nanotechnology in the Agri-Food Industry*, Vol. 3. pp. 1–36. DOI: 10.1016/B978-0-12-804306-6.00001-5.

Kabalnov, A. (2001) Ostwald ripening and related phenomena. *Journal of Dispersion Science and Technology* 22(1), 1–12. DOI: 10.1081/DIS-100102675.

Kaderides, K., Goula, A.M. and Adamopoulos, K.G. (2015) A process for turning pomegranate peels into a valuable food ingredient using ultrasound-assisted extraction and encapsulation. *Innovative Food Science & Emerging Technologies* 31, 204–215. DOI: 10.1016/j.ifset.2015.08.006.

Karthik, P., Ezhilarasi, P.N. and Anandharamakrishnan, C. (2017) Challenges associated in stability of food grade nanoemulsions. *Critical Reviews in Food Science and Nutrition* 57(7), 1435–1450. DOI: 10.1080/10408398.2015.1006767.

Kasapoglu, E.D., Kahraman, S., and Fatih Tornuk, F. (2020) Apricot juice processing byproducts as sources of value-added compounds for food industry. *European Food Science and Engineering* 1, 18–23.

Khan, M.R., Sadiq, M.B. and Mehmood, Z. (2020) Development of edible gelatin composite films enriched with polyphenol loaded nanoemulsions as chicken meat packaging material. *CyTA - Journal of Food* 18(1), 137–146. DOI: 10.1080/19476337.2020.1720826.

Koroleva, M.Y. and Yurtov, E.V. (2021) Ostwald ripening in macro- and nanoemulsions. *Russian Chemical Reviews* 90(3), 293–323. DOI: 10.1070/RCR4962.

Kralova, I. and Sjöblom, J. (2009) Surfactants used in food industry: a review. *Journal of Dispersion Science and Technology* 30(9), 1363–1383. DOI: 10.1080/01932690902735561.

Lane, K.E., Li, W., Smith, C. and Derbyshire, E. (2014) The bioavailability of an omega-3-rich algal oil is improved by nanoemulsion technology using yogurt as a food vehicle. *International Journal of Food Science & Technology* 49(5), 1264–1271. DOI: 10.1111/ijfs.12455.

Lavelli, V., Sri Harsha, P.S.C. and Spigno, G. (2016) Modelling the stability of maltodextrin-encapsulated grape skin phenolics used as a new ingredient in apple puree. *Food Chemistry* 209, 323–331. DOI: 10.1016/j.foodchem.2016.04.055.

Li, L., Chen, H., Lü, X., Gong, J. and Xiao, G. (2022) Effects of papain concentration, coagulation temperature, and coagulation time on the properties of model soft cheese during ripening. *LWT—Food Science and Technology* 161, 113404. DOI: 10.1016/j.lwt.2022.113404.

Liu, J., Bi, J., Liu, X., Zhang, B., Wu, X. *et al.* (2019) Effects of high pressure homogenization and addition of oil on the carotenoid bioaccessibility of carrot juice. *Food & Function* 10(1), 458–468. DOI: 10.1039/C8FO01925H.

Maali, A. and Hamed Mosavian, M.T. (2013) Preparation and application of nanoemulsions in the last decade (2000–2010). *Journal of Dispersion Science and Technology* 34(1), 92–105. DOI: 10.1080/01932691.2011.648498.

Madene, A., Jacquot, M., Scher, J. and Desobry, S. (2006) Flavour encapsulation and controlled release - a review. *International Journal of Food Science and Technology* 41(1), 1–21. DOI: 10.1111/j.1365-2621.2005.00980.x.

Manach, C., Williamson, G., Morand, C., Scalbert, A. and Rémésy, C. (2005) Bioavailability and bioefficacy of polyphenols in humans. I. Review of 97 bioavailability studies. *American Journal of Clinical Nutrition* 81(Suppl 1), 230–242. DOI: 10.1093/ajcn/81.1.230S.

Mason, T.G., Wilking, J.N., Meleson, K., Chang, C.B. and și Graves, S.M. (2006) Nanoemulsions: formation, structure, and physical properties. *Journal of Physics Condensed Matter* 18, 635–666. DOI: 10.1088/0953-8984/18/41/R01.

McClements, D.J. (2004) *Food Emulsions. Principles, Practices, and Techniques*, 2nd edn. CRC Press, Boca Raton, FL.

McClements, D.J. (2016) *Food Emulsions. Principles, Practices and Techniques*, 3rd edn, CRC Press, Boca Raton, FL.

McClements, D.J. and Rao, J. (2011) Food-grade nanoemulsions: formulation, fabrication, properties, performance, biological fate, and potential toxicity. *Critical Reviews in Food Science and Nutrition* 51(4), 285–330. DOI: 10.1080/10408398.2011.559558.

McClements, D.J. and Yan, L. (2010) Review of *in vitro* digestion models for rapid screening of emulsion-based systems. *Food & Function* 1(1), 32–59. DOI: 10.1039/c0fo00111b.

McClements, D.J., Decker, E.A. and Weiss, J. (2007) Emulsion-based delivery systems for lipophilic bioactive components. *Journal of Food Science* 72(8), 109–124. DOI: 10.1111/j.1750-3841.2007.00507.x.

McClements, D.J., Decker, E.A., Park, Y. and Weiss, J. (2009) Structural design principles for delivery of bioactive components in nutraceuticals and functional foods. *Critical Reviews in Food Science and Nutrition* 49(6), 577–606. DOI: 10.1080/10408390902841529.

McCowen, K.C., Ling, P.R., Decker, E., Djordjevic, D., Roberts, R.F. *et al.* (2010) A simple method of supplementation of omega-3 polyunsaturated fatty acids: use of fortified yogurt in healthy volunteers. *Nutrition in Clinical Practice* 25(6), 641–645. DOI: 10.1177/0884533610385699.

Mehanna, N.S., Hassan, F.A.M., El-Messery, T.M. and Mohamed, A.G. (2017) Production of functional processed cheese by using tomato juice. *International Journal of Dairy Science* 12(2), 155–160. DOI: 10.3923/ijds.2017.155.160.

Mehmood, T., Ahmed, A. and Ahmed, Z. (2021) Food-grade nanoemulsions for the effective delivery of β-carotene. *Langmuir* 37(10), 3086–3092. DOI: 10.1021/acs.langmuir.0c03399.

Milanović, V., Sabbatini, R., Garofalo, C., Cardinali, F., Pasquini, M. *et al.* (2021) Evaluation of the inhibitory activity of essential oils against spoilage yeasts and their potential application in yogurt. *International Journal of Food Microbiology* 341, 109048. DOI: 10.1016/j.ijfoodmicro.2021.109048.

Mohammed, N.K., Muhialdin, B.J. and Hussin, A.S.M. (2020) Characterization of nanoemulsion of *Nigella sativa* oil and its application in ice cream. *Food Science & Nutrition* 8(6), 2608–2618. DOI: 10.1002/fsn3.1500.

Monti, A. and Alexopoulou, E. (2013) *Kenaf: A Multi-Purpose Crop for Several Industrial Applications: New Insights from the Biokenaf Project*. Springer, London. DOI: 10.1007/978-1-4471-5067-1.

Mustafa, R., He, Y., Shim, Y.Y. and Reaney, M.J.T. (2018) Aquafaba, wastewater from chickpea canning, functions as an egg replacer in sponge cake. *International Journal of Food Science & Technology* 53(10), 2247–2255. DOI: 10.1111/ijfs.13813.

Noori, S., Zeynali, F. and Almasi, H. (2018) Antimicrobial and antioxidant efficiency of nanoemulsion-based edible coating containing ginger (*Zingiber officinale*) essential oil and its effect on safety and quality attributes of chicken breast fillets. *Food Control* 84, 312–320. DOI: 10.1016/j.foodcont.2017.08.015.

Norcino, L.B., Mendes, J.F., Natarelli, C.V.L., Manrich, A., Oliveira, J.E. *et al.* (2020) Pectin films loaded with copaiba oil nanoemulsions for potential use as bio-based active packaging. *Food Hydrocolloids* 106, 105862. DOI: 10.1016/j.foodhyd.2020.105862.

Ozogul, Y., Karsli, G.T., Durmuş, M., Yazgan, H., Oztop, H.M., *et al.* (2022) Recent developments in industrial applications of nanoemulsions. *Advances in Colloid and Interface Science* 304, 102685. DOI: 10.1016/j.cis.2022.102685.

Ozogul, Y., Yuvka, İ., Ucar, Y., Durmus, M., Kösker, A.R. *et al.* (2017) Evaluation of effects of nanoemulsion based on herb essential oils (rosemary, laurel, thyme and sage) on sensory, chemical and microbiological quality of rainbow trout (*Oncorhynchus mykiss*) fillets during ice storage. *LWT—Food Science and Technology* 75, 677–684. DOI: 10.1016/j.lwt.2016.10.009.

Öztürk, B. (2017) Nanoemulsions for food fortification with lipophilic vitamins: production challenges, stability, and bioavailability. *European Journal of Lipid Science and Technology* 118, 1500539. DOI: 10.1002/ejlt.201500539.

Pan, K., Chen, H., Davidson, P.M. and Zhong, Q. (2014) Thymol nanoencapsulated by sodium caseinate: physical and antilisterial properties. *Journal of Agricultural and Food Chemistry* 62(7), 1649–1657. DOI: 10.1021/jf4055402.

Panghal, A., Chhikara, N., Anshid, V., Charan, M.V.S., Surendran, V. *et al.* (2019) Nanoemulsions: a promising tool for dairy sector. In: Prasad, R., Kumar, V., Kumar, M. and Choudhary, D. (eds) *Nanobiotechnology in Bioformulations*. Springer, Cham, Switzerland, pp. 99–117.

Pérez-Soto, E., Cenobio-Galindo, A. de J., Espino-Manzano, S.O., Franco-Fernández, M.J., Ludeña-Urquizo, F.E. *et al.* (2021) The addition of microencapsulated or nanoemulsified bioactive compounds influences the antioxidant and antimicrobial activities of a fresh cheese. *Molecules* 26(8), 2170. DOI: 10.3390/molecules26082170.

Puttasiddaiah, R., Lakshminarayana, R., Somashekar, N.L., Gupta, V.K., Inbaraj, B.S. *et al.* (2022) Advances in nanofabrication technology for nutraceuticals: new insights and future trends. *Bioengineering* 9(9), 478. DOI: 10.3390/bioengineering9090478.

Qazi, H.J., Ye, A., Acevedo-Fani, A. and Singh, H. (2021) *In vitro* digestion of curcumin-nanoemulsion-enriched dairy protein matrices: impact of the type of gel structure on the bioaccessibility of curcumin. *Food Hydrocolloids* 117, 106692. DOI: 10.1016/j.foodhyd.2021.106692.

Rocha, G.A., Fávaro-Trindade, C.S. and Grosso, C.R.F. (2012) Microencapsulation of lycopene by spray drying: characterization, stability and application of microcapsules. *Food and Bioproducts Processing* 90(1), 37–42. DOI: 10.1016/j.fbp.2011.01.001.

Romulo, A., Meindrawan, B. and Marpietylie (2021) Effect of dairy and non-dairy ingredients on the physical characteristic of ice cream: review. *IOP Conference Series: Earth and Environmental Science* 794(1), 012145. DOI: 10.1088/1755-1315/794/1/012145.

Ron, N., Zimet, P., Bargarum, J. and Livney, Y.D. (2010) Beta-lactoglobulin–polysaccharide complexes as nanovehicles for hydrophobic nutraceuticals in non-fat foods and clear beverages. *International Dairy Journal* 20(10), 686–693. DOI: 10.1016/j.idairyj.2010.04.001.

Ruiz-Montañez, G., Ragazzo-Sanchez, J.A., Picart-Palmade, L., Calderón-Santoyo, M. and Chevalier-Lucia, D. (2017) Optimization of nanoemulsions processed by high-pressure homogenization to protect a bioactive extract of jackfruit (*Artocarpus heterophyllus* Lam). *Innovative Food Science & Emerging Technologies* 40, 35–41. DOI: 10.1016/j.ifset.2016.10.020.

Salama, H.H., El-Sayed, H.S., Kholif, A.M.M. and Edris, A.E. (2021) Essential oils nanoemulsion for the flavoring of functional stirred yogurt: manufacturing, physicochemical, microbiological, and sensorial investigation. *Journal of the Saudi Society of Agricultural Sciences* 21(6), 372–382. DOI: 10.1016/j.jssas.2021.10.001.

Salem, M.A. and Ezzat, S.M. (2018) Nanoemulsions in food industry. In: Milani, J.M. (ed.) *Some New Aspects of Colloidal Systems in Foods*. InTech Open. Available at: https://www.intechopen.com/chapters/63457

Salvia-Trujillo, L., Soliva-Fortuny, R., Rojas-Graü, M.A., McClements, D.J. and Martín-Belloso, O. (2017) Edible nanoemulsions as carriers of active ingredients: a review. *Annual Review of Food Science and Technology* 8, 439–466. DOI: 10.1146/annurev-food-030216-025908.

Sanchez, L.T., Pinzon, M.I. and Villa, C.C. (2022) Development of active edible films made from banana starch and curcumin-loaded nanoemulsions. *Food Chemistry* 371, 131121. DOI: 10.1016/j.foodchem.2021.131121.

Sari, T.P., Mann, B., Kumar, R., Singh, R.R.B., Sharma, R. *et al.* (2015) Preparation and characterization of nanoemulsion encapsulating curcumin. *Food Hydrocolloids* 43, 540–546. DOI: 10.1016/j.foodhyd.2014.07.011.

Severino, R., Ferrari, G., Vu, K.D., Donsì, F., Salmieri, S. *et al.* (2015) Antimicrobial effects of modified chitosan based coating containing nanoemulsion of essential oils, modified atmosphere packaging and gamma irradiation against *Escherichia coli* O157:H7 and *Salmonella typhimurium* on green beans. *Food Control* 50, 215–222. DOI: 10.1016/j.foodcont.2014.08.029.

Shahidi, F. and Zhong, Y. (2010) Lipid oxidation and improving the oxidative stability. *Chemical Society Reviews* 39(11), 4067–4079. DOI: 10.1039/b922183m.

Shariffa, Y.N., Tan, T.B., Uthumporn, U., Abas, F., Mirhosseini, H. *et al.* (2017) Producing a lycopene nanodispersion: formulation development and the effects of high pressure homogenization. *Food Research International* 101, 165–172. DOI: 10.1016/j.foodres.2017.09.005.

Sharifi, A., Golestan, L. and Sharifzadeh Baei, M. (2013) Studying the enrichment of ice cream with alginate nanoparticles including Fe and Zn salts. *Journal of Nanoparticles* 2013, 1–5. DOI: 10.1155/2013/754385.

Shehzad, Q., Rehman Slara, A., Ali, A., and Khan, S.*et al.* (2020) Preparation and characterization of resveratrol loaded nanoemulsions. *International Journal of Agriculture Innovations and Research* 8, 300–310.

Silva, H.D., Cerqueira, M.Â. and Vicente, A.A. (2012) Nanoemulsions for food applications: development and characterization. *Food and Bioprocess Technology* 5(3), 854–867. DOI: 10.1007/s11947-011-0683-7.

Silva, H.D., Cerqueira, M.A. and Vicente, A.A. (2015) Influence of surfactant and processing conditions in the stability of oil-in-water nanoemulsions. *Journal of Food Engineering* 167, 89–98. DOI: 10.1016/j.jfoodeng.2015.07.037.

Singhal, A., Karaca, A.C., Tyler, R. and Nickerson, M. (2016) Pulse proteins: from processing to structure-function relationships. In: Goyal, A.K. (ed.) *Grain Legumes*. InTech, Rijeka, Croatia, pp. 55–78. DOI: 10.5772/61382.

Solans, C. and Solé, I. (2012) Nano-emulsions: formation by low-energy methods. *Current Opinion in Colloid & Interface Science* 17(5), 246–254. DOI: 10.1016/j.cocis.2012.07.003.

Solans, C., Izquierdo, P., Nolla, J., Azemar, N. and Garcia-Celma, M.J. (2005) Nano-emulsions. *Current Opinion in Colloid & Interface Science* 10(3–4), 102–110. DOI: 10.1016/j.cocis.2005.06.004.

Sonneville-Aubrun, O., Simonnet, J.-T. and L'Alloret, F. (2004) Nanoemulsions: a new vehicle for skin-care products. *Advances in Colloid and Interface Science* 108–109, 145–149. DOI: 10.1016/j.cis.2003.10.026.

Stang, M., Schuchmann, H. and Schubert, H. (2001) Emulsification in high-pressure homogenizers. *Engineering in Life Sciences* 1(4), 151–157. DOI: 10.1002/1618-2863(200110)1:4<151::AID-ELSC151>3.0.CO;2-D.

Sugumar, S., Singh, S., Mukherjee, A. and Chandrasekaran, N. (2015) Nanoemulsion of orange oil with non ionic surfactant produced emulsion using ultrasonication technique: evaluating against food spoilage yeast. *Applied Nanoscience* 6(1), 113–120. DOI: 10.1007/s13204-015-0412-z.

Tabilo-Munizaga, G., Villalobos-Carvajal, R., Herrera-Lavados, C., Moreno-Osorio, L., Jarpa-Parra, M. *et al.* (2019) Physicochemical properties of high-pressure treated lentil protein-based nanoemulsions. *LWT—Food Science and Technology* 101, 590–598. DOI: 10.1016/j.lwt.2018.11.070.

Tadros, T., Izquierdo, P., Esquena, J. and Solans, C. (2004) Formation and stability of nano-emulsions. *Advances in Colloid and Interface Science* 108–109, 303–318. DOI: 10.1016/j.cis.2003.10.023.

Terjung, N., Löffler, M., Gibis, M., Hinrichs, J. and Weiss, J. (2011) Influence of droplet size on the efficacy of oil-in-water emulsions loaded with phenolic antimicrobials. *Food & Function* 3(3), 290–301. DOI: 10.1039/c2fo10198j.

Tesch, S., Freudig, B. and Schubert, H. (2003) Production of emulsions in high-pressure homogenizers – Part I: disruption and stabilization of droplets. *Chemical Engineering & Technology* 26(5), 569–573. DOI: 10.1002/ceat.200390086.

Tipchuwong, N., Chatraporn, C., Ngamchuachit, P. and Tansawat, R. (2017) Increasing retention of vitamin D 3 in vitamin D 3 fortified ice cream with milk protein emulsifier. *International Dairy Journal* 74, 74–79. DOI: 10.1016/j.idairyj.2017.01.003.

Tongnuanchan, P., Benjakul, S. and Prodpran, T. (2013) Physico-chemical properties, morphology and antioxidant activity of film from fish skin gelatin incorporated with root essential oils. *Journal of Food Engineering* 117(3), 350–360. DOI: 10.1016/j.jfoodeng.2013.03.005.

Ullah, R., Nadeem, M. and Imran, M. (2017) Omega-3 fatty acids and oxidative stability of ice cream supplemented with olein fraction of chia (*Salvia hispanica* L.) oil. *Lipids in Health and Disease* 16(1), 34. DOI: 10.1186/s12944-017-0420-y.

van der Hee, R.M., Miret, S., Slettenaar, M., Duchateau, G.S.M.J.E., Rietveld, A.G. *et al.* (2009) Calcium absorption from fortified ice cream formulations compared with calcium absorption from milk. *Journal of the American Dietetic Association* 109(5), 830–835. DOI: 10.1016/j.jada.2009.02.017.

Waglewska, E. and Bazylińska, U. (2021) Biodegradable amphoteric surfactants in titration-ultrasound formulation of oil-in-water nanoemulsions: rational design, development, and kinetic stability. *International Journal of Molecular Sciences* 22(21), 11776. DOI: 10.3390/ijms222111776.

Wang, W., Zhao, D., Xiang, Q., Li, K., Wang, B. *et al.* (2021) Effect of cinnamon essential oil nanoemulsions on microbiological safety and quality properties of chicken breast fillets during refrigerated storage. *LWT—Food Science and Technology* 152, 112376. DOI: 10.1016/j.lwt.2021.112376.

Widyastuti, Y. and Febrisiantosa, A. (2013) Milk and different types of milk products. In: Visakh, P.M., Iturriaga, L.B., and Ribotta, P.D. (eds.) *Advances in Food Science and Nutrition*. Scrivener Publishing, Beverly, MA, pp. 49–68.

Windhab, E.J., Dressler, M., Feigl, K., Fischer, P. and Megias-Alguacil, D. (2005) Emulsion processing—from single-drop deformation to design of complex processes and products. *Chemical Engineering Science* 60(8–9), 2101–2113. DOI: 10.1016/j.ces.2004.12.003.

Witthayapanyanon, A., Acosta, E.J., Harwell, J.H. and Sabatini, D.A. (2006) Formulation of ultralow interfacial tension systems using extended surfactants. *Journal of Surfactants and Detergents* 9(4), 331–339. DOI: 10.1007/s11743-006-5011-2.

Wu, C., Wang, L., Hu, Y., Chen, S., Liu, D. *et al.* (2016) Edible coating from citrus essential oil-loaded nanoemulsions: physicochemical characterization and preservation performance. *RSC Advances* 6(25), 20892–20900. DOI: 10.1039/C6RA00757K.

Xiong, Y., Li, S., Warner, R.D. and Fang, Z. (2020) Effect of oregano essential oil and resveratrol nanoemulsion loaded pectin edible coating on the preservation of pork loin in modified atmosphere packaging. *Food Control* 114, 107226. DOI: 10.1016/j.foodcont.2020.107226.

Xue, J., Davidson, P.M. and Zhong, Q. (2017) Inhibition of *Escherichia coli* O157:H7 and *Listeria monocytognes* growth in milk and cantaloupe juice by thymol nanoemulsions prepared with gelatin and lecithin. *Food Control* 73(Part B), 1499–1506. DOI: 10.1016/j.foodcont.2016.11.015.

Yadav, V., Alka Sharma, A. and Singh, S.K. (2015) Microencapsulation techniques applicable to food flavours research and development: a comprehensive review. *International Journal of Food and Nutritional Sciences* 4, 119–124. Available at: https://www.researchgate.net/publication/280028817

Yu, H. and Huang, Q. (2012) Improving the oral bioavailability of curcumin using novel organogel-based nanoemulsions. *Journal of Agricultural and Food Chemistry* 60(21), 5373–5379. DOI: 10.1021/jf300609p.

Zheng, B. and McClements, D.J. (2020) Formulation of more efficacious curcumin delivery systems using colloid science: enhanced solubility, stability, and bioavailability. *Molecules* 25(12), 2791. DOI: 10.3390/molecules25122791.

Zheng, B., Peng, S., Zhang, X. and McClements, D.J. (2018) Impact of delivery system type on curcumin bioaccessibility: comparison of curcumin-loaded nanoemulsions with commercial curcumin supplements. *Journal of Agricultural and Food Chemistry* 66(41), 10816–10826. DOI: 10.1021/acs.jafc.8b03174.

Zhong, J., Yang, R., Cao, X., Liu, X. and Qin, X. (2018) Improved physicochemical properties of yogurt fortified with fish oil/γ-oryzanol by nanoemulsion technology. *Molecules* 23(1), 56. DOI: 10.3390/molecules23010056.

4 Smart Nanodelivery Systems for Transporting Chemicals and DNA into Plants

Khairul Anwar Ishak[1,2]* and Rauzah Hashim[1]

[1]*Centre for Fundamental and Frontier Sciences in Nanostructure Self-Assembly, Department of Chemistry, Faculty of Science, Universiti Malaya, 50603 Kuala Lumpur, Malaysia; [2]Institute of Biological Sciences, Faculty of Science, University of Malaya, 50603 Kuala Lumpur, Malaysia*

Abstract

Bibliometric analysis using the Web of Science has shown a dramatic increase in the number of studies relating to plant nanotechnology research from 2010 to the present, demonstrating that researchers have recognized the potential of using nanotechnology for agriculture advancement. The focus areas of research encompass the development of nanoparticles for plant stress detection, bioimaging, fertilization, soil modification, and other applications. Furthermore, the use of nanoparticles as smart nanodelivery systems has piqued the interest of many researchers, as demonstrated by a significant annual increase in publications on this topic. The basic principles of nanoparticle engineering as a smart nanodelivery system will be reviewed in this chapter. Their potential application to deliver chemical compounds and genetic materials into plants will be discussed with examples.

Keywords: nanotechnology, nanobiotechnology, nanoparticles, smart nanodelivery system

4.1 Introduction

In human civilization, crops such as paddy, wheat, maize, and barley have been more than sustenance providers; they have served as emblems of a nation's economic power. For centuries, they have been the source of great pride and celebration. These photoautotrophic organisms have been subtly shaping the political and social structures of nations since the dawn of civilization (Tauger, 2010). Mesopotamia, Ancient Egypt, Ancient China, and the Indus Valley arose as a direct result of major advancements in agriculture where humans no longer depended on hunting and gathering activities for survival.

Intercropping and crop rotation are two examples of ancient practices that are used to restore soil fertility. More recent times have seen the introduction of synthetic agrochemicals that are sprayed or broadcast over crops to provide fertilizer and/or pest control to increase crop yield. When applied directly, these agrochemicals have poor efficacy or insufficiently reach the crop's target sites and are often wasted, resulting in levels far below the minimal effective concentration needed for healthy plant growth. Some factors that cause these losses are chemical

*Corresponding author: ka.ishak@um.edu.my

© CAB International 2023. *Nanoformulations for Sustainable Agriculture and Environmental Risk Mitigation* (eds Z. Khan and N.A. Ṣuṭan)
DOI: 10.1079/9781800623095.0004

leaching, evaporation, photolysis, hydrolysis, and microbiological deterioration, among other factors (Yang *et al.*, 2016; Wang *et al.*, 2019). Additionally, issues such as overapplication and surface runoff have an impact on this traditional application (Xia *et al.*, 2020). Recently, nanotechnology has gained more interest in agriculture, particularly the use of nanoparticles (NPs) as vehicles to deliver agrochemicals to plants in an effective way (Solanki *et al.*, 2015; An *et al.*, 2022). NPs help the substances to achieve full biological competency while reducing their loss and negative effects, through effective delivery and controlled release of chemical compounds in sufficient and necessary amounts (An *et al.*, 2022).

Converging nanotechnology with biotechnology opens new avenues for plant gene modification or engineering (Cunningham *et al.*, 2018; Lv *et al.*, 2020). The delivery of bioactive molecules such as nucleotides, proteins, and activators may be accomplished using NPs as novel vehicles or vectors (Jat *et al.*, 2020). Traditional approaches such as electroporation, particle bombardment (gene gun-mediated), polyethylene glycol (PEG)-mediated, and *Agrobacterium*-mediated delivery have a number of disadvantages, including the potential to harm cells, reduce cell viability, transport DNA with low integrity, and be ineffective against many different plant species (Demirer and Landry, 2017). Nanobiotechnology may offer high transformation efficiency and good biocompatibility, as well as excellent preservation of nucleic acids and plant viability (Jat *et al.*, 2020). However, there are many obstacles that need to be overcome before NPs can be widely used to deliver genes into plants.

Nanotechnology in agriculture has now progressively advanced from laboratory trials to real-world applications (Kumar *et al.*, 2019). A significant amount of research has been reported in recent decades on the engineering of NPs or nanomaterials to have specific desirable properties, with a particular emphasis on nanodelivery systems (Fig. 4.1). This chapter outlines the technological development of NPs as smart nanodelivery (SND) systems and their application in delivering chemicals, DNA, and other compounds into plants.

4.2 Development of Smart Nanodelivery Systems

SND systems are responsive and functional NPs or nanomaterials that are engineered to carry and deliver active components (cargo) specifically to targeted sites (Fig. 4.2). The potential applications of SND systems are subjected to much research in humans, such as cancer and genetic disease therapy and targeted drug delivery, among others (Basak, 2020; Shah *et al.*, 2021). In general, similar principles of SND systems for humans can be applied to plant biological systems. This section outlines the basic aspects of SND system development for chemicals and DNA delivery used for humans and plants.

4.2.1 NP variety and production

As shown in Fig. 4.2a, there are various types of NP. Metal elements that are commonly used to produce metallic NPs (MNPs) include gold (Au), silver (Ag), palladium (Pd), platinum (Pt), copper (Cu), zinc (Zn), and titanium (Ti). They are either pristine (Au, Ag, Pd, Pt, Cu) or oxides (ZnO, TiO_2). MNPs are created top down using a size-reduction technique that breaks down bulk materials into small components. This can be achieved by employing ultrasonic generators, high-energy pulse homogenizers, microwaves, and other methods (Jamkhande *et al.*, 2019). Bulk metallic material can also be dispersed into NPs using deposition techniques such as sputtering, vapor deposition, and pulsed electrochemical etching. Bottom-up assembly involves the fabrication of nanostructures atom by atom or particle by particle. This can be accomplished through flame spraying, laser pyrolysis, and other methods (Jamkhande *et al.*, 2019). Chemical reactions or processes can also be used to create MNPs. The sol–gel process is an example of a chemical method for fabricating various metal oxide NPs (Sui and Charpentier, 2012). Strong chemical additives such as hydrazine, dimethyl formamide, and sodium borohydride are used as reducing agents in some chemical reactions to prepare MNPs (Phan and Nguyen, 2017). Alternatives that are friendlier to the environment can be used in place of these

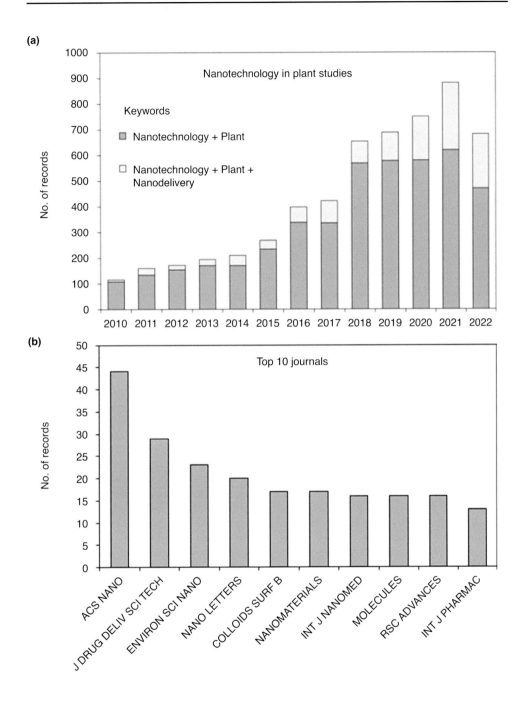

Fig. 4.1. Web of Science survey on (a) nanotechnology application in plant studies; (b) top ten journals that publish articles related to nanodelivery system for plant in the present year.

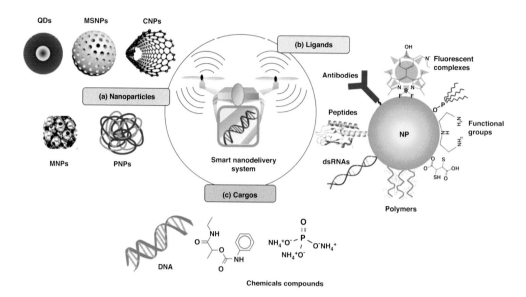

Fig. 4.2. Illustration of a smart nanodelivery (SND) system composed of basic nanoparticles (NPs) (a) as the core, functionalized with ligands (b) for targeting delivery of DNA and other agrochemicals (c). The examples shown are only a few of the many available. CNPs, carbon NPs; ds RNA, double-stranded RNA; MSNPs, mesoporous silica NPs; MNPs, metallic NPs; PNPs, polymeric NPs; QDs, quantum dots.

harsh chemical reagents, such as extracts from leaves, stems, roots, and other plant parts (Rai and Yadav, 2013). The plant extract contains secondary metabolites such as amino acids, vitamins, and many more that can act as reducing and stabilizing agents during the synthesis of metal NPs (Ishak *et al.*, 2019).

Nonmetal elements such as silica and carbon are also used in the manufacture of NPs. To form NPs, these elements are assembled into various nanostructures. Mesoporous silica NPs (MSNPs), for example, can be found in lamellar, hexagonal, and cubic structural arrangements with varying pore sizes (Kankala *et al.*, 2020). There are several methods for producing MSNPs such as the sol–gel process and the Stöber method. Surfactants may be used as modulators to form nanoporous silica particles (Vazquez *et al.*, 2017; Chen *et al.*, 2020). Carbon NPs (CNPs) come in a variety of nanostructured forms (Zaytseva and Neumann, 2016), including cylindrical nanotubes (single- and multiwalled), graphene sheets (single- and multilayered), and spherical fullerene (C60, C70, and higher order). In general, CNPs are synthesized using similar physico-chemical processes such as laser ablation, arc-discharge

techniques, and chemical vapor deposition (Zaytseva and Neumann, 2016; Manawi *et al.*, 2018). However, the processing conditions used to create the various nanostructures may differ.

Quantum dots (QDs), also known as 'artificial atoms,' are nanoscale crystals made up of semiconductor elements (groups II–VI or III–V); examples include cadmium selenide (CdSe), cadmium telluride (CdTe), zinc sulfide (ZnS), and carbon-based QDs. They have distinctive optical (luminescence) and electronic properties, excellent photostability, and a large Stokes shift (Bera *et al.*, 2010; Molaei, 2019). While QDs such as CdSe, CdTe, and ZnS are common in nanosensor and bioimaging applications, carbon-based QDs are commonly studied for chemical and DNA delivery into plants (Molaei, 2019; Wang *et al.*, 2020). Organic compounds such as polysaccharides, organic acids, and others have been investigated as carbon dot precursors via microwave treatments, thermal decomposition, and solvothermal/hydrothermal methods. Electrochemical etching, arc discharge, and laser ablation are some examples that can be used to 'cut' carbonic materials such as carbon nanotubes, graphite, and others into carbon dots (Liu *et al.*, 2019).

Polymeric NPs (PNPs) can be divided into three categories according to their structural organization: nanocapsules, nanospheres, and polymerosomes (Crucho and Barros, 2017). Biodegradable PNPs are commonly made from natural polysaccharides such as chitosan, starch, alginates, pectins and many more. Biodegradable synthetic polymers include polycaprolactone, polylactides, poly(lactide-co-glycolide), and PEG. PNPs can be produced through nanoprecipitation, which involves mixing polymer solution (in a water-miscible solvent) with chemicals/drugs in a stirred aqueous solution that contains a stabilizer or surfactant. The diffusion of the solvent into an aqueous medium followed by evaporation will lead to polymer aggregation/precipitation (Crucho and Barros, 2017). In the polymer-coating method, nano-sized particles (e.g., oil-swollen micelles, MNPs and CNPs) are used as templates in the incubation process with the polymer solution under predetermined conditions (Crucho and Barros, 2017). Recently, the production of PNPs via phase-inversion emulsification (Ishak and Annuar, 2017) has been reported, with the surfactant ratio (Ishak et al., 2017), polymer molecular weight (Ishak and Annuar, 2016), and temperature (Ishak and Annuar, 2018) being controlled. In the chemical coacervation process, the interaction of oppositely charged polyelectrolytes to form NP complexes is involved. The synthesized PNPs are usually subjected to additional cross-linking steps to increase their rigidity. Amphiphilic block copolymers, such as triblock copolymers of poly(caprolactone)-poly(ethylene-glycol)-poly(caprolactone), can self-assemble into polymersomes, a type of PNP with properties similar to liposomes and niosomes (Lee and Feijen, 2012).

4.2.2 NP surface modification and chemical/DNA loading

Surface modification is one of the most important fields in nanotechnology, along with the design and fabrication of NPs (Sperling and Parak, 2010). NP surface modification with a variety of ligands (Fig. 4.2b) is necessary for NP stabilization and preventing agglomeration. It is also important to ensure good biocompatibility,

adhesion and wettability properties of NPs with reduced toxicity prior to their application as SND systems.

The surface modification of inorganic NPs (i.e., MNPs, MSNPs, CNPs, and QDs) is typically carried out by adherence of ligands possessing functional groups such as a thiol group, carboxylic acids, amines, phosphines, and so on (Neouze and Schubert, 2008). For example, MNPs or QDs stabilized with mercaptocarboxylic acids (e.g., dimercaptosuccinic acid) have been demonstrated to have both thiol and carboxylic acid groups (Algar and Krull, 2007). Moreover, these functional groups can subsequently be transformed into different functional groups, for example, carboxylic to amide bonds via a condensation reaction with primary amines. CdSe, ZnS and other QDs are often stabilized by trioctylphosphine or trioctylphosphine oxide to increase their solubility in aqueous solutions (Moreels et al., 2007).

As well as small ligands, the surface of NPs is often coated or modified with polymers to improve their properties (Grubbs, 2007). Uncharged polymers such as PEG are suitable for stabilizing NPs in high-salt buffers and biological environments (Shah et al., 2012). PEG is a biocompatible, hydrophilic, inert polymer that uses steric effects to repel other molecules such as proteins and other cellular components. Coating the surface of NPs with polyethylene oxide has become common to increase their biocompatibility and solubility in water, and thus provide a longer circulation time in vivo (Aqil et al., 2008). To provide NPs with resistance to acid, hyperbranched polyethyleneimine can be used (Prabhakar et al., 2016). Polyacrylic acid is a weak polyelectrolyte that can be applied to stabilize NPs in aqueous medium (Jia et al., 2013). Many polyacrylic acid derivatives have a number of reactive functional groups that could provide a high binding capacity of NPs with biomolecules, making this very appealing for biological applications.

Conjugation of inorganic NPs to biomolecules as the targeting ligand would lead to high specificity in the interaction between NPs and biological systems (Sapsford et al., 2013; Yoo et al., 2019). Nature provides a variety of organic molecules that differ in composition, size, and complexity, which are used to give biological systems and organisms structure and function. On the one hand, there are small molecules such as lipids, vitamins, amino acids, and sugars, and on the other hand, there are larger molecules such as proteins, enzymes, DNA, and RNA, which are all natural polymers. Examples of using these ligands for targeting delivery are provided in Section 4.4.

NPs offer benefits for efficient chemical and DNA delivery (Fig. 4.2c) due to their high surface area and fast mass transfer (An *et al.*, 2022). For these features, active components (e.g., agrochemicals, DNA) are incorporated/loaded into the engineered NPs via a variety of mechanisms, including absorption, adsorption, ionic and/or weak intermolecular bonding, encapsulation, and entrapment within the NP matrix (Xin *et al.*, 2020). For instance, carbon nanotubules can absorb different types of organic compounds, including pyrethroid insecticides, organophosphorus/organochlorine pesticides, and many more (Jakubus *et al.*, 2017). The presence of a thiol group in the 5′ or 3′ position of DNA suggests the possibility of forming a covalent bond with the surface of AuNPs, which could increase colloidal stability and functionalize these metallic NPs for gene delivery (Carnerero *et al.*, 2017). PNPs with a hydrophobic interior have been used to encapsulate lipid-soluble compounds (e.g., coumarin 6) to provide drug stability and controlled release (Xin *et al.*, 2018). MSNPs have been used to entrap and protect active molecules such as prochloraz and fenoxanil within the porous matrix for delivery into different plant species (Zhao *et al.*, 2018; Zhu *et al.*, 2018).

4.2.3 Physico-chemical characterization of NP

Engineered NPs have distinct physico-chemical properties from their bulk counterparts, which is due to their nanoscale size, shape, and chemical and surface structure. Therefore, characterizing these NPs for various physico-chemical properties is critical prior to their application (Mourdikoudis *et al.*, 2018). Particle size and size distribution are the most important pieces of information gathered at the beginning of NP characterization because they determine their suitability for a particular application. The most common techniques for determining

particle size are laser diffraction and photon correlation spectroscopy. Both methods (Filella *et al.*, 1997; Ma *et al.*, 2000) measure the light scattering effect that the NPs cause in order to determine their size. The distribution of particle size populations within a given sample is represented by the polydispersity index. This describes the homogeneity of NP solutions with values ranging from 0 (highly homogeneous) to 1 (highly heterogeneous).

A Zetasizer characterizes the NP surface charge (also known as zeta potential, normally ranging from +100 to −100 mV) to predict colloidal suspension stability. It refers to the electrical potential that exists on a shear plane, the imaginary surface that separates the moving solid phase from the thin liquid layer (Kirby and Hasselbrink, 2004). NPs with a higher zeta potential have higher stability in liquid suspensions as they repel each other to avoid coagulation (Heurtault *et al.*, 2003).

Scanning or tunneling electron microscopy can be used to obtain information about the surface morphology and structure of NPs (Liu, 2005; Puchalski *et al.*, 2007). However, the major constraint to using biological associated samples (e.g., nucleic acid-coated NPs) is the complex and time-consuming sample preparation methods (Low *et al.*, 2011). Low-quality electron micrographs may be obtained when NPs are coated with biomolecules. Dehydration, which occurs during sample preparation, causes the samples to shrink and aggregate as they dry, which can lead to the appearance of unexpected artifacts.

The chemical elements of NPs can be determined using energy-dispersive X-ray analysis (Puchalski *et al.*, 2007). This is based on incident beam electrons causing characteristic X-rays to be produced in specimen atoms. When atoms transition to their ground state from excitation, they will emit distinctive X-rays, where the potential energy is determined by the difference in two orbitals. The X-ray photon energy is very specific to the element atomic number in which the interaction occurs. As a result, the X-rays emitted from a specimen contain information about its chemical composition.

X-ray diffraction is a versatile technique used to characterize NPs and obtain accurate information about their composition and crystalline structure. It works best with powdered samples that have been freshly prepared by drying colloidal solutions of the samples. When the X-rays hit the NPs, the spinning of electrons in the atom makes the X-rays scatter. Interference patterns are made when the scattered X-rays are reflected in different directions. An X-ray diffraction instrument compares the intensity and peak pattern produced by the sample with a diffraction database to determine the composition and crystalline structure of the NPs (Upadhyay et al., 2016).

Fourier-transform infrared (FTIR) spectroscopy is another method that complements crystallographic analysis. It is extremely sensitive to the chemical surface of NPs and is used to identify functional groups in the material. Different chemical bonds vibrate at specific frequencies upon absorbing wavelength energy (Berthomieu and Hienerwadel, 2009). This will produce a unique infrared spectrum, which can then be compared to known signatures of identified materials. FTIR is also capable of analyzing intermolecular interactions within a colloidal system as well as monitoring functional groups in a mixed suspension (Vieira et al., 2011).

For many NPs, Raman spectroscopy has emerged as one of the preferred characterization techniques (Guo et al., 2017). The main reason for this is that Raman spectroscopy can differentiate the structural arrangement of the same type of NPs—for example, between single- and multiwalled carbon nanotubes (Costa et al., 2008). When light is scattered at a different wavelength from the laser source, Raman scattering occurs, which provides useful information about the composition and structural arrangement of the NPs.

4.3 Chemicals and DNA Delivery Into Plants

The delivery of engineered NPs containing agrochemicals or DNA (i.e., the SND system) into the plant can be influenced by different factors, including the NP properties (e.g., size, surface charge, chemical composition), the application mode (e.g., foliar liquid and aerosol spraying, lamina injection/infiltration, drenching, seed priming), interactions of the NPs with environmental conditions (e.g., soil texture

and moisture, microbiota), and the restrictions imposed by the complex anatomy or physiology of different plant species.

4.3.1 SND–plant primary interaction

The SND system can essentially be applied to either the vegetative parts of the plant or the plant roots. To better understand the dynamics of NP–plant interactions, a number of plant anatomical and physiological aspects must be considered.

Openings such as stomata and hydathodes at the shooting surface (foliar entry, as shown in Fig. 4.3) and other vegetative parts such as stigma and bark texture (Eichert and Goldbach, 2008; Eichert et al., 2008; Kurepa et al., 2010) can passively uptake the SND system. Shoot surfaces are naturally covered by a cuticle composed of macromolecules (e.g., cutin, lignin and suberin). Although NPs have been shown to be capable of producing holes in the cuticle, this physical barricade is usually nearly impassable to NPs (Larue et al., 2014; Schwab et al., 2016). However, trichomes on the shooting surface can influence plant surface dynamics by entrapping NPs and thus could increase the amount of time that the NPs stay on the tissues. Damage and wounds to the plant's aerial and hypogeal sections may also provide other viable routes for NP uptake (Al-Salim et al., 2011). Other delivery strategies, such as leaf lamina infiltration, have been proven to successfully infuse SND systems into plants (Giraldo et al., 2014).

While the upper parts of the plant may be impermeable due to the existence of the lipophilic barrier, the rhizodermis provides easy root entry for NPs (Fig. 4.3), particularly near the root tip (Chichiricò and Poma, 2015). However, the mechanism of NP uptake in soil seems to be more complex than in plant aerial parts. Plant mucilage (gel-like substances), organic matters, and symbiotic microorganisms in soil may all have an impact on the availability of NPs. Root mucilage and exudates, for example, could support the attachment of NPs to the root surface, which in turn may increase the NP uptake rate but at the same time may entrap and aggregate NPs (Avellan et al., 2017). In addition, the particle's charge may also

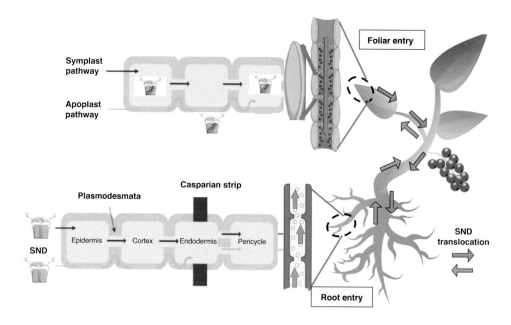

Fig. 4.3. Translocation of a smart nanodelivery (SND) system in the plant upon entry.

influence NP aggregation (Milewska-Hendel *et al.*, 2017). Recent findings using enhanced dark-field microscopy have shown that, regardless of the particle's charge, root mucilage tends to accumulate AuNPs. However, NPs with a negative charge were shown not to accumulate in the root mucilage of thale cress and were able to move directly into root tissue (Avellan *et al.*, 2017).

4.3.2   SND translocation in plants

Once they have successfully entered the plant, NPs can be mobilized through two different routes—the symplastic and apoplastic pathways (Fig. 4.3).The former involves the transport of solutes between the cytoplasm of adjacent cells via plasmodesmata and sieve plate pores, while the latter transports solutes through the cell wall and extracellular spaces. Eventually, NPs reach the xylem and phloem vessels for further translocation to other parts of the plant tissue and organs (Fig. 4.3). Organs such as fruits, seeds, and flowers are usually the end point for NP accumulation.

Through the apoplastic pathway, NPs may be transported via radial movement and eventually reach the vascular tissues, where they are transported further to the aerial regions (Larue *et al.*, 2014; Sun *et al.*, 2014). However, the Casparian strip may prevent NPs from completing this radial movement (Lv *et al.*, 2019). To avoid this natural barrier, NPs may switch from the apoplastic to the symplastic pathway. The rigid cell wall, however, poses a physical barricade for NP entry, making symplastic transport much more difficult. The cell wall is a porous multilayered structure made mostly of cellulose/hemicellulose and scaffold proteins. Its filter-like structure can sieve NPs, with pores ranging in size from 10 to 20nm (Carpita *et al.*, 1979). Different types of NP, such as carbon nanotubes with a diameter of less than 50nm, have been demonstrated to easily infiltrate the cell wall in a number of plant species (Lin *et al.*, 2009; Kurepa *et al.*, 2010; Lin *et al.*, 2010; Chang *et al.*, 2013; Etxeberria *et al.*, 2016).

At some point, the symplastic pathway requires the NPs to enter the cytoplasm (Palocci *et al.*, 2017). Endocytosis, membrane translocation, pore formation, and facilitated diffusion through carrier proteins are examples of processes that uptake NPs into cells (Lin *et al.*, 2010; Wang *et al.*, 2012; Serag *et al.*, 2013). Carbon nanotubules have been shown to enter periwinkle plant cells through an endosome-mediated uptake mechanism (Serag *et al.*, 2011). Once the NPs are in the cytoplasm, they can move between cells via the plasmodesmata, a narrow thread of cytoplasm with a diameter of 20–50 nm that connects adjacent cells in plant tissues. NPs with different sizes have been shown to move between cells through plasmodesmata in different plant species (Lin *et al.*, 2009; Geisler-Lee *et al.*, 2013; Zhai *et al.*, 2014).

In addition to application to the entire plant, seeds and protoplasts are also potential routes for SND application. Seed priming is a technique that involves pre-soaking the seeds in a functional NP solution prior to germination. MNPs have been shown to penetrate seeds without significantly affecting seed viability, rate of germination, or shoot development (Racuciu and Creanga, 2009; Pokhrel and Dubey, 2013). These NPs were able to translocate into germinated seedlings. In general, the same idea can be applied to protoplasts (plant cells without a cell wall) (Silva *et al.*, 2010; Lew *et al.*, 2018). They are exposed to a functional NP solution prior to regeneration via plant tissue culture. This is commonly used in plant genetic engineering, whereby the NPs infiltrate the cell membrane and deliver the gene of interest into the cell.

4.3.3   SND *in situ* detection

There are several techniques for detecting NPs in biological specimens. Although some of these were not deliberately used for plant research, the principle of NP detection and analysis remains applicable.

Synchrotron-based X-ray fluorescence microscopy is a unique way to look at the element distribution in biological specimens *in situ* (Lombi and Susini, 2009; Wang *et al.*, 2014; Zhao *et al.*, 2014). For example, endocytosis of lanthanum particles in plant cells has been studied using high spatial resolution (Wang *et al.*, 2014). Synchrotron scanning transmission X-ray microscopy was used to look at the three-dimensional distribution of AgNPs inside a human monocyte with a lateral resolution of 60 nm (Wang *et al.*, 2015). Synchrotron X-ray absorption spectroscopy can be used to measure chemical transformation in cells and plant tissues in real time (Wang *et al.*, 2015).

Single-particle inductively coupled plasma mass spectrometry has recently been developed to determine the size, distribution, and concentration of AuNPs in tomato tissues (Dan *et al.*, 2015). Meanwhile, field flow fractionation and inductively couple plasma mass spectrometry have been jointly used to differentiate and characterize NPs and other colloidal particles in environmental systems and food samples (Baalousha *et al.*, 2011; von der Kammer *et al.*, 2011).

Although nanoscale secondary ion mass spectrometry (nano-SIMS) has only been utilized to analyze the presence of NPs in animal cells (Kempen *et al.*, 2013; Sekine *et al.*, 2017), it should also be applicable to plant studies. By using a high-energy primary ion beam to erode the sample surface and release secondary ion particles, nano-SIMS can acquire nanoscale resolution (down to 50 nm) of elemental distribution in samples. It can measure most of the elements in the periodic table (Wirtz *et al.*, 2015).

4.4   Reported Studies

4.4.1   Application of NPs as SND systems

Any nanodelivery system should ideally have a large loading capacity as well as high biocompatibility. None the less, they are considered 'smart' when they possess special characteristics such as the ability to overcome several physical barriers in order to penetrate the plant's biological system and deliver active components to the desired location (targeted delivery). Furthermore, when exposed to external stimuli, they can release active components in a controlled manner. Currently, SND systems are designed to have multiple functions, which could facilitate other processes such as bioimaging.

Because of their high surface area, biocompatibility, and other attributes, MNPs can be utilized as nanocarriers for pesticides, fertilizers, and other chemical compounds.

For instance, graphene oxide–iron NPs have been shown to be good phosphate carriers with controlled release as an attribute compared with commercial fertilizer granules (Andelkovic *et al.*, 2018). Significant increases in stem and root length were seen in spinach due to enhanced delivery of iron using iron oxide (Fe_2O_3) NPs (Jeyasubramanian *et al.*, 2016). As well as delivering chemical compounds, MNPs could also provide an attractive tool to deliver proteins and genes into plants. For example, one study has shown enhanced DNA delivery inside the epidermis cells of onion plants using gold-functionalized MSNPs (Martin-Ortigosa *et al.*, 2012). In another study, dual-functional AuNP was shown to be useful for delivering synthetic plant hormone (auxin) and facilitating particle detection inside tobacco plants using surface-enhanced Raman spectroscopy and X-ray photoelectron spectroscopy (Nima *et al.*, 2014; Jia *et al.*, 2016).

Because of their high stability and bio-compatibility, low-priced silica NPs are among the most effective for agrochemical delivery (Sun *et al.*, 2014). MSNPs have been utilized to deliver fungicides and pesticides to plants with sustained release behavior, demonstrating the great potential of MSNPs as a controlled-release system. For instance, amino-functionalized MSNPs have been used as nanocarriers for azoxystrobin to improve fungicidal activity against late blight fungus and reduce the adverse effects of pesticides on tomato plants (Xu *et al.*, 2018). As well as MSNPs, CNPs have also been actively used as nanodelivery systems, typically in plant genetic engineering. For example, peptide-conjugated carbon nanotubes provided targeted delivery of a DNA plasmid to the chloroplast of thale cress plants with no permanent chloroplast membrane damage and an absence of impact on photosystem (II) quantum yield (Santana *et al.*, 2022). Polyethyleneimine-passivated carbon QDs were demonstrated to effectively enter cucumber plants and deliver double-stranded RNA 50-fold more effectively than naked dsRNA, indicating the 'smart' attribute of this NP (Delgado-Martín *et al.*, 2022).

Given that most PNPs are ecofriendly and biodegradable, it should be easier to overcome regulatory barriers and gain public acceptance. Overall, PNPs have huge potential in nanodelivery applications with controlled-release

properties to improve crop production. For example, chitosan-based PNPs showed high biocompatibility with cherry tomato plants, resulting in enhanced delivery of gibberellic acid, which promoted fourfold more fruit production (Pereira *et al.*, 2019). In other study, it was demonstrated that PNP derived from PEG and poly(lactide-co-glycolide) could penetrate the seedcoat and deliver recombinant activator protein (fluorescein isothiocyanate-conjugated PEaT1 protein) to improve wheat seedling growth (Gao *et al.*, 2018). By using fluorescent-conjugated poly(phenylene ethynylene) PNP as a nanocarrier, the delivery and uptake of small-interfering (or silencing) RNA (siRNA) by tobacco cells was readily observed (Silva *et al.*, 2010). This study demonstrated successful siRNA delivery through effective silencing of the *NtCesA-1a* and *NtCesA-1b* genes involved in cell wall production. Table 4.1 summarizes other examples of NP applications in chemical and DNA delivery into plants.

4.4.2 Phytotoxicity of NPs

Applications of NPs in DNA and chemical delivery applications have shown tremendous potential for agriculture advancement. However, it is also important to determine their possible toxicity toward plants through nanotoxicological studies. The toxicity effects of NPs on plants can be indicated by NP accumulation levels in plant cells, changes in plant physical growth parameters (e.g., leaf size or length), changes in enzymatic and photosynthetic activities, as well as the levels of phenols, flavonoids and malondialdehyde (Tarrahi *et al.*, 2021).

There are several properties that contribute to the phytotoxicity of NPs, which are highly related to their physico-chemical characteristics—size, shape, surface characteristics, and concentration. The impact of titanium dioxide NPs (TiO_2NPs) on aquatic organisms and plants, for example, showed a particular pattern of responses based on the crystal structure and size of the NPs. When compared with micro-sized particles, TiO_2NPs had the highest toxicity, and their rutile form was less toxic than their anatase form (Clément *et al.*, 2013). The ligands used to modify the surface of the NPs also have crucial

Table 4.1. Different applications of nanoparticles (NPs) as a smart nanodelivery system in plants.

NP	Ligand	Cargo	Plant	Smart attributes/remarks	Reference
Gold NP	PEG with functional thiol and amino group	Plasmid DNA and fluorescein isothiocyanate	Canola protoplast	Successful GUS gene delivery and fluorescence colour for facile imaging (dual function)	Hao *et al.* (2013)
Gold-silver NP	Thiolated PEG	Growth regulator 2,4-D	Tobacco callus	Effective NP uptake and negligible phytotoxicity observed (high biocompatibility)	Nima *et al.* (2014)
MSNP	3-Aminopropyl-trimethoxysilane	AmCyan1 and DsRed2 plasmids and Cre protein	Maize cells	Stable co-delivery of DNA and protein (high loading capacity)	Martin-Ortigosa *et al.* (2014)
MSNP	Poly(*N*-isopropylacrylamide-co-methacrylic acid)	Thiamethoxam (an insecticide)	Rice seedlings	The NPs provided a controlled compound release upon temperature changes and were stable for a long time in the plant biological system	Gao *et al.* (2020)
Carbon nanotube	Polyethylenimine	Plasmid DNA	Arugula, wheat, cotton	The NPs provided an enhanced gene delivery for stable protein expression and high biocompatibility	Demirer *et al.* (2019)

Continued

Table 4.1. Continued

NP	Ligand	Cargo	Plant	Smart attributes/ remarks	Reference
Carbon nanotube	Chitosan	Plasmid DNA	Arugula, watercress	Targeted transgene delivery to chloroplast and transient expression in the mature plant were achieved	Kwak et al. (2019)
Carbon quantum dot	Polyethylenimine	siRNA	Nicotiana benthamiana, tomato	The NPs successfully infiltrated the plant cell wall and released siRNA upon pH changes between cellular compartments	Schwartz et al. (2020)
Polyglutamic acid NP	Chitosan	Gibberellic acid	Common bean	The NPs provided sustained release of hormone (prolonging availability), as well as protection against degradation	Pereira et al. (2017)
PEG NP	Single guide RNA	Cas9 ribonucleoprotein	Petunia protoplast	Direct delivery of purified recombinant protein targeting specific nitrate reductase gene locus	Subburaj et al. (2016)

2,4-D, 2,4-dichlorophenoxyacetic acid; GUS, β-glucuronidase; MSNP, mesoporous silica NP; PEG, polyethylene glycol; siRNS, small interfering RNA.

effects on the toxicity outcomes observed in nanotoxicological studies. For instance, coating copper oxide NPs (CuONPs) with polystyrene resulted in a severe toxic effect in *Chlamydomonas reinhardtii* (a single-celled green algae) in contrast to naked CuONPs (Saison *et al.*, 2010). In addition, the plant species may also contribute to NP phytotoxicity. Cerium oxide NPs, for example, have been shown to inhibit the growth of head lettuce roots, but have negligible toxic effects on wheat, cucumber, and radish (Zhang *et al.*, 2015). In other cases, under similar experimental conditions, graphene NPs cause significant toxic effects on the physiology and morphology of tomato, spinach and cabbage seedlings but not to lettuce seedling (Begum *et al.*, 2011).

Several toxicity mechanisms have been proposed that could potentially harm plant health. Many studies have proposed that the toxicity mechanism in plant cells is primarily due to the production of reactive oxygen species (ROS) (Khan *et al.*, 2019). As a way of detoxifying NPs, plants produce hydrogen peroxide, hydroxyl radical, superoxide, and other types of ROS. The accumulation of high levels of ROS in a plant cell lead to higher lipid peroxidation, DNA degradation, and electrolyte leakage, all of which lead to cell apoptosis (Dev *et al.*, 2018). Furthermore, NPs might release ions that could cause toxic effects in plant cells. For instance, research on the toxicity of CuONPs on duckweed revealed that the toxicity was primarily caused by Cu^{2+} release from the NPs (Perreault *et al.*, 2014). NP aggregation in the cell wall could cause pore blockage, thus reducing the water uptake (apoplastic flow) efficiency of the plant cell (Chen *et al.*, 2010). Furthermore, NPs could have adverse interactions with plasma membranes, chloroplasts, mitochondria, chromosomes, and other organelles, which may disrupt their functions (Mehrian and de Lima, 2016). It has also been proposed that NPs may unnecessarily stimulate a number of signal transduction pathways in the cell, such as calcium flux, ROS generation, and phosphorylation of mitogen-activated protein kinase (Kohan-Baghkheirati and Geisler-Lee, 2015; Marslin *et al.*, 2017). As a result, important plant biochemical activities that are essential for primary growth could be interrupted.

4.5 Conclusions

In contemporary agriculture, the use of NPs as SND systems is undoubtedly growing. For instance, SNDs may be used to deliver agrochemicals to crops on demand in a smart farming system, where everything is precisely managed through the implementation of the Internet of Things and automation. This agricultural development is viewed as a potential future solution to the food security problem, as it has the ability to effectively increase crop productivity. None the less, it is important to consider how novel NPs might be harmful to the environment, to plants and for human consumption. To ensure the safety of their application in agriculture, NPs should be developed in accordance with the 'safety-by-design' principle and nanodelivery systems inspired by nanomedicine.

Acknowledgment

The authors acknowledge financial support by Universiti Malaya through a UM International Collaboration grant (ST021-2022).

References

Algar, W.R. and Krull, U.J. (2007) Luminescence and stability of aqueous thioalkyl acid capped CdSe/ZnS quantum dots correlated to ligand ionization. *ChemPhysChem* 8(4), 561–568. DOI: 10.1002/cphc.200600686.

Al-Salim, N., Barraclough, E., Burgess, E., Clothier, B., Deurer, M. *et al.* (2011) Quantum dot transport in soil, plants, and insects. *The Science of the Total Environment* 409(17), 3237–3248. DOI: 10.1016/j.scitotenv.2011.05.017.

An, C., Sun, C., Li, N., Huang, B., Jiang, J. *et al.* (2022) Nanomaterials and nanotechnology for the delivery of agrochemicals: strategies towards sustainable agriculture. *Journal of Nanobiotechnology* 20(1), 11. DOI: 10.1186/s12951-021-01214-7.

Andelkovic, I.B., Kabiri, S., Tavakkoli, E., Kirby, J.K., McLaughlin, M.J. *et al.* (2018) Graphene oxide-Fe(III) composite containing phosphate – A novel slow release fertilizer for improved agriculture management. *Journal of Cleaner Production* 185, 97–104. DOI: 10.1016/j.jclepro.2018.03.050.

Aqil, A., Vasseur, S., Duguet, E., Passirani, C., Benoît, J.P. *et al.* (2008) PEO coated magnetic nanoparticles for biomedical application. *European Polymer Journal* 44(10), 3191–3199. DOI: 10.1016/j.eurpolymj.2008.07.011.

Avellan, A., Schwab, F., Masion, A., Chaurand, P., Borschneck, D. *et al.* (2017) Nanoparticle uptake in plants: gold nanomaterial localized in roots of *Arabidopsis thaliana* by X-ray computed nanotomography and hyperspectral imaging. *Environmental Science & Technology* 51(15), 8682–8691. DOI: 10.1021/acs.est.7b01133.

Baalousha, M., Stolpe, B. and Lead, J.R. (2011) Flow field-flow fractionation for the analysis and characterization of natural colloids and manufactured nanoparticles in environmental systems: a critical review. *Journal of Chromatography A* 1218(27), 4078–4103. DOI: 10.1016/j.chroma.2011.04.063.

Basak, S. (2020) The age of multistimuli-responsive nanogels: the finest evolved nano delivery system in biomedical sciences. *Biotechnology and Bioprocess Engineering* 25(5), 655–669. DOI: 10.1007/s12257-020-0152-0.

Begum, P., Ikhtiari, R. and Fugetsu, B. (2011) Graphene phytotoxicity in the seedling stage of cabbage, tomato, red spinach, and lettuce. *Carbon* 49(12), 3907–3919. DOI: 10.1016/j.carbon.2011.05.029.

Bera, D., Qian, L., Tseng, T.-K. and Holloway, P.H. (2010) Quantum dots and their multimodal applications: a review. *Materials* 3(4), 2260–2345. DOI: 10.3390/ma3042260.

Berthomieu, C. and Hienerwadel, R. (2009) Fourier transform infrared (FTIR) spectroscopy. *Photosynthesis Research* 101(2–3), 157–170. DOI: 10.1007/s11120-009-9439-x.

Carnerero, J.M., Jimenez-Ruiz, A., Castillo, P.M. and Prado-Gotor, R. (2017) Covalent and non-covalent dna–gold-nanoparticle interactions: new avenues of research. *ChemPhysChem* 18(1), 17–33. DOI: 10.1002/cphc.201601077.

Carpita, N., Sabularse, D., Montezinos, D. and Delmer, D.P. (1979) Determination of the pore size of cell walls of living plant cells. *Science* 205(4411), 1144–1147. DOI: 10.1126/science.205.4411.1144.

Chang, F.-P., Kuang, L.-Y., Huang, C.-A., Jane, W.-N., Hung, Y. *et al.* (2013) A simple plant gene delivery system using mesoporous silica nanoparticles as carriers. *Journal of Materials Chemistry B* 1(39), 5279–5287. DOI: 10.1039/c3tb20529k.

Chen, R., Ratnikova, T.A., Stone, M.B., Lin, S., Lard, M, *et al.* (2010) Differential uptake of carbon nanoparticles by plant and mammalian cells. *Small* 6(5), 612–617. DOI: 10.1002/smll.200901911.

Chen, Z., Peng, B., Xu, J.-Q., Xiang, X.-C., Ren, D.-F. *et al.* (2020) A non-surfactant self-templating strategy for mesoporous silica nanospheres: beyond the Stöber method. *Nanoscale* 12(6), 3657–3662. DOI: 10.1039/c9nr10939k.

Chichiriccò, G. and Poma, A. (2015) Penetration and toxicity of nanomaterials in higher plants. *Nanomaterials* 5(2), 851–873. DOI: 10.3390/nano5020851.

Clément, L., Hurel, C. and Marmier, N. (2013) Toxicity of TiO2 nanoparticles to cladocerans, algae, rotifers and plants - effects of size and crystalline structure. *Chemosphere* 90(3), 1083–1090. DOI: 10.1016/j.chemosphere.2012.09.013.

Costa, S., Borowiak-Palen, E., Kruszynska, M., Bachmatiuk, A., and Kalenczuk, R. (2008) Characterization of carbon nanotubes by Raman spectroscopy. *Materials Science—Poland* 26, 433–441.

Crucho, C.I.C. and Barros, M.T. (2017) Polymeric nanoparticles: a study on the preparation variables and characterization methods. *Materials Science & Engineering: C* 80, 771–784. DOI: 10.1016/j.msec.2017.06.004.

Cunningham, F.J., Goh, N.S., Demirer, G.S., Matos, J.L. and Landry, M.P. (2018) Nanoparticle-mediated delivery towards advancing plant genetic engineering. *Trends in Biotechnology* 36(9), 882–897. DOI: 10.1016/j.tibtech.2018.03.009.

Dan, Y., Zhang, W., Xue, R., Ma, X., Stephan, C. *et al.* (2015) Characterization of gold nanoparticle uptake by tomato plants using enzymatic extraction followed by single-particle inductively coupled plasma–mass spectrometry analysis. *Environmental Science & Technology* 49(5), 3007–3014. DOI: 10.1021/es506179e.

Delgado-Martín, J., Delgado-Olidén, A. and Velasco, L. (2022) Carbon dots boost dsRNA delivery in plants and increase local and systemic siRNA production. *International Journal of Molecular Sciences* 23(10), 5338. DOI: 10.3390/ijms23105338.

Demirer, G.S. and Landry, M.P. (2017) Delivering genes to plants. *Chemical Engineering Progress* 113, 40–45.

Demirer, G.S., Zhang, H., Matos, J.L., Goh, N.S., Cunningham, F.J. *et al.* (2019) High aspect ratio nanomaterials enable delivery of functional genetic material without DNA integration in mature plants. *Nature Nanotechnology* 14(5), 456–464. DOI: 10.1038/s41565-019-0382-5.

Dev, A., Srivastava, A.K. and Karmakar, S. (2018) Nanomaterial toxicity for plants. *Environmental Chemistry Letters* 16(1), 85–100. DOI: 10.1007/s10311-017-0667-6.

Eichert, T. and Goldbach, H.E. (2008) Equivalent pore radii of hydrophilic foliar uptake routes in stomatous and astomatous leaf surfaces--further evidence for a stomatal pathway. *Physiologia Plantarum* 132(4), 491–502. DOI: 10.1111/j.1399-3054.2007.01023.x.

Eichert, T., Kurtz, A., Steiner, U. and Goldbach, H.E. (2008) Size exclusion limits and lateral heterogeneity of the stomatal foliar uptake pathway for aqueous solutes and water-suspended nanoparticles. *Physiologia Plantarum* 134(1), 151–160. DOI: 10.1111/j.1399-3054.2008.01135.x.

Etxeberria, E., Gonzalez, P., Bhattacharya, P., Sharma, P. and Ke, P.C. (2016) Determining the size exclusion for nanoparticles in citrus leaves. *HortScience* 51(6), 732–737. DOI: 10.21273/HORTSCI.51.6.732.

Filella, M., Zhang, J., Newman, M.E. and Buffle, J. (1997) Analytical applications of photon correlation spectroscopy for size distribution measurements of natural colloidal suspensions: capabilities and limitations. *Colloids and Surfaces A: Physicochemical and Engineering Aspects* 120(1–3), 27–46. DOI: 10.1016/S0927-7757(96)03677-1.

Gao, H., Qin, Y., Guo, R., Wu, Y., Qiu, D. *et al.* (2018) Enhanced plant growth promoting role of mPEG-PLGA-based nanoparticles as an activator protein PeaT1 carrier in wheat (*Triticum aestivum L* .) . *Journal of Chemical Technology & Biotechnology* 93(11), 3143–3151. DOI: 10.1002/jctb.5668.

Gao, Y., Xiao, Y., Mao, K., Qin, X., Zhang, Y. *et al.* (2020) Thermoresponsive polymer-encapsulated hollow mesoporous silica nanoparticles and their application in insecticide delivery. *Chemical Engineering Journal* 383, 123169. DOI: 10.1016/j.cej.2019.123169.

Geisler-Lee, J., Wang, Q., Yao, Y., Zhang, W., Geisler, M. *et al.* (2013) Phytotoxicity, accumulation and transport of silver nanoparticles by *Arabidopsis thaliana*. *Nanotoxicology* 7(3), 323–337. DOI: 10.3109/17435390.2012.658094.

Giraldo, J.P., Landry, M.P., Faltermeier, S.M., McNicholas, T.P., Iverson, N.M. *et al.* (2014) Plant nanobionics approach to augment photosynthesis and biochemical sensing. *Nature Materials* 13(4), 400–408. DOI: 10.1038/nmat3890.

Grubbs, R.B. (2007) Roles of polymer ligands in nanoparticle stabilization. *Polymer Reviews* 47(2), 197–215. DOI: 10.1080/15583720701271245.

Guo, H., He, L. and Xing, B. (2017) Applications of surface-enhanced Raman spectroscopy in the analysis of nanoparticles in the environment. *Environmental Science: Nano* 4(11), 2093–2107. DOI: 10.1039/C7EN00653E.

Hao, Y., Yang, X., Shi, Y., Song, S., Xing, J. *et al.* (2013) Magnetic gold nanoparticles as a vehicle for fluorescein isothiocyanate and DNA delivery into plant cells. *Botany* 91(7), 457–466. DOI: 10.1139/cjb-2012-0281.

Heurtault, B., Saulnier, P., Pech, B., Proust, J.-E. and Benoit, J.-P. (2003) Physico-chemical stability of colloidal lipid particles. *Biomaterials* 24(23), 4283–4300. DOI: 10.1016/s0142-9612(03)00331-4.

Ishak, K.A. and Annuar, M.S.M. (2016) Phase inversion of medium-chain-length poly-3-hydroxyalkanoates (mcl-PHA)-incorporated nanoemulsion: effects of mcl-PHA molecular weight and amount on its mechanism. *Colloid and Polymer Science* 294(12), 1969–1981. DOI: 10.1007/s00396-016-3957-9.

Ishak, K.A. and Annuar, M.S.M. (2017) Facile formation of medium-chain-length poly-3-hydroxyalkanoates (mcl-PHA)-incorporated nanoparticle using combination of non-ionic surfactants. *Journal of Surfactants and Detergents* 20(2), 341–353. DOI: 10.1007/s11743-017-1928-x.

Ishak, K.A. and Annuar, M.S.M. (2018) Temperature-induced three-phase equilibrium of medium-chain-length poly-3-hydroxyalkanoates-incorporated emulsion system for production of polymeric nanoparticle. *Journal of Dispersion Science and Technology* 39(3), 375–383. DOI: 10.1080/01932691.2017.1320563.

Ishak, K.A., Annuar, M.S.M. and Ahmad, N. (2017) Optimization of water/oil/surfactant system for preparation of medium-chain-length poly-3-hydroxyalkanoates (mcl-PHA)-incorporated nanoparticles via
nanoemulsion templating technique. *Applied Biochemistry and Biotechnology* 183(4), 1191–1208.
DOI: 10.1007/s12010-017-2492-6.

Ishak, N.M., Kamarudin, S.K. and Timmiati, S.N. (2019) Green synthesis of metal and metal oxide
nanoparticles via plant extracts: an overview. *Materials Research Express* 6(11), 112004. DOI:
10.1088/2053-1591/ab4458.

Jakubus, A., Paszkiewicz, M. and Stepnowski, P. (2017) Carbon nanotubes application in the extraction techniques of pesticides: a review. *Critical Reviews in Analytical Chemistry* 47(1), 76–91. DOI:
10.1080/10408347.2016.1209105.

Jamkhande, P.G., Ghule, N.W., Bamer, A.H. and Kalaskar, M.G. (2019) Metal nanoparticles synthesis: an
overview on methods of preparation, advantages and disadvantages, and applications. *Journal of
Drug Delivery Science and Technology* 53, 101174. DOI: 10.1016/j.jddst.2019.101174.

Jat, S.K., Bhattacharya, J. and Sharma, M.K. (2020) Nanomaterial based gene delivery: a promising
method for plant genome engineering. *Journal of Materials Chemistry B* 8(19), 4165–4175. DOI:
10.1039/d0tb00217h.

Jeyasubramanian, K., Thoppey, U.U.G., Hikku, G.S., Selvakumar, N., Subramania, A. *et al*. (2016)
Enhancement in growth rate and productivity of spinach grown in hydroponics with iron oxide
nanoparticles. *RSC Advances* 6(19), 15451–15459. DOI: 10.1039/C5RA23425E.

Jia, X., Yin, J., He, D., He, X., Wang, K. *et al*. (2013) Polyacrylic Acid Modified Upconversion Nanoparticles
for Simultaneous pH-Triggered Drug Delivery and Release Imaging. *Journal of Biomedical
Nanotechnology* 9(12), 2063–2072. DOI: 10.1166/jbn.2013.1764.

Jia, J.-L., Jin, X.-Y., Liu, Q.-L., Liang, W.-L., Lin, M.-S. *et al*. (2016) Preparation, characterization and
intracellular imaging of 2,4-dichlorophenoxyacetic acid conjugated gold nanorods. *Journal of
Nanoscience and Nanotechnology* 16(5), 4936–4942. DOI: 10.1166/jnn.2016.12096.

Kankala, R.K., Han, Y.-H., Na, J., Lee, C.-H., Sun, Z, *et al*. (2020) Nanoarchitectured structure and surface
biofunctionality of mesoporous silica nanoparticles. *Advanced Materials* 32(23), e1907035. DOI:
10.1002/adma.201907035.

Kempen, P.J., Hitzman, C., Sasportas, L.S., Gambhir, S.S. and Sinclair, R. (2013) Advanced characterization techniques for nanoparticles for cancer research: applications of SEM and NanoSIMS for locat-
ing Au nanoparticles in cells. *Materials Research Society Symposia Proceedings. Materials Research
Society* 1569, 157–163. DOI: 10.1557/opl.2013.613.

Khan, Z., Shahwar, D., Yunus Ansari, M.K. and Chandel, R. (2019) Toxicity assessment of anatase (TiO_2)
nanoparticles: a pilot study on stress response alterations and DNA damage studies in *Lens culinaris*
Medik. *Heliyon* 5(7), e02069. DOI: 10.1016/j.heliyon.2019.e02069.

Kirby, B.J. and Hasselbrink, E.F. Jr. (2004) Zeta potential of microfluidic substrates: 1. Theory, experimental techniques, and effects on separations. *Electrophoresis* 25(2), 187–202. DOI: 10.1002/
elps.200305754.

Kohan-Baghkheirati, E. and Geisler-Lee, J. (2015) Gene expression, protein function and pathways of
Arabidopsis thaliana responding to silver nanoparticles in comparison to silver ions, cold, salt,
drought, and heat. *Nanomaterials* 5(2), 436–467. DOI: 10.3390/nano5020436.

Kumar, A., Gupta, K., Dixit, S., Mishra, K. and Srivastava, S. (2019) A review on positive and negative impacts of nanotechnology in agriculture. *International Journal of Environmental Science and
Technology* 16(4), 2175–2184. DOI: 10.1007/s13762-018-2119-7.

Kurepa, J., Paunesku, T., Vogt, S., Arora, H., Rabatic, B.M. *et al*. (2010) Uptake and distribution of
ultrasmall anatase TiO2 Alizarin red S nanoconjugates in *Arabidopsis thaliana*. *Nano Letters* 10(7),
2296–2302. DOI: 10.1021/nl903518f.

Kwak, S.-Y., Lew, T.T.S., Sweeney, C.J., Koman, V.B., Wong, M.H. *et al*. (2019) Chloroplast-selective gene
delivery and expression in planta using chitosan-complexed single-walled carbon nanotube carriers.
Nature Nanotechnology 14(5), 447–455. DOI: 10.1038/s41565-019-0375-4.

Larue, C., Castillo-Michel, H., Sobanska, S., Cécillon, L., Bureau, S. *et al*. (2014) Foliar exposure of the
crop *Lactuca sativa* to silver nanoparticles: evidence for internalization and changes in Ag speciation. *Journal of Hazardous Materials* 264, 98–106. DOI: 10.1016/j.jhazmat.2013.10.053.

Lee, J.S. and Feijen, J. (2012) Polymersomes for drug delivery: design, formation and characterization.
Journal of Controlled Release 161(2), 473–483. DOI: 10.1016/j.jconrel.2011.10.005.

Lew, T.T.S., Wong, M.H., Kwak, S.Y., Sinclair, R., Koman, V.B. *et al.* (2018) Rational design principles for the transport and subcellular distribution of nanomaterials into plant protoplasts. *Small* 14(44), e1802086. DOI: 10.1002/smll.201802086.

Lin, S., Reppert, J., Hu, Q., Hudson, J.S., Reid, M.L. *et al.* (2009) Uptake, translocation, and transmission of carbon nanomaterials in rice plants. *Small* 5 NA–NA. DOI: 10.1002/smll.200801556.

Lin, J., Zhang, H., Chen, Z. and Zheng, Y. (2010) Penetration of lipid membranes by gold nanoparticles: insights into cellular uptake, cytotoxicity, and their relationship. *ACS Nano* 4(9), 5421–5429. DOI: 10.1021/nn1010792.

Liu, J. (2005) Scanning transmission electron microscopy and its application to the study of nanoparticles and nanoparticle systems. *Journal of Electron Microscopy* 54(3), 251–278. DOI: 10.1093/jmicro/dfi034.

Liu, M.L., Chen, B.B., Li, C.M. and Huang, C.Z. (2019) Carbon dots: synthesis, formation mechanism, fluorescence origin and sensing applications. *Green Chemistry* 21(3), 449–471. DOI: 10.1039/C8GC02736F.

Lombi, E. and Susini, J. (2009) Synchrotron-based techniques for plant and soil science: opportunities, challenges and future perspectives. *Plant and Soil* 320(1–2), 1–35. DOI: 10.1007/s11104-008-9876-x.

Low, M., Yu, S., Han, M.Y. and Su, X. (2011) Investigative study of nucleic acid-gold nanoparticle interactions using laser-based techniques, electron microscopy, and resistive pulse sensing with a nanopore. *Australian Journal of Chemistry* 64(9), 1229–1234. DOI: 10.1071/CH11200.

Lv, J., Christie, P. and Zhang, S. (2019) Uptake, translocation, and transformation of metal-based nanoparticles in plants: recent advances and methodological challenges. *Environmental Science* 6(1), 41–59. DOI: 10.1039/C8EN00645H.

Lv, Z., Jiang, R., Chen, J. and Chen, W. (2020) Nanoparticle-mediated gene transformation strategies for plant genetic engineering. *Plant Journal* 104, 880–891. DOI: 10.1111/tpj.14973.

Ma, Z., Merkus, H.G., de Smet, J.G.A.E., Heffels, C. and Scarlett, B. (2000) New developments in particle characterization by laser diffraction: size and shape. *Powder Technology* 111(1–2), 66–78. DOI: 10.1016/S0032-5910(00)00242-4.

Manawi, Y.M., Samara, A., Al-Ansari, T., Atieh, M.A. and Ihsanullah (2018) A review of carbon nanomaterials' synthesis via the chemical vapor deposition (CVD) method. *Materials* 11(5), 822. DOI: 10.3390/ma11050822.

Marslin, G., Sheeba, C.J. and Franklin, G. (2017) Nanoparticles alter secondary metabolism in plants via ROS burst. *Frontiers in Plant Science* 8, 832. DOI: 10.3389/fpls.2017.00832.

Martin-Ortigosa, S., Peterson, D.J., Valenstein, J.S., Lin, V.S.-Y., Trewyn, B.G. *et al.* (2014) Mesoporous silica nanoparticle-mediated intracellular cre protein delivery for maize genome editing via loxP site excision. *Plant Physiology* 164(2), 537–547. DOI: 10.1104/pp.113.233650.

Martin-Ortigosa, S., Valenstein, J.S., Lin, V.S.-Y., Trewyn, B.G. and Wang, K. (2012) Gold functionalized mesoporous silica nanoparticle mediated protein and DNA codelivery to plant cells via the biolistic method. *Advanced Functional Materials* 22(17), 3576–3582. DOI: 10.1002/adfm.201200359.

Mehrian, S.K. and de Lima, R. (2016) Nanoparticles cyto and genotoxicity in plants: mechanisms and abnormalities. *Environmental Nanotechnology, Monitoring & Management* 6, 184–193. DOI: 10.1016/j.enmm.2016.08.003.

Milewska-Hendel, A., Zubko, M., Karcz, J., Stróż, D. and Kurczyńska, E. (2017) Fate of neutral-charged gold nanoparticles in the roots of the Hordeum vulgare L. cultivar Karat. *Scientific Reports* 7(1), 3014. DOI: 10.1038/s41598-017-02965-w.

Molaei, M.J. (2019) A review on nanostructured carbon quantum dots and their applications in biotechnology, sensors, and chemiluminescence. *Talanta* 196, 456–478. DOI: 10.1016/j.talanta.2018.12.042.

Moreels, I., Martins, J.C. and Hens, Z. (2007) Solution NMR techniques for investigating colloidal nanocrystal ligands: a case study on trioctylphosphine oxide at InP quantum dots. *Sensors and Actuators B: Chemical* 126(1), 283–288. DOI: 10.1016/j.snb.2006.12.026.

Mourdikoudis, S., Pallares, R.M. and Thanh, N.T.K. (2018) Characterization techniques for nanoparticles: comparison and complementarity upon studying nanoparticle properties. *Nanoscale* 10(27), 12871–12934. DOI: 10.1039/c8nr02278j.

Neouze, M.-A. and Schubert, U. (2008) Surface modification and functionalization of metal and metal oxide nanoparticles by organic ligands. *Monatshefte Für Chemie - Chemical Monthly* 139(3), 183–195. DOI: 10.1007/s00706-007-0775-2.

Nima, Z.A., Lahiani, M.H., Watanabe, F., Xu, Y., Khodakovskaya, M.V. et al. (2014) Plasmonically active nanorods for delivery of bio-active agents and high-sensitivity SERS detection in planta. RSC Advances 4(110), 64985–64993. DOI: 10.1039/C4RA10358K.

Palocci, C., Valletta, A., Chronopoulou, L., Donati, L., Bramosanti, M. et al. (2017) Endocytic pathways involved in PLGA nanoparticle uptake by grapevine cells and role of cell wall and membrane in size selection. Plant Cell Reports 36(12), 1917–1928. DOI: 10.1007/s00299-017-2206-0.

Pereira, A.E.S., Sandoval-Herrera, I.E., Zavala-Betancourt, S.A., Oliveira, H.C., Ledezma-Pérez, A.S. et al. (2017) γ-Polyglutamic acid/chitosan nanoparticles for the plant growth regulator gibberellic acid: characterization and evaluation of biological activity. Carbohydrate Polymers 157, 1862–1873. DOI: 10.1016/j.carbpol.2016.11.073.

Pereira, A.E.S., Oliveira, H.C. and Fraceto, L.F. (2019) Polymeric nanoparticles as an alternative for application of gibberellic acid in sustainable agriculture: a field study. Scientific Reports 9(1), 7135. DOI: 10.1038/s41598-019-43494-y.

Perreault, F., Samadani, M. and Dewez, D. (2014) Effect of soluble copper released from copper oxide nanoparticles solubilisation on growth and photosynthetic processes of Lemna gibba L. Nanotoxicology 8(4), 374–382. DOI: 10.3109/17435390.2013.789936.

Phan, C.M. and Nguyen, H.M. (2017) Role of capping agent in wet synthesis of nanoparticles. The Journal of Physical Chemistry A 121(17), 3213–3219. DOI: 10.1021/acs.jpca.7b02186.

Pokhrel, L.R. and Dubey, B. (2013) Evaluation of developmental responses of two crop plants exposed to silver and zinc oxide nanoparticles. The Science of the Total Environment 452–453, 321–332. DOI: 10.1016/j.scitotenv.2013.02.059.

Prabhakar, N., Zhang, J., Desai, D., Casals, E., Gulin-Sarfraz, T. et al. (2016) Stimuli-responsive hybrid nanocarriers developed by controllable integration of hyperbranched PEI with mesoporous silica nanoparticles for sustained intracellular siRNA delivery. International Journal of Nanomedicine 11, 6591–6608. DOI: 10.2147/IJN.S120611.

Puchalski, M., Dąbrowski, P., Olejniczak, W., Krukowski, P., Kowalczyk, P., and Polański, K. (2007) The study of silver nanoparticles by scanning electron microscopy, energy dispersive X-ray analysis and scanning tunnelling microscopy. Materials Science—Poland 25, 473–478.

Racuciu, M. and Creanga, D.-E. (2009) Cytogenetical changes induced by β-cyclodextrin coated nanoparticles in plant seeds. Romanian Journal of Physics 54, 2.

Rai, M. and Yadav, A. (2013) Plants as potential synthesiser of precious metal nanoparticles: progress and prospects. IET Nanobiotechnology 7(3), 117–124. DOI: 10.1049/iet-nbt.2012.0031.

Saison, C., Perreault, F., Daigle, J.-C., Fortin, C., Claverie, J. et al. (2010) Effect of core-shell copper oxide nanoparticles on cell culture morphology and photosynthesis (photosystem II energy distribution) in the green alga, Chlamydomonas reinhardtii. Aquatic Toxicology 96(2), 109–114. DOI: 10.1016/j.aquatox.2009.10.002.

Santana, I., Jeon, S.-J., Kim, H.-I., Islam, M.R., Castillo, C. et al. (2022) Targeted carbon nanostructures for chemical and gene delivery to plant chloroplasts. ACS Nano 16(8), 12156–12173. DOI: 10.1021/acsnano.2c02714.

Sapsford, K.E., Algar, W.R., Berti, L., Gemmill, K.B., Casey, B.J. et al. (2013) Functionalizing nanoparticles with biological molecules: developing chemistries that facilitate nanotechnology. Chemical Reviews 113(3), 1904–2074. DOI: 10.1021/cr300143v.

Schwab, F., Zhai, G., Kern, M., Turner, A., Schnoor, J.L. et al. (2016) Barriers, pathways and processes for uptake, translocation and accumulation of nanomaterials in plants--Critical review. Nanotoxicology 10(3), 257–278. DOI: 10.3109/17435390.2015.1048326.

Schwartz, S.H., Hendrix, B., Hoffer, P., Sanders, R.A. and Zheng, W. (2020) Carbon dots for efficient small interfering RNA delivery and gene silencing in plants. Plant Physiology 184(2), 647–657. DOI: 10.1104/pp.20.00733.

Sekine, R., Moore, K.L., Matzke, M., Vallotton, P., Jiang, H. et al. (2017) Complementary imaging of silver nanoparticle interactions with green algae: dark-field microscopy, electron microscopy, and nanoscale secondary ion mass spectrometry. ACS Nano 11(11), 10894–10902. DOI: 10.1021/acsnano.7b04556.

Serag, M.F., Kaji, N., Gaillard, C., Okamoto, Y., Terasaka, K. et al. (2011) Trafficking and subcellular localization of multiwalled carbon nanotubes in plant cells. ACS Nano 5(1), 493–499. DOI: 10.1021/nn102344t.

Serag, M.F., Kaji, N., Habuchi, S., Bianco, A. and Baba, Y. (2013) Nanobiotechnology meets plant cell biology: carbon nanotubes as organelle targeting nanocarriers. *RSC Advances* 3(15), 4856–4862. DOI: 10.1039/c2ra22766e.

Shah, A., Aftab, S., Nisar, J., Ashiq, M.N. and Iftikhar, F.J. (2021) Nanocarriers for targeted drug delivery. *Journal of Drug Delivery Science and Technology* 62, 102426. DOI: 10.1016/j.jddst.2021.102426.

Shah, N.B., Vercellotti, G.M., White, J.G., Fegan, A., Wagner, C.R. *et al.* (2012) Blood-nanoparticle interactions and in vivo biodistribution: impact of surface PEG and ligand properties. *Molecular Pharmaceutics* 9(8), 2146–2155. DOI: 10.1021/mp200626j.

Silva, A.T., Nguyen, A., Ye, C., Verchot, J. and Moon, J.H. (2010) Conjugated polymer nanoparticles for effective siRNA delivery to tobacco BY-2 protoplasts. *BMC Plant Biology* 10, 291. DOI: 10.1186/1471-2229-10-291.

Solanki, P., Bhargava, A., Chhipa, H., Jain, N., and Panwar, J. (2015) Nano-fertilizers and their smart delivery system. In: Rai, M., Ribeiro, C., Mattoso, L., and Duran, N. (eds) *Nanotechnologies in Food and Agriculture*. Springer. Cham, Switzerland, pp. 81–101.

Sperling, R.A. and Parak, W.J. (2010) Surface modification, functionalization and bioconjugation of colloidal inorganic nanoparticles. *Philosophical Transactions of the Royal Society A: Mathematical, Physical and Engineering Sciences* 368(1915), 1333–1383. DOI: 10.1098/rsta.2009.0273.

Subburaj, S., Chung, S.J., Lee, C., Ryu, S.-M., Kim, D.H. *et al.* (2016) Site-directed mutagenesis in Petunia × hybrida protoplast system using direct delivery of purified recombinant Cas9 ribonucleoproteins. *Plant Cell Reports* 35(7), 1535–1544. DOI: 10.1007/s00299-016-1937-7.

Sui, R. and Charpentier, P. (2012) Synthesis of metal oxide nanostructures by direct sol-gel chemistry in supercritical fluids. *Chemical Reviews* 112(6), 3057–3082. DOI: 10.1021/cr2000465.

Sun, D., Hussain, H.I., Yi, Z., Siegele, R., Cresswell, T. *et al.* (2014) Uptake and cellular distribution, in four plant species, of fluorescently labeled mesoporous silica nanoparticles. *Plant Cell Reports* 33(8), 1389–1402. DOI: 10.1007/s00299-014-1624-5.

Tarrahi, R., Mahjouri, S. and Khataee, A. (2021) A review on in vivo and in vitro nanotoxicological studies in plants: a headlight for future targets. *Ecotoxicology and Environmental Safety* 208, 111697. DOI: 10.1016/j.ecoenv.2020.111697.

Tauger, M.B. (2010) *Agriculture in World History*. Routledge, London. DOI: 10.4324/9780203847480.

Upadhyay, S., Parekh, K. and Pandey, B. (2016) Influence of crystallite size on the magnetic properties of Fe3O4 nanoparticles. *Journal of Alloys and Compounds* 678, 478–485. DOI: 10.1016/j.jallcom.2016.03.279.

Vazquez, N.I., Gonzalez, Z., Ferrari, B. and Castro, Y. (2017) Synthesis of mesoporous silica nanoparticles by sol–gel as nanocontainer for future drug delivery applications. *Boletín de La Sociedad Española de Cerámica y Vidrio* 56(3), 139–145. DOI: 10.1016/j.bsecv.2017.03.002.

Vieira, A.P., Berndt, G., de Souza Junior, I.G., Di Mauro, E., Paesano, A, *et al.* (2011) Adsorption of cysteine on hematite, magnetite and ferrihydrite: FT-IR, Mössbauer, EPR spectroscopy and X-ray diffractometry studies. *Amino Acids* 40(1), 205–214. DOI: 10.1007/s00726-010-0635-y.

von der Kammer, F., Legros, S., Hofmann, T., Larsen, E.H. and Loeschner, K. (2011) Separation and characterization of nanoparticles in complex food and environmental samples by field-flow fractionation. *Trends in Analytical Chemistry* 30(3), 425–436. DOI: 10.1016/j.trac.2010.11.012.

Wang, B., Huang, J., Zhang, M., Wang, Y., Wang, H. *et al.* (2020) Carbon dots enable efficient delivery of functional DNA in plants. *ACS Applied Bio Materials* 3(12), 8857–8864. DOI: 10.1021/acsabm.0c01170.

Wang, L., Li, J., Zhou, Q., Yang, G., Ding, X.L. *et al.* (2014) Rare earth elements activate endocytosis in plant cells. *Proceedings of the National Academy of Sciences* 111(35), 12936–12941. DOI: 10.1073/pnas.1413376111.

Wang, L., Zhang, T., Li, P., Huang, W., Tang, J. *et al.* (2015) Use of synchrotron radiation-analytical techniques to reveal chemical origin of silver-nanoparticle cytotoxicity. *ACS Nano* 9(6), 6532–6547. DOI: 10.1021/acsnano.5b02483.

Wang, R., Min, J., Kronzucker, H.J., Li, Y. and Shi, W. (2019) N and P runoff losses in China's vegetable production systems: loss characteristics, impact, and management practices. *Science of The Total Environment* 663, 971–979. DOI: 10.1016/j.scitotenv.2019.01.368.

Wang, T., Bai, J., Jiang, X. and Nienhaus, G.U. (2012) Cellular uptake of nanoparticles by membrane penetration: a study combining confocal microscopy with FTIR spectroelectrochemistry. *ACS Nano* 6(2), 1251–1259. DOI: 10.1021/nn203892h.

Wirtz, T., Philipp, P., Audinot, J.-N., Dowsett, D. and Eswara, S. (2015) High-resolution high-sensitivity elemental imaging by secondary ion mass spectrometry: from traditional 2D and 3D imaging to correlative microscopy. *Nanotechnology* 26(43), 434001. DOI: 10.1088/0957-4484/26/43/434001.

Xia, Y., Zhang, M., Tsang, D.C.W., Geng, N., Lu, D. *et al.* (2020) Recent advances in control technologies for non-point source pollution with nitrogen and phosphorous from agricultural runoff: current practices and future prospects. *Applied Biological Chemistry* 63(1), 8. DOI: 10.1186/s13765-020-0493-6.

Xin, X., He, Z., Hill, M.R., Niedz, R.P., Jiang, X. *et al.* (2018) Efficiency of biodegradable and pH-responsive polysuccinimide nanoparticles (PSI-NPs) as Smart nanodelivery systems in grapefruit: *in vitro* cellular investigation. *Macromolecular Bioscience* 18(7), e1800159. DOI: 10.1002/mabi.201800159.

Xin, X., Judy, J.D., Sumerlin, B.B. and He, Z. (2020) Nano-enabled agriculture: from nanoparticles to smart nanodelivery systems. *Environmental Chemistry* 17(6), 413–425. DOI: 10.1071/EN19254.

Xu, C., Cao, L., Zhao, P., Zhou, Z., Cao, C. *et al.* (2018) Emulsion-based synchronous pesticide encapsulation and surface modification of mesoporous silica nanoparticles with carboxymethyl chitosan for controlled azoxystrobin release. *Chemical Engineering Journal* 348, 244–254. DOI: 10.1016/j.cej.2018.05.008.

Yang, H., Xu, M., Koide, R.T., Liu, Q., Dai, Y. *et al.* (2016) Effects of ditch-buried straw return on water percolation, nitrogen leaching and crop yields in a rice-wheat rotation system. *Journal of the Science of Food and Agriculture* 96(4), 1141–1149. DOI: 10.1002/jsfa.7196.

Yoo, J., Park, C., Yi, G., Lee, D. and Koo, H. (2019) Active targeting strategies using biological ligands for nanoparticle drug delivery systems. *Cancers* 11(5), 640. DOI: 10.3390/cancers11050640.

Zaytseva, O. and Neumann, G. (2016) Carbon nanomaterials: production, impact on plant development, agricultural and environmental applications. *Chemical and Biological Technologies in Agriculture* 3(1), 1–26. DOI: 10.1186/s40538-016-0070-8.

Zhai, G., Walters, K.S., Peate, D.W., Alvarez, P.J.J. and Schnoor, J.L. (2014) Transport of gold nanoparticles through plasmodesmata and precipitation of gold ions in woody poplar. *Environmental Science & Technology Letters* 1(2), 146–151. DOI: 10.1021/ez400202b.

Zhang, P., Ma, Y., Zhang, Z., He, X., Li, Y. *et al.* (2015) Species-specific toxicity of ceria nanoparticles to *Lactuca* plants. *Nanotoxicology* 9(1), 1–8. DOI: 10.3109/17435390.2013.855829.

Zhao, F.-J., Moore, K.L., Lombi, E. and Zhu, Y.-G. (2014) Imaging element distribution and speciation in plant cells. *Trends in Plant Science* 19(3), 183–192. DOI: 10.1016/j.tplants.2013.12.001.

Zhao, P., Cao, L., Ma, D., Zhou, Z., Huang, Q. *et al.* (2018) Translocation, distribution and degradation of prochloraz-loaded mesoporous silica nanoparticles in cucumber plants. *Nanoscale* 10(4), 1798–1806. DOI: 10.1039/C7NR08107C.

Zhu, F., Liu, X., Cao, L., Cao, C., Li, F. *et al.* (2018) Uptake and distribution of fenoxanil-loaded mesoporous silica nanoparticles in rice plants. *International Journal of Molecular Sciences* 19(10), 2854. DOI: 10.3390/ijms19102854.

5 Use of Nanoformulations and Nano-Enabled Products in Mitigating the Risk Associated with the Current Use of Agrochemicals

Kotuwegoda Guruge Kaushani and Gayan Priyadarshana*

Department of Materials and Mechanical Technology, Faculty of Technology, University of Sri Jayewardenepura, Homagama, Sri Lanka

---

Abstract

The economies of self-sufficient and emerging countries have always been driven primarily by agriculture. It supplies the necessary raw materials for human and livestock existence. The extensive use of conventional agrochemicals and their method of application have created numerous problems for the agricultural sector, including uncontrolled release, soil fertility issues, nutrient deficiencies, water scarcity, toxicity to nontarget organisms, decline of biological matter in the soil, and decreased crop yield. The application of nanoformulations and nano-enabled products can considerably improve the efficacy of crop production and provide a vital approach ensuring agricultural sustainability while promoting food safety. The potential use of nanoscale formulations, such as nanofertilizers, nanopesticides, and nano-enabled products, including nano-additives, nanosensors, and nanocleansers, has made them more ecologically friendly and sustainable. This chapter provides an insight into recent advancements and implications in nanoformulations and nano-enabled products in mitigating the risks associated with the use of conventional agrochemicals and their ecological impact.

Keywords: nanoformulations, nano-enabled products, agrochemicals, agriculture, risks

5.1 Introduction

Over the last 50 years, traditional agricultural methods have seen remarkable gains in efficiency and productivity, resulting in significant increases in food supply, ranging from 70 to 90% (Ur Rahim *et al.*, 2021). However, the agricultural industry is currently confronted with a variety of problems, including nutrient deficiency in the soil, declining crop yields, water scarcity, climatic changes, the decline of biological matter in the growing medium, pest infestations, and inadequate awareness of genetically modified organisms (Singh *et al.*, 2021). These may be caused by some conventional agronomic practices, such as intensive agricultural crop production, commercialization of hybrid crops of high productivity, and extensive use of additives such as water, pesticides, and fertilizers, which raise a variety of ecological and socio-economic concerns that involve an evolution to more sustainable food production and agriculture. Ecological problems include the

*Corresponding author: gayanp@sjp.ac.lk

© CAB International 2023. *Nanoformulations for Sustainable Agriculture and Environmental Risk Mitigation* (eds Z. Khan and N.A. Șuțan)
DOI: 10.1079/9781800623095.0005

continuous degradation of surface soil due to the loss of soil nutrients, the decline in soil structure, and salinity, which results in soil infertility and a decrease in crop yields. In addition, excessive use of inorganic fertilizers and agrochemicals increases greenhouse-gas emissions and causes eutrophication of underground, marine, and freshwater ecosystems, which results in the development of toxic algal blooms, the destruction of large quantities of fish, and risks to food safety and human health. The application of synthetic conventional pesticide formulations such as soluble liquid concentrates, emulsifiable concentrates, wettable powders, dust or dustable powder, herbicides, or antibiotics may result in the contamination of food, water, and soil due to their inefficient release and absorption in crops (Ur Rahim *et al.*, 2021). This will cause pesticide resistance in pathogens and the destruction of beneficial organisms such as soil microbiota and pollinators, leading to significant ecosystem damage and food shortages (Xin *et al.*, 2020).

Minimizing the risks and challenges associated with the current usage of agrochemicals can be achieved by using an eco-friendly strategy that reduces the number of active ingredients used and provides targeted delivery to enhance efficacy. The use of nanotechnology can greatly improve the efficacy of modern agricultural inputs and provide a vital remedy for ensuring the development of sustainable agriculture. Through advancements in crop productivity, precision agriculture, water consumption efficiency, and crop management against pathogens by using nanoformulations and nano-enabled products, nanotechnology could indeed change the food production systems while preserving environmental integrity and ecological sustainability. Nanotechnology involves the generation, characterization, and utilization of nanoparticles (NPs), which are extremely small, between 1 and 100 nm (Madkour, 2019). The surface area:volume ratio increases noticeably as the size decreases, making NPs extremely reactive, with distinct and novel properties (structural, chemical, optical, magnetic, and electrical) to develop sustainability. Specifically, this will target them in the soil or plant and provide controlled delivery (Singh *et al.*, 2021). Nanotechnology has introduced nano-based formulations such as nanopesticides, nanofertilizers, and nano-enabled products to increase crop yields and production (Usman *et al.*, 2020). The aim of incorporating nanoformulations is to minimize the premature degradation of active components through their controlled delivery and release. Furthermore, they improve antibiotic solubility, reduce damage to non-target species through targetable delivery, and decrease chemical residues through biological degradation, while enhancing pest control efficiency. Nano-enabled products can enhance plant resistance to environmental factors, detect plant infections, and improve plant nutrition as a defense against infections. This chapter presents the implications of nanotechnological tools in sustainable agriculture, the most recent advancements in research on nanoformulations and nano-enabled products for enhancing crop protection and plant nutrition, nano-remediation techniques for contaminated soil and water, the use of nanosensors for plant quality, health, and safety monitoring, and their future prospects.

5.2 Nanotechnology in Sustainable Agriculture

Annually, about 3 billion t of a diverse range of crop production is achieved worldwide for consumption (e.g. wheat, cereals, legumes, maize), fuel, and textiles, demanding large quantities of pesticides, fertilizers, water, and energy. Agrochemicals often help to keep the crops protected from various insect pests and maintain food safety (Ur Rahim *et al.*, 2021). However, the majority of agrochemicals have extremely low usage effectiveness and high overall losses. These constraints have triggered intense study into the development of novel nanoscale formulations. The modern agricultural sector could be greatly improved by using nanotechnology to conserve crop supplies, livestock farming, and aquaculture. The application of nanotechnology in agriculture may be advantageous for increasing crop yields through proper nutrient control, assuring better water quality management and targeted application of fertilizers and pesticides to improve soil health and quality (Zehra *et al.*, 2021). Furthermore, the morphological and physiological structures of plants are enhanced by the delivery of nutrients encapsulated with

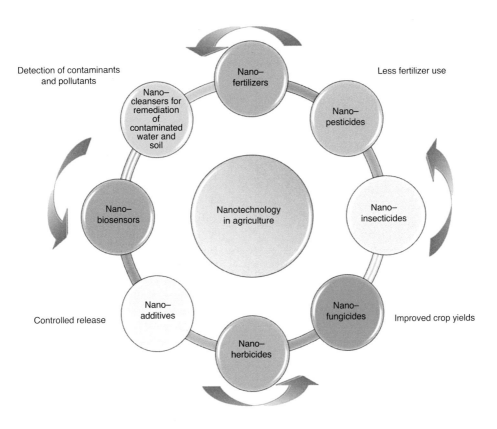

Fig. 5.1. Applications of nanotechnology in sustainable agriculture.

nanomaterials (NMs) in the soil structure (Sabir *et al.*, 2014). By removing hazardous substances from the soil and so improving its natural quality, soil and water remediation with NMs will significantly contribute to the maintenance of sustainability (Singh *et al.*, 2021). Moreover, to detect toxic chemicals, heavy metals, fungal pathogens, and environmental contaminants in soil media, electrochemical and optical biosensors based on NMs have been utilized. In this section, we have emphasized various advancements and innovations in the fields of nanoformulations including nanopesticides, nano-insecticides, nanofungicides, nanofertilizers, nano-enabled products including nano-additives, nanobiosensors, and nanocleansers for remediation of contaminated water, and nano-enabled products for soil remediation using nanotechnology (Fig. 5.1, Table 5.1).

5.2.1 Nanoformulations

Crop protection through nanopesticides

Commercial agriculture frequently employs the use of pesticides. Only 0.1% of the pesticides used reach their intended targets; 99.9% harm the environment, negatively affecting the food chain and living organisms (Carriger *et al.*, 2006). Pesticide-resistant plants, insects, and infections have emerged as a result of the widespread use of pesticides in the environment, in addition to their negative effects on non-target and beneficial species and soil biodiversity (Ur Rahim *et al.*, 2021). Nanopesticides can provide slow decomposition, controllable delivery of active components, and long-term pest control (Bratovcic *et al.*, 2021). Nanopesticides, which are biocompatible, environmentally friendly,

Table 5.1. Recent developments of nanomaterials as nanopesticides, nanobactericides, nanoherbicides, and nanofungicides for agricultural sustainability

Nanomaterial/active component	Type of nanoproduct	Crop/disease	Target plant pathogen	Significance	Reference
Nanosilica (SiO$_2$)	Nano-insecticide	Zea mays	Rhyzopertha dominica, Sitophilus oryzae, Oryzaephilus surinamensis, Tribolium castaneum	Demonstrated efficiency against the insects found in stored goods, with a 100% mortality rate	El-Naggar et al. (2020)
Hydrophilic SiO$_2$NPs and zinc oxide NPs	Nano-insecticide	Triticum aestivum	Tribolium castaneum Herbst, Sitophilus oryzae, Callosobruchus maculatus	Highly efficient insecticide activity	Haroun et al. (2020)
Titanium dioxide with zinc and silver NPs	Nano-insecticide	Tomato spot	Xanthomonas perforans	Significantly higher bacteria reduction and photocatalytic activity	Paret et al. (2013)
Silver NPs/ silver–chitosan nanocomposite	Nanobactericide	Bacterial canker on fruit trees	Pseudomonas syringae	Indicated that synthesized AgNPs can function as effective antibacterial agents	Shahryari et al. (2020)
Silver–graphene oxide composite	Nanobactericide	Bacterial spot on tomatoes	Xanthomonas perforans	Effective suppression of bacterial growth with no phytotoxicity	Ocsoy et al. (2013)
Silver and copper NPs	Nanofungicide	Woody trees	Phytophthora cactorum, Rhizoctonia solani, Grifola frondosa, Fistulina hepatica	AgNPs and CuNPs may be used to protect nursery seedlings, wood, and trees by being toxic to some diseases and wood-decay fungi	Aleksandrowicz-Trzcińska et al. (2018)
Copper NPs	Nanofungicide	–	Fusarium spp.	Effective antifungal activity against Fusarium equiseti followed by Fusarium oxysporum	Bramhanwade et al. (2016)

Continued

Table 5.1. Continued

Nanomaterial/active component	Type of nanoproduct	Crop/disease	Target plant pathogen	Significance	Reference
Solid lipid NPs loaded with *Zataria multiflora* essential oil	Nanofungicide	–	*Aspergillus ochraceus, Aspergillus solani, Aspergillus niger, Aspergillus flavus, Rhizopus stolonifera, Rhizopus solani*	Solid lipid NPs act as effective carriers of *Z. multiflora* essential oil to suppress fungal infections	Nasseri *et al.* (2016)
Imazapyr- and imazapic-loaded chitosan NPs	Nanoherbicide	*Allium cepa*	*Bidens pilosa*	Improved mode of action and reduction in toxicity by encapsulation of the herbicide	Maruyama *et al.* (2016)
Nano-encapsulation of *Satureja hortensis* L. essential oil	Nanoherbicide	Tomato (*Lycopersicon esculentum* Mill.) and amaranth (*Amaranthus retroflexus* L.)	–	Improved herbicidal activity of encapsulated savory essential oil compared with non-nano essential oil emulsion	Taban *et al.* (2020)

NP, nanoparticle.

and safe to use, include nano-insecticides, nano-weedicides, nanofungicides, and nanobactericides. They are used for effective delivery systems and are inexpensive and easier to apply (Baker *et al.*, 2019). Numerous formulations, including solid lipid NPs, polymeric NPs, inorganic NPs (e.g. alumina (Al_2O_3NPs), silica (SiNPs), titanium dioxide (TiO_2NPs), copper (CuNPs), silver (AgNPs)), carbon materials (e.g. carbon nanotubes (CNTs), graphene), nano-emulsions/nanodispersions, and nanogels, have been proven to deliver targeted, effective nanopesticides due to their insecticidal, fungicidal, and bactericidal properties (Kah *et al.*, 2013; Usman *et al.*, 2020).

Copper-based NPs, a commercially available nanopesticide, are widely produced and used in agriculture, water treatment, and food preservation. Due to the copper ions' antifungal and antimicrobial properties, which allow the control of a wide range of pathogens, they are widely used in agriculture (Mehta *et al.*, 2021). Broad-spectrum active AgNPs are effective against a variety of phytopathogens, including *Botrytis cinerea*, *Biploaris sorokiniana*, *Fusarium culmorum*, *Colletotrichum gloeosporioides*, *Phoma* spp., *Magnaporthe grisea*, *Pythium ultimum*, and *Trichoderma* spp. (Gajbhiye *et al.*, 2009; Sharon *et al.*, 2010). Tebuconazole entrapped in nanofibrillated cellulose/silica nanocomposites has been shown to achieve a 50% improvement in the release rate of pesticides to prevent leaf senescence and decay against rust and mold (Mattos and Magalhães, 2016).

NANO-INSECTICIDES. Some insecticides must be dissolved in organic solvents because they are only partially soluble in water. This makes the pesticide more expensive and hazardous (Herrmann and Guillard, 2000; Lushchak *et al.*, 2018). However, NPs have the potential to minimize toxicity while improving the solubility of pesticides. Recently, conventional insecticides and bioactive substances with insecticidal effects have been loaded on to a variety of NPs such as lipids, silica, chitosan, and inorganic NPs to increase the stability, slow down the release of the active compounds, reduce volatilization, improve low water solubility, and reduce the toxicity of the insecticide (Rajput *et al.*, 2021). Hydrophobic insecticides such as bioinspired mussel avermectin with improved polydopamine

NPs, which were applied to cucumber and broccoli, exhibited promising inhibition of aphids (Liang *et al.*, 2018). Excellent insecticidal effects were demonstrated by nanostructured alumina against leaf-cutting ants, and against Dipteran (*Ceratitis capitata*), and Coleoptera (*Oryzaephilus surinamensis*, *Sitophilus oryzae*) pests, developing sustainable approaches in pest management (López-García *et al.*, 2018). Nanostructured alumina particles that are nontoxic, biocompatible, and environmentally friendly may be able to remove the insect cuticle, leading to death by dehydration (Stadler *et al.*, 2018). Another study examined the efficacy of applying nanosilica to the soil to treat it as a pesticide to control storage insects such as *Rhyzopertha dominica*, *S. oryzae*, *O. surinamensis*, and *Tribolium castaneum* in maize (El-Naggar *et al.*, 2020). Furthermore, nanosilica applied as a carrier for enhanced bioactivity of chlorfenapyr was evaluated as an effective pesticide against *Helicoverpa armigera* and *Plutella xylostella* on Chinese cabbage (Song *et al.*, 2012). Inorganic NPs, including zinc oxide NPs (ZnONPs), AgNPs, and TiO_2NPs, have been widely applied, along with insecticidal preparations to protect crops against pulse beetles, mosquito vectors, and white grubs. Most importantly, AgNPs are more successful in the management of agricultural pests than conventional pesticides (Gahukar and Das, 2020; Pho *et al.*, 2021).

NANOFUNGICIDES. Developing and producing efficient nanofungicides that can reduce fungal infections, in particular by suppressing pathogenic fungi, can help solve problems related to plant fungal diseases. Chitosan, silica, and polymer combinations have been well studied, using their NPs as fungicide carriers. In addition, the efficacy of some metals (copper (Cu), Ag, and Zn) and their oxide NPs (e.g. ZnONPs and copper oxide NPs (CuONPs)) as fungicides have been analysed (Jampílek and Kráľová, 2015). Applying NP-based fungicides might provide a solution to issues associated with conventional insecticides, including their limited water solubility, poor stability, and high volatility (Worrall *et al.*, 2018), which can interfere with their prolonged release. In a study by Sharma *et al.* (2017), three plant-pathogenic fungi, *Colletotrichum gloeosporioides*, *Fusarium oxysporum*, and *Dematophora necatrix*, were successfully evaluated for the antimycotic activity of nickel

(NiFe$_2$O$_4$) and cobalt ferrite (CoFe$_2$O$_4$) NPs, and were found to be effective fungicides on various crops to control plant infections. Carbendazim, a low-water-soluble fungicide, was encapsulated in chitosan-pectin nanocapsules and tested against two fungal species, *Aspergillus parasiticus* and *F. oxysporum* for bioefficacy, and against three types of seeds, *Zea mays*, *Cucumis sativus*, and *Lycopersicum esculentum* for phytotoxicity studies (Kumar *et al.*, 2017). The nanoformulation of carbendazim demonstrated better efficiency at lower concentrations against target species when compared with pure carbendazim and commercial formulations, and was shown to be safer for seed germination and root growth in the three types of seeds. Prochloraz-loaded mesoporous SiNPs (MSNPs) were synthesized by Zhao *et al.* (2018) to study the translocation, distribution, and degradation of the target pesticide in cucumber cultivars. The analysis revealed that MSNPs can inhibit prochloraz breakdown on the surface of cucumber leaves. Furthermore, prochloraz-loaded MSNPs demonstrated a lower risk of pesticide accumulation in cucumber fruit, which is the edible component of the plant. Because of their good adhesion to bacterial and fungal cell surfaces and their high adherence to bacterial and fungal cell walls, well-stabilized Ag and CuNPs can function as effective broad-spectrum fungicides (Jampílek and Kráľová, 2015). Furthermore, NPs containing Zn and ZnO can be used as low-dose, economical, antimicrobials against the oomycete *Peronospora tabacina*, which causes root rot and seedling and foliar blight on tobacco leaves (Wagner *et al.*, 2016).

NANOHERBICIDES. The development of nano-weedicides and nanoherbicides is specifically intended to target particular weeds and herbs that invade agricultural environments and trigger nutrient depletion in crops (Singh *et al.*, 2021). Herbicides have lower non-target toxicity than insecticide and fungicide nanocarriers. Nano-active herbicides can increase the harvest potential by reducing the need for synthetic herbicides. Herbicide encapsulation (specifically ametryn, atrazine, clomazone, simazine, diuron, paraquat, and pendimethalin) in montmorillonite, poly(lactic-co-glycolic acid), and polymers including poly-V-caprolactone, alginate, and chitosan have demonstrated excellent effectiveness against targeted weeds (Grillo *et al.*, 2014; Pereira *et al.*, 2014). Sodium tripolyphosphate and chitosan polymeric NPs encapsulated with paraquat were proven to be effective on target plants such as maize and mustard (*Brassica* spp.) with enhanced herbicidal activity and lower toxicity than the pure compound (Grillo *et al.*, 2014). The herbicidal activity of nano-atrazine against *Bidens pilosa* L. and *Amaranthus viridis* L. was assessed, and it was found that even at lower concentrations, nano-atrazine had the same inhibitory function as commercial atrazine on the growth of shoots and roots (Sousa *et al.*, 2018). Nano-sized rice husk nanosorbents were loaded with 2,4-dichlorophenoxyacetic acid (2,4-D) in a study by Evy *et al.* (2016) and evaluated for their effective herbicidal activity. It was concluded that nanoparticle-loaded 2,4-D had better herbicidal activity than 2,4-D alone against the targeted *Brassica* spp.

Plant nutrition through nanofertilizers

According to Singh *et al.* (2021), nanofertilizers are 'smart carriers' of NPs or nano-emulsions delivering one or more micro- or macronutrients. Examples include nano-apatite, nano-urea, nano-potash, nanozeolite, and nano-iron-containing fertilizers that encourage the accessibility of the active ingredients to the plants, resulting in higher crop yields. The use of nanofertilizer increases agricultural sustainability by replacing conventional fertilizer usage, thereby minimizing production costs. Besides providing multiple nutrients simultaneously in a slow-release, targeted delivery system, having higher diffusion and solubility, and avoiding nutrient losses through leaching and volatilization, NPs can also assist in firmly retaining the useful nutrients and improving nutrient-use efficiency (Chhipa, 2017; Guha *et al.*, 2020). For these reasons, nanofertilizers are recommended over conventional fertilizers because they reduce agrochemical requirements and are environmentally safe (Elemike *et al.*, 2019). When compared with conventional fertilizers, nanofertilizers reduce the rate of nutrient release while increasing their stability through aggregation or adsorption without producing changes in chemical diversification due to pH-dependent mechanisms, redox potential, and ligand availability in the soil (Raliya *et al.*, 2018).

For soil and foliage application, the most commonly designed components utilized in nano-enabled fertilizers are hydroxyapatite, calcite, zeolites, mesoporous silica, chitosan, metal oxide NPs (CuONPs, iron oxide NPs (FeONPs), ZnONPs, TiO_2NPs), CNTs, nanoclays, and magnetite (Guo et al., 2018; Zehra et al., 2021). Several reports have been conducted on the delayed and prolonged release of nitrogen (N) in urea–hydroxyapatite NPs to boost plant N agronomic efficiency. In comparison with other commonly used nitrogenous fertilizers, hydroxyapatite has a larger surface area and interacts more with urea, resulting in a delayed and controlled release of N and phosphorus (P) from urea over 60 days (Kottegoda et al., 2011). In a study by Raguraj et al. (2020), the slow-release fertilizer urea–hydroxyapatite nanohybrid significantly improved the N and P content in Camellia sinensis (L.) Kuntze (tea), as well as the P content in the soil, and had a positive impact on adverse climatic conditions. Furthermore, a controlled-release potassium chloride (KCl) fertilizer was developed by mixing in a 1:1 ratio with conventional KCl fertilizers, and it was reported that the developed nanofertilizer was able to improve K-use efficiency, improve the soil fertility by delaying K release, provide continuous nutrient absorption, promote crop yields, and minimize production costs (Li et al., 2020). ZnONPs have been used successfully in the development of nanofertilizers and achieved an increase in root dry weight and gluten content in cucumber (C. sativus), improvement in the plant development when used at 20 ppm in mung beans (Vigna radiata) and 1 ppm in chickpeas (Cicer arietinum), improvement of their shoot dry weight (Mahajan et al., 2011; Burman et al., 2013), and elongation of the roots of rape seeds (Brassica napus) when applied at a concentration of 1–2000 ppm (Lin and Xing, 2007). Table 5.2 provides an overview of a comprehensive list of nanofertilizers and their effects on various crops.

5.2.2 Nano-enabled products

Nano-additives

Nano-engineered materials can potentially be used as additives to improve plant growth,

nutrient utilization, and plant germination, as well as to mitigate the negative effects of traditional fertilizer applications. Silicon dioxide NPs (SiO_2NPs) have been successfully applied to reduce the detrimental impact of drought conditions on wheat plants (Triticum aestivum L.), specifically at rates of 30 and 60 ppm. Soil application of NPs was able to improve relative water content (84%) while improving the crop yield by 18–25% under drought stress (Behboudi et al., 2018). In a study by Khodakovskaya et al. (2009), it was shown that CNTs are effective additions for increasing the growth rate of seedlings and seed germination of tomatoes (Solanum lycopersicum L.). The effectiveness of K- and P-incorporated nanofertilizer composed of nanozeolite was evaluated in Ipomoea aquatica L. (kalmi) pot culture tests (Rajonee et al., 2017). There was an increase in K and P deposition in crops grown with the nanofertilizer and they exhibited better moisture content, pH, and cation exchange capacity compared with plants treated with conventional fertilizer. Additionally, certain NMs have the potential to reduce pollution residues in plants. Application of SiNPs and TiO_2NPs in crops, particularly in rice (Oryza sativa) grown on cadmium (Cd)-contaminated soil, was able to improve biomass (51% total biomass) and photosynthesis, and minimize the toxicity that occurred due to Cd accumulation (Rizwan et al., 2019). This was mostly due to the ability of the NPs to suppress oxidative bursts and strengthen the antioxidant defense system of the plants.

Nanosensors for detection of pathogens and diseases, and monitoring of crop health

Nanosensors can significantly improve crop yields by managing agricultural soil and water, as well as monitoring soil nutrient absorption and moisture, and detecting pollutants such as pesticide residues, excessive fertilizer, plant pathogens (bacteria, viruses, and fungi), gas molecules, and heavy-metal ions (Fig. 5.2) (Singh et al., 2021). Nanosensors are immobilized bioreceptor probes that provide real-time detection, greater sensitivity, stability, and selectivity, and fast response times for target identification, and are thus becoming more advantageous than traditional sensors (Xin et al., 2020). A broad variety of nanobiosensors have been developed

Table 5.2. Nanofertilizers that help plant growth and development.

Nanofertilizer	Crop	Concentration (ppm)	Significance	Reference(s)
Carbon-based NPs (carbon nanotubes, graphene)	Wheat, tomato, soybean	5–500	Improvement in root cell elongation, increase in dehydrogenase activity, higher biomass production, and reduction of stress and heavy-metal toxicity	Wang et al. (2012); Khodakovskaya et al. (2013); Lahiani et al. (2013)
Iron oxide (Fe_2O_3) NPs	Watermelon, wheat, soybean	1.5–4000	Increase in the germination of seeds and seedling growth, improvement in grain yield, nutritional quality, and biomass, and an increase in the amount of chlorophyll	Ghafari and Razmjoo (2013); Li et al. (2013)
Silica (Si, SiO, SiO_2) NPs	Tomato, wheat, Larix olgensis	8–15	Enhancement of seed germination, growth, and quality, and improvement in protein and chlorophyll content	Bao-shan et al. (2004); Haghighi and Pessaraki (2013); Sun et al. (2016)
Urea-modified silica NPs	–	–	Confirmed t extended-release behaviour of nitrogen controlled by the rate of diffusion of urea	Siriwardena et al. (2017)
Titanium dioxide NPs	Spinach (Spinacia oleracea), mung bean (Vigna radiata)	0.25–4	About 73% improvement in plant dry weight, and improvement in nitrogen gas fixation, plant nutrient content and growth	Zheng et al. (2005)
Copper NPs	Lettuce (Lactuca sativa)	130–600	Increase in root and shoot length	Shah and Belozerova (2009)
Copper oxide NP	Maize (Zea mays)	10	Increase in plant growth up to 51% by regulating the different enzyme activities	Adhikari et al. (2016)
Magnedium oxide NPs	Cluster bean (Cyamopsis tetragonoloba L.)	15	An increase in chlorophyll content in the leaves may help improve light absorption for photosynthesis	Raliya et al. (2014)
Urea-modified hydroxyapatite nanohybrids	Maize (Z. mays)	–	Enhanced biomass, germination of seeds by efficient delivery of phosphorus and nitrogen, and an increase in the number of roots, root length, and maximum plant width	Pabodha et al. (2022)

Continued

Table 5.2. Continued

Nanofertilizer	Crop	Concentration (ppm)	Significance	Reference(s)
Urea-modified calcium carbonate nanohybrids	–	–	Demonstrated controlled nitrogen release over conventional nitrogen fertilizer systems	Rathnaweera *et al.* (2019)
Zinc-doped urea–hydroxyapatite nanohybrids	Maize (*Z. mays*)	–	Had a good potential of delivering nitrogen, phosphorus, and zinc at the seedling stage and thereby increased the growth rate of *Z. mays* seeds by 69% and the seed germination rate by 19%	Abeywardana *et al.* (2021)

NP, nanoparticle.

using NMs, such as carbonaceous NMs, magnetic NPs, CNTs, quantum dots (QDs), and gold NPs (AuNPs). Nanomaterials can distinguish nutrients including nitrite, nitrate, urea, urease, and ammonia in the soil, as well as pesticide residues such as carbamates, atrazine, neonicotinoids, organophosphates, and glyphosate, which have a toxic impact on the environment and on public health at higher concentrations (Ahmed *et al.*, 2017; Mahna *et al.*, 2021). Several studies have

used nanobiosensors to detect these nitrogenous nutrients in soil or water bodies using AuNPs (Mura *et al.*, 2015).

The potential application of AgNPs as colorimetric sensors for ammonium detection in wastewater samples has been investigated (Singh *et al.*, 2021). The ability of metal oxide NPs, polymeric NPs, and QDs to detect microbial infections in the vicinity of plants has been investigated. Additionally, metal NPs, including magnetic NPs, AgNPs, and AuNPs, have been used to detect bacteria such as *Listeria monocytogenes*, *Salmonella* Typhimurium, *Salmonella enterica*, *Staphylococcus aureus*, and *Escherichia coli* optically. For instance, immunosensors based on AuNPs have demonstrated the potential for quickly detecting chlorosis virus in bell peppers (*Capsicum annuum* L.) (Sharma *et al.*, 2019). For the detection of a wide range of toxic heavy-metal ions including lead (Pb^{2+}), Cd^{3+}, and mercury (Hg^{2+}) in soil and water bodies, different types of NPs, including AgNPs, AuNPs, graphene, QDs, and CNTs, have been used in fluorescent, photometric, electrochemical, and surface-enhanced Raman scattering sensors. Despite these distinctive benefits for application, the majority of nanosensors have been produced only on a laboratory scale, and more work is needed to develop these sensing

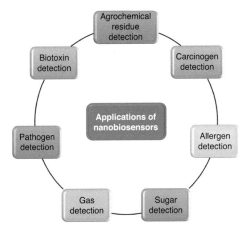

Fig. 5.2. Graphical representation of the applications of nanobiosensors in agriculture.

Table 5.3. Applications of nanosensors in promoting sustainability in agriculture.

Nanobiosensor	Nanomaterial used	Sensor type	Applications	Reference
Smart nanobiosensor	Zinc oxide and copper, and iron NPs	Electrochemical	Improved growth and germination of tomato, chili, and cucumber	Negrete (2020)
Salicylic acid sensors	Copper and gold NPs	Electrochemical	Detection of the fungal pathogen *Sclerotinia sclerotiorum* in infected oilseed rape	Wang *et al.* (2010)
Gold NPs deposited on CNT modified glassy carbon electrode surface	Au-coated CNT	Electrochemical	Detection of the pesticide triazophos in vegetables and food samples	Li *et al.* (2012)
Fluorescence paper-based sensor	Cadmium telluride quantum dots and zincNPs	Optical	Visual detection of three carbamate pesticides (carbofuran, metolcarb, and carbaryl)	Chen *et al.* (2020)

CNT, carbon nanotube; NP, nanoparticle.

systems for in-field agricultural applications (Table 5.3).

Nanocleansers for remediation of contaminated water

Nanotechnology can purify and allow the reuse of low-quality contaminated irrigation water (saltwater and brackish water) for crops and vegetation (Xin *et al.*, 2020). Nanotechnological applications provide many advantages in water treatment over conventional methods due to their cost effectiveness, fast action, and increased long-term resistance to pollution and biocorrosion (Semra and Nadaroglu, 2015). Some of the most widely explored nanowastewater treatments consist of nanofilters (e.g. magnetic NPs, CNTs, and nanoporous ceramics), nano-adsorbents (e.g. magnesium oxide (MgO), ZnO, MnO, CNTs, Fe), nanomembranes (e.g. polymers, TiO_2, Fe_3O_4), and nanophotocatalysts (e.g. cadmium sulfide (CdS), TiO_2, ZnO) (Xin *et al.*, 2020). CNTs can exhibit semiconductor or metallic properties and can be utilized to filter out pollutants and nano-sized toxic particulates from contaminated water (Semra and Nadaroglu, 2015). For instance, multi-walled CNTs can completely remove organophosphorus pesticides (e.g. diazinon) from irrigation water after 15 min of contact (Dehghani *et al.*, 2019). A synthetic clay substance called hydrotalcite is used to remove arsenic (As (III)) and arsenate (As (V)) from soil and underground water (Gillman, 2006). The cost of production and distribution is decreased as it is simple to transform 'used' hydrotalcite into profitable phosphatic fertilizer that has no impact on soil As levels. A novel and recyclable Ag/Fe_2O_3 graphene oxide nano-adsorbent was successfully used to eliminate toxic lead (Pb(II)) from wastewater, and optimum removal of Pb(II) (93%) was achieved via FeNPs in 40 min (Haerizade *et al.*, 2018). Furthermore, a nanocurcumin-incorporated polyethersulfone membrane was successfully developed to enhance anti-biofouling properties against *E. coli* and *Pseudomonas aeruginosa* in water treatment (Sathish Kumar and Arthanareeswaran, 2019).

Nano-enabled products for soil remediation

Soil is a very precious naturally-occurring resource that cannot be renewed, making careful protection necessary to prevent unacceptable losses. Soil helps to provide about 99% of our food, and also makes wood, fibre, and industrially valuable raw materials available. Additionally, it stores carbon, controls infections, recycles trash, controls greenhouse-gas emissions, and performs several other functions (Ur Rahim *et al.*, 2021). However, contamination caused by human activities can result in prolonged soil poisoning as a result of the discharge of either organic (e.g. plastics, biphenyls, pesticides, and fertilizers) or inorganic pollutants, such as metalloids and heavy metals (including Cu, Cd, Zn, arsenic (As), Pb, Hg), along with emergent pollutants (e.g. pharmaceuticals, nano/microplastics, personal hygiene products, flame retardants, detergents) (Choi *et al.*, 2021). Traditional bioremediation techniques may be constrained in their efficacy due to prolonged treatment periods, poor pollutant availability, limited remediation effectiveness in severely polluted soils induced by the toxicity of contaminants to microbial pathogens (e.g. bacteria, fungi, plants), and the development of toxic byproducts (Ur Rahim *et al.*, 2021). Nanotechnology provides cost-effective ways to decontaminate harmful pollutants from the soil through nano-assisted bioremediation by reducing clean-up times and costs, avoiding complex processes necessary for contaminated soil treatment and disposal, and reducing concentrations of specific contaminants to almost undetectable levels (Singh *et al.*, 2021). The commonly employed NMs for agricultural soil remediation are graphene-based, carbon-based NMs (biochar, single- and multi-walled CNTs, activated carbon), metal oxides (TiO_2, MnO, FeO, and MgO), chemical reductant-based reducing materials (iron sulfide (FeS), nano-zerovalent iron, molybdenum sulfides (MoS_2), thiosulfates ($S_2O_3^{-2}$), Zn, Mn, hydrogen peroxide (H_2O_2)), and minerals (Ur Rahim *et al.*, 2021). To remediate arsenic-contaminated soils under anoxic conditions, the utilization of three economical Fe/Al-based materials—sponge iron filters, red soil, and Al-based water treatment sludge—were examined. Compared with the other two materials, sponge iron filters were shown to have the most potential for As-contaminated soil remediation (Hou *et al.*, 2020). Furthermore, As in contaminated soil was successfully stabilized by biogenic MnO. The findings also revealed that biogenic MnO, compared with nonbiogenic MnO, is a far more effective, economical, and environmentally safe material for As stabilization in contaminated soil and showed a favourable impact on the biodiversity of the soil bacterial community (Wang *et al.*, 2020). NMs can also be employed to oxidize antibiotics in polluted soils and degrade organic contaminants via catalytic mineralization (Jain *et al.*, 2020). NMs have made significant strides in soil biodegradation of organic contaminants such as dichlorodiphenyldichloroethylene, triclosan, polyclorinated biphenyls, diphenyl ethers, and lindane (Bebić *et al.*, 2020; Nguyen *et al.*, 2020). NMs may promote bioremediation by decreasing the toxicity of pollutants to biocontrol agents. Nickel (Ni)/Fe bimetallic NPs were studied for the translocation of polybrominated diphenyl ether in contaminated soil and its phytotoxicity on Chinese cabbage. The combined effect of NMs and bioremediation was found to instantly decrease the toxicity of contaminants in the environment and NMs in plants (Wu *et al.*, 2016). Even though there is adequate information concerning their useful applications in bioremediation, the broad and useful utilization of NMs in contaminated water and soil is still not well understood. Therefore, to provide beneficial information and prevent harmful exposure, an appropriate assessment system for the application of NMs in the agriculture industry must be devised.

5.3 Commercial Applications of Nanoformulations in Agriculture

Nanoformulations have to face many challenges of cost and efficacy to be used as alternatives to the already established market of chemical pesticides. Some commercially available nanoformulations that fulfil the conditions of high efficiency and environment safety are described in Table 5.4.

5.4 Ecological Impact of Nanoformulations and Nano-Enabled Products

A decrease in the quantity of pesticides required to ensure crop protection is one of the main factors driving the adoption of nanotechnology. According to various studies, nanoformulations can reduce the rate of release and improve pest targeting. Nanoformulations can shield against a variety of

degradation processes, such as photolysis, hydrolysis, and soil destruction (Bratovcic *et al.*, 2021). The potential of NMs to support the transition to more resilient and environmentally friendly agriculture requires a thorough and sufficient risk analysis to prevent adverse ecological impacts. Although agrochemical nanoformulation is an important and novel technology, release of NMs into the field could have an impact on human health and the environment. Most novel NMs are claimed to be highly dispersible and reactive upon release into the environment, resulting in a potential risk to human health. Regarding their possible risks and hazards, it has been noted that some metals, such as Ag, Cu, and metal oxides (e.g. FeO and ZnO), may disperse quickly, whereas others, such as graphene, CNTs, SiO_2, and TiO_2, are considerably more persistent (Ur Rahim *et al.*, 2021). TiO_2NPs have been widely used in industrial applications and consumer products owing to their strong catalytic properties, and are predicted to accumulate in sewage sludge and have substantial potential for accumulation in agricultural soil (Khan *et al.*, 2022). Consequently, hazard identification and characterization, as well as risk evaluation, are required to define the risks and hence to take appropriate risk management measures. Additionally, to prevent false expectations related to the use of nanotechnology in agriculture, nano-agrochemicals should be evaluated against products already available in the marketplace. Toxicological issues affecting organelles, biomolecules, and macromolecules must also be studied to establish an accurate and responsible risk assessment of agrochemicals. This assessment must take into account effects at the organism, population, and dose-response levels. In addition, integrating life cycle assessment into risk assessment approaches may help to minimize other unintended consequences. These techniques may serve as a framework for developing procedures and guidelines that will enable the adoption of eco-friendly, sustainable, novel, and effective NMs for use in environmental and agricultural applications.

5.5 Conclusions and Future Prospects

Numerous applications of nanotechnology have been found in the agricultural industry,

Table 5.4. Commercially available nanoformulations used in sustainable agriculture.

Type of nanomaterial (active component)	Commercial/trade name	Modified functions	Reference
Nanosilver	NSPW-L30SS, silver–silica-based antimicrobial additive from NanoSilva, LLC., Georgia	The growth of bacteria, molds, and fungi that generate foul odors and discoloration is suppressed by the use of nanoforms made of silver–silica, which are used to protect polymers and polymer-based products; the allowed limit for silver is 30 ppm	Isman et al. (2011)
Nanoclay	Aegis® OXCE Barrier Nylon Resin, from Honeywell International Inc., USA	Provides an effective barrier in comparison with the performance of a glass bottle, suitable for co-injection due to its processing temperature being equal to that of poly(ethylene terephthalate)	Fortunati et al. (2019)
Nanoclay	Plantic® R1 Tray, from Plantic Technologies Ltd, Australia	Fabricated with resources that are sustainable and renewable, environmental non-toxicity after usage, biodegradable, and superior mechanical and rheological characteristics	Fortunati et al. (2019)
Nanosilver	Anson Nano Freshness-Keeping Film, Storage Bag, Anson Nano Silver Fresh Containers from Anson Nano-Biotechnology (Zhuhai) Co., Ltd, China	US Food and Drug Administration standard, keeping the food fresh for longer; safe food storage	Odetayo et al. (2022)
Copper hydroxide (Cu(OH)$_2$)	Kocide®3000 (fungicide), from Bharat Certis AgriScience Ltd, Japan	Improved effectiveness against a variety of fungal and bacterial diseases in different crops and offers better, longer-lasting disease control	Li et al. (2019)
Nanometal sulfates (iron, aluminum, cobalt, nickel, magnesium, silver, and manganese)	Nano-Gro™ from Agro Nanotechnology Corp, USA	Improve the plant's natural defenses to withstand stress, enhance germination and strong plant structures, improve crop production per plant, and lead to higher protein/sugar content	

such as the usage of nanopesticides, nanofertilizers, nano-additives, nanobiosensors, and nanocleansers for environmental remediation, which are crucial in addressing the different agronomic and ecological challenges, especially in minimizing the risks associated with the current use of agrochemicals. Agriculture may benefit from nanotechnology by delivering novel technologies such as nanofertilizer, micronutrient, and nanopesticide delivery, and agrochemical-encapsulated nanocarrier systems. Nanotechnology can minimize the amount of active substances used due to their improved bioavailability and resistance to degradation, resulting in lower dose-dependent toxicity for non-target species and less environmental impact. However, these nanoproducts must be thoroughly evaluated in terms of their release behaviour before application. The results

of studies in this field are primarily limited to laboratories, and the documentation of relevant rules and regulations is insufficient. Therefore, further research is required to examine the exposure levels, behaviour, mechanistic applications, and ecotoxicity assessment of nano-engineered products to create effective, versatile, stable, economical, and ecologically friendly NMs. The characterization and optimization of NMs for various species of plants must be thoroughly investigated before their commercial usage, as their response might differ depending on the plant species. Therefore, continuous advancement in the development of new and enhanced synthesis techniques would be very beneficial in improving their efficacy. More expert knowledge of in-field experiments would also be extremely beneficial for large-scale nano-based techniques in sustainable agriculture.

References

Abeywardana, L., de Silva, M., Sandaruwan, C., Dahanayake, D., Priyadarshana, G. *et al.* (2021) Zinc-doped hydroxyapatite–urea nanoseed coating as an efficient macro–micro plant nutrient delivery agent. *ACS Agricultural Science & Technology* 1(3), 230–239. DOI: 10.1021/acsagscitech.1c00033.

Adhikari, T., Sarkar, D., Mashayekhi, H. and Xing, B. (2016) Growth and enzymatic activity of maize (*Zea mays* L.) plant: solution culture test for copper dioxide nano particles. *Journal of Plant Nutrition* 39(1), 99–115. DOI: 10.1080/01904167.2015.1044012.

Ahmed, M., Rauf, M., Mukhtar, Z. and Saeed, N.A. (2017) Excessive use of nitrogenous fertilizers: an unawareness causing serious threats to environment and human health. *Environmental Science and Pollution Research International* 24(35), 26983–26987. DOI: 10.1007/s11356-017-0589-7.

Aleksandrowicz-Trzcińska, M., Szaniawski, A., Olchowik, J. and Drozdowski, S. (2018) Effects of copper and silver nanoparticles on growth of selected species of pathogenic and wood-decay fungi *in vitro*. *Forestry Chronicle* 94(2), 109–116. DOI: 10.5558/tfc2018-017.

Baker, S., Satish, S., Prasad, N., and Chouhan, R.S. (2019) Nano-agromaterials: influence on plant growth and crop protection. In: Thomas, S., Grohens, Y. and Pottathara, Y.B. (eds.) *Industrial Applications of Nanomaterials*. Elsevier, Amsterdam, pp. 341–363.

Bao-shan, L., shao-qi, d., Chun-hui, L., Li-jun, F., Shu-chun, Q. *et al.* (2004) Effect of TMS (nanostructured silicon dioxide) on growth of Changbai larch seedlings. *Journal of Forestry Research* 15(2), 138–140. DOI: 10.1007/BF02856749.

Bebić, J., Banjanac, K., Ćorović, M., Milivojević, A., Simović, M. *et al.* (2020) Immobilization of laccase from *Myceliophthora thermophila* on functionalized silica nanoparticles: optimization and application in lindane degradation. *Chinese Journal of Chemical Engineering* 28(4), 1136–1144. DOI: 10.1016/j.cjche.2019.12.025.

Behboudi, F., Sarvestani, Z.T., Kassaee, M.Z., Modaes Sanavi, S.A.M.M., and Sorooshzadeh, A. (2018) Improving growth and yield of wheat under drought stress via application of SiO_2 nanoparticles. *Journal of Agricultural Science and Technology* 20, 1479–1492.

Bramhanwade, K., Shende, S., Bonde, S., Gade, A. and Rai, M. (2016) Fungicidal activity of Cu nanoparticles against *Fusarium* causing crop diseases. *Environmental Chemistry Letters* 14(2), 229–235. DOI: 10.1007/s10311-015-0543-1.

Bratovcic, A., Hikal, W.M., Said-Al Ahl, H.A.H., Tkachenko, K.G., Baeshen, R.S. *et al.* (2021) Nanopesticides and nanofertilizers and agricultural development: scopes, advances and applications. *Open Journal of Ecology* 11(04), 301–316. DOI: 10.4236/oje.2021.114022.

Burman, U., Saini, M. and Kumar, P. (2013) Effect of zinc oxide nanoparticles on growth and antioxidant system of chickpea seedlings. *Toxicological & Environmental Chemistry* 95(4), 605–612. DOI: 10.1080/02772248.2013.803796.

Carriger, J.F., Rand, G.M., Gardinali, P.R., Perry, W.B., Tompkins, M.S. *et al.* (2006) Pesticides of potential ecological concern in sediment from South Florida canals: an ecological risk prioritization for aquatic arthropods. *Soil and Sediment Contamination* 15(1), 21–45. DOI: 10.1080/15320380500363095.

Chen, H., Hu, O., Fan, Y., Xu, L., Zhang, L. *et al.* (2020) Fluorescence paper-based sensor for visual detection of carbamate pesticides in food based on CdTe quantum dot and nano ZnTPyP. *Food Chemistry* 327, 127075. DOI: 10.1016/j.foodchem.2020.127075.

Chhipa, H. (2017) Nanofertilizers and nanopesticides for agriculture. *Environmental Chemistry Letters* 15(1), 15–22. DOI: 10.1007/s10311-016-0600-4.

Choi, Y., Lee, J.-H., Kim, K., Mun, H., Park, N. *et al.* (2021) Identification, quantification, and prioritization of new emerging pollutants in domestic and industrial effluents, Korea: application of LC-HRMS based suspect and non-target screening. *Journal of Hazardous Materials* 402, 123706. DOI: 10.1016/j.jhazmat.2020.123706.

Dehghani, M.H., Kamalian, S., Shayeghi, M., Yousefi, M., Heidarinejad, Z. *et al.* (2019) High-performance removal of diazinon pesticide from water using multi-walled carbon nanotubes. *Microchemical Journal* 145, 486–491. DOI: 10.1016/j.microc.2018.10.053.

Elemike, E.E., Uzoh, I.M., Onwudiwe, D.C. and Babalola, O.O. (2019) The role of nanotechnology in the fortification of plant nutrients and improvement of crop production. *Applied Sciences* 9(3), 499. DOI: 10.3390/app9030499.

El-Naggar, M.E., Abdelsalam, N.R., Fouda, M.M.G., Mackled, M.I., Al-Jaddadi, M.A.M. *et al.* (2020) Soil application of nano silica on maize yield and its insecticidal activity against some stored insects after the post-harvest. *Nanomaterials* 10(4), 739. DOI: 10.3390/nano10040739.

Evy, A.A.M., Melvin, S.M. and Ramalingam, C. (2016) Application of rice husk nanosorbents containing 2,4-dichlorophenoxyacetic acid herbicide to control weeds and reduce leaching from soil. *Journal of the Taiwan Institute of Chemical Engineers* 63, 318–326. DOI: 10.1016/j.jtice.2016.03.024.

Fortunati, E., Mazzaglia, A. and Balestra, G.M. (2019) Sustainable control strategies for plant protection and food packaging sectors by natural substances and novel nanotechnological approaches. *Journal of the Science of Food and Agriculture* 99(3), 986–1000. DOI: 10.1002/jsfa.9341.

Gahukar, R.T. and Das, R.K. (2020) Plant-derived nanopesticides for agricultural pest control: challenges and prospects. *Nanotechnology for Environmental Engineering* 5(1), 3. DOI: 10.1007/s41204-020-0066-2.

Gajbhiye, M., Kesharwani, J., Ingle, A., Gade, A. and Rai, M. (2009) Fungus-mediated synthesis of silver nanoparticles and their activity against pathogenic fungi in combination with fluconazole. *Nanomedicine: Nanotechnology, Biology and Medicine* 5(4), 382–386. DOI: 10.1016/j.nano.2009.06.005.

Ghafari, H. and Razmjoo, J. (2013) Effect of foliar application of nano-iron oxidase, iron chelate and iron sulphate rates on yield and quality of wheat. *International Journal of Agronomy and Plant Production* 4, 2997–3003.

Gillman, G.P. (2006) A simple technology for arsenic removal from drinking water using hydrotalcite. *The Science of the Total Environment* 366(2–3), 926–931. DOI: 10.1016/j.scitotenv.2006.01.036.

Grillo, R., Pereira, A.E.S., Nishisaka, C.S., de Lima, R., Oehlke, K. *et al.* (2014) Chitosan/tripolyphosphate nanoparticles loaded with paraquat herbicide: an environmentally safer alternative for weed control. *Journal of Hazardous Materials* 278, 163–171. DOI: 10.1016/j.jhazmat.2014.05.079.

Guha, T., Gopal, G., Kundu, R. and Mukherjee, A. (2020) Nanocomposites for delivering agrochemicals: a comprehensive review. *Journal of Agricultural and Food Chemistry* 68(12), 3691–3702. DOI: 10.1021/acs.jafc.9b06982.

Guo, H., White, J.C., Wang, Z. and Xing, B. (2018) Nano-enabled fertilizers to control the release and use efficiency of nutrients. *Current Opinion in Environmental Science & Health* 6, 77–83. DOI: 10.1016/j.coesh.2018.07.009.

Haerizade, B.N., Ghavami, M., Koohi, M., Darzi, S.J., Rezaee, N., *et al.* (2018) Green removal of toxic Pb (II) from water by a novel and recyclable ag/γ-fe2o3 @ r-GO nanocomposite. *Iranian Journal of Chemistry and Chemical Engineering* 37, 29–37.

Haghighi, M. and Pessarakli, M. (2013) Influence of silicon and nano-silicon on salinity tolerance of cherry tomatoes (*Solanum lycopersicum* L.) at early growth stage. *Scientia Horticulturae* 161, 111–117. DOI: 10.1016/j.scienta.2013.06.034.

Haroun, S.A., Elnaggar, M.E., Zein, D.M. and Gad, R.I. (2020) Insecticidal efficiency and safety of zinc oxide and hydrophilic silica nanoparticles against some stored seed insects. *Journal of Plant Protection Research* 60, 77–85. DOI: 10.24425/jppr.2020.132211.

Herrmann, J.M. and Guillard, C. (2000) Photocatalytic degradation of pesticides in agricultural used waters. *Comptes Rendus de l'Académie Des Sciences - Series IIc - Chemistry* 3(6), 417–422. DOI: 10.1016/S1387-1609(00)01137-3.

Hou, Q., Han, D., Zhang, Y., Han, M., Huang, G. *et al.* (2020) The bioaccessibility and fractionation of arsenic in anoxic soils as a function of stabilization using low-cost Fe/Al-based materials: a long-term experiment. *Ecotoxicology and Environmental Safety* 191, 110210. DOI: 10.1016/j.ecoenv.2020.110210.

Isman, M.B., Miresmailli, S. and Machial, C. (2011) Commercial opportunities for pesticides based on plant essential oils in agriculture, industry and consumer products. *Phytochemistry Reviews* 10(2), 197–204. DOI: 10.1007/s11101-010-9170-4.

Jain, M., Mudhoo, A., Ramasamy, D.L., Najafi, M., Usman, M. *et al.* (2020) Adsorption, degradation, and mineralization of emerging pollutants (pharmaceuticals and agrochemicals) by nanostructures: a comprehensive review. *Environmental Science and Pollution Research* 27(28), 34862–34905. DOI: 10.1007/s11356-020-09635-x.

Jampílek, J. and Kráľová, K. (2015) Applications of nanoformulations in agricultural production and their impact on food and human health. *Proceedings of ECOpole* 9, 14–16. DOI: 10.2429/proc.2015.9(2)054.

Kah, M., Beulke, S., Tiede, K. and Hofmann, T. (2013) Nanopesticides: state of knowledge, environmental fate, and exposure modeling. *Critical Reviews in Environmental Science and Technology* 43(16), 1823–1867. DOI: 10.1080/10643389.2012.671750.

Khan, Z., Shahwar, D. and Khatoon, B. (2022) Trans-generational response of TiO_2 nanoparticles in inducing variability and changes in biochemical pool of lentil F_2 progenies. *Journal of Biosciences* 47(3), 35. DOI: 10.1007/s12038-022-00268-5.

Khodakovskaya, M., Dervishi, E., Mahmood, M., Xu, Y., Li, Z. *et al.* (2009) Carbon nanotubes are able to penetrate plant seed coat and dramatically affect seed germination and plant growth. *ACS Nano* 3(10), 3221–3227. DOI: 10.1021/nn900887m.

Khodakovskaya, M.V., Kim, B.-S., Kim, J.N., Alimohammadi, M., Dervishi, E. *et al.* (2013) Carbon nanotubes as plant growth regulators: effects on tomato growth, reproductive system, and soil microbial community. *Small* 9(1), 115–123. DOI: 10.1002/smll.201201225.

Kocira, A., Kocira, S., Złotek, U., Kornas, R. and Świeca, M. (2015) Effects of nano-gro preparation applications on yield components and antioxidant properties of common bean (*Phaseolus vulgaris* L.). *Fresenius Environmental Bulletin* 24, 4034–4041.

Kottegoda, N., Munaweera, I., Madusanka, N. and Karunaratne, V. (2011) A green slow-release fertilizer composition based on urea-modified hydroxyapatite nanoparticles encapsulated wood. *Current Science* 73–78. Available at: https://www.jstor.org/stable/24077865

Kumar, S., Kumar, D. and Dilbaghi, N. (2017) Preparation, characterization, and bio-efficacy evaluation of controlled release carbendazim-loaded polymeric nanoparticles. *Environmental Science and Pollution Research International* 24(1), 926–937. DOI: 10.1007/s11356-016-7774-y.

Lahiani, M.H., Dervishi, E., Chen, J., Nima, Z., Gaume, A. *et al.* (2013) Impact of carbon nanotube exposure to seeds of valuable crops. *ACS Applied Materials & Interfaces* 5(16), 7965–7973. DOI: 10.1021/am402052x.

Li, H., Xie, C., Li, S. and Xu, K. (2012) Electropolymerized molecular imprinting on gold nanoparticle-carbon nanotube modified electrode for electrochemical detection of triazophos. *Colloids and Surfaces B: Biointerfaces* 89, 175–181. DOI: 10.1016/j.colsurfb.2011.09.010.

Li, J., Rodrigues, S., Tsyusko, O.V. and Unrine, J.M. (2019) Comparing plant–insect trophic transfer of Cu from lab-synthesised nano-$Cu(OH)_2$ with a commercial nano-$Cu(OH)_2$ fungicide formulation. *Environmental Chemistry* 16(6), 411. DOI: 10.1071/EN19011.

Li, J., Chang, P.R., Huang, J., Wang, Y., Yuan, H., *et al.* (2013) Physiological effects of magnetic iron oxide nanoparticles towards watermelon. *Journal of Nanoscience and Nanotechnology* 13(8), 5561–5567. DOI: 10.1166/jnn.2013.7533.

Li, Z., Liu, Z., Zhang, M., Li, C., Li, Y.C. *et al.* (2020) Long-term effects of controlled-release potassium chloride on soil available potassium, nutrient absorption and yield of maize plants. *Soil and Tillage Research* 196, 104438. DOI: 10.1016/j.still.2019.104438.

Liang, J., Yu, M., Guo, L., Cui, B., Zhao, X. *et al.* (2018) Bioinspired development of P(St-MAA)-avermectin nanoparticles with high affinity for foliage to enhance folia retention. *Journal of Agricultural and Food Chemistry* 66(26), 6578–6584. DOI: 10.1021/acs.jafc.7b01998.

Lin, D. and Xing, B. (2007) Phytotoxicity of nanoparticles: inhibition of seed germination and root growth. *Environmental Pollution* 150(2), 243–250. DOI: 10.1016/j.envpol.2007.01.016.

López-García, G.P., Buteler, M. and Stadler, T. (2018) Testing the insecticidal activity of nanostructured alumina on *Sitophilus oryzae* (L.) (Coleoptera: Curculionidae) under laboratory conditions using galvanized steel containers. *Insects* 9(3), 87. DOI: 10.3390/insects9030087.

Lushchak, V.I., Matviishyn, T.M., Husak, V.V., Storey, J.M. and Storey, K.B. (2018) Pesticide toxicity: a mechanistic approach. *EXCLI Journal* 17, 1101–1136. DOI: 10.17179/excli2018-1710.

Madkour, L.H. (2019) Introduction to nanotechnology (NT) and nanomaterials (NMs). In: *Nanoelectronic Materials. Advanced Structured Materials*, Vol. 116. Springer, Cham, pp. 1–47.

Mahajan, P., Dhoke, S.K. and Khanna, A.S. (2011) Effect of nano-ZnO particle suspension on growth of mung (*Vigna radiata*) and gram (*Cicer arietinum*) seedlings using plant agar method. *Journal of Nanotechnology* 2011, 1–7. DOI: 10.1155/2011/696535.

Mahna, D., Puri, S. and Sharma, S. (2021) DNA methylation modifications: mediation to stipulate pesticide toxicity. *International Journal of Environmental Science and Technology* 18(2), 531–544. DOI: 10.1007/s13762-020-02807-9.

Maruyama, C.R., Guilger, M., Pascoli, M., Bileshy-José, N., Abhilash, P.C. *et al.* (2016) Nanoparticles based on chitosan as carriers for the combined herbicides imazapic and imazapyr. *Scientific Reports* 6(1), 19768. DOI: 10.1038/srep19768.

Mattos, B.D. and Magalhães, W.L.E. (2016) Biogenic nanosilica blended by nanofibrillated cellulose as support for slow-release of tebuconazole. *Journal of Nanoparticle Research* 18(9), 274. DOI: 10.1007/s11051-016-3586-8.

Mehta, M.R., Mahajan, H.P. and Hivrale, A.U. (2021) Green synthesis of chitosan capped-copper nano biocomposites: synthesis, characterization, and biological activity against plant pathogens. *BioNanoScience* 11(2), 417–427. DOI: 10.1007/s12668-021-00823-8.

Mura, S., Greppi, G., Roggero, P.P., Musu, E., Pittalis, D. *et al.* (2015) Functionalized gold nanoparticles for the detection of nitrates in water. *International Journal of Environmental Science and Technology* 12(3), 1021–1028. DOI: 10.1007/s13762-013-0494-7.

Nasseri, M., Golmohammadzadeh, S., Arouiee, H., Jaafari, M.R. and Neamati, H. (2016) Antifungal activity of zataria multiflora essential oil-loaded solid lipid nanoparticles *in-vitro* condition. *Iranian Journal of Basic Medical Sciences* 19, 1231. DOI: 10.22038/ijbms.2016.7824.

Negrete, J.C. (2020) Nanotechnology an option in Mexican agriculture. *Journal of Biotechnology & Bioinformatics Research* 2, 1–3. DOI: 10.47363/JBBR/2020(2)106.

Nguyen, T.H., Nguyen, T.T.L., Pham, T.D. and Le, T.S. (2020) Removal of lindane from aqueous solution using aluminum hydroxide nanoparticles with surface modification by anionic surfactant. *Polymers* 12(4), 960. DOI: 10.3390/polym12040960.

Ocsoy, I., Paret, M.L., Ocsoy, M.A., Kunwar, S., Chen, T. *et al.* (2013) Nanotechnology in plant disease management: DNA-directed silver nanoparticles on graphene oxide as an antibacterial against *Xanthomonas perforans* . *ACS Nano* 7(10), 8972–8980. DOI: 10.1021/nn4034794.

Odetayo, T., Tesfay, S. and Ngobese, N.Z. (2022) Nanotechnology-enhanced edible coating application on climacteric fruits. *Food Science & Nutrition* 10(7), 2149–2167. DOI: 10.1002/fsn3.2557.

Pabodha, D., Abeywardana, L., Sandaruwan, C., Herath, L. and Priyadarshana, G. (2022) Urea-hydroxyapatite-polymer nanohybrids as seed coatings for enhanced germination. *Vidyodaya Journal of Science* 25(1), 50–64. DOI: 10.31357/vjs.v25i01.5924.

Paret, M.L., Vallad, G.E., Averett, D.R., Jones, J.B. and Olson, S.M. (2013) Photocatalysis: effect of light-activated nanoscale formulations of TiO(2) on *Xanthomonas perforans* and control of bacterial spot of tomato. *Phytopathology* 103(3), 228–236. DOI: 10.1094/PHYTO-08-12-0183-R.

Pereira, A.E.S., Grillo, R., Mello, N.F.S., Rosa, A.H. and Fraceto, L.F. (2014) Application of poly(epsilon-caprolactone) nanoparticles containing atrazine herbicide as an alternative technique to control weeds and reduce damage to the environment. *Journal of Hazardous Materials* 268, 207–215. DOI: 10.1016/j.jhazmat.2014.01.025.

Pho, Q.H., Escriba-Gelonch, M., Losic, D., Rebrov, E.V., Tran, N.N. *et al.* (2021) Survey of synthesis processes for n-doped carbon dots assessed by green chemistry and circular and ecoscale metrics. *ACS Sustainable Chemistry & Engineering* 9(13), 4755–4770. DOI: 10.1021/acssuschemeng.0c09279.

Raguraj, S., Wijayathunga, W.M.S., Gunaratne, G.P., Amali, R.K.A., Priyadarshana, G. *et al.* (2020) Urea–hydroxyapatite nanohybrid as an efficient nutrient source in *Camellia sinensis* (L.) Kuntze (tea) . *Journal of Plant Nutrition* 43(15), 2383–2394. DOI: 10.1080/01904167.2020.1771576.

Rajonee, A.A., Zaman, S. and Huq, S.M.I. (2017) Preparation, characterization and evaluation of efficacy of phosphorus and potassium incorporated nano fertilizer. *Advances in Nanoparticles* 06(02), 62–74. DOI: 10.4236/anp.2017.62006.

Rajput, V.D., Singh, A., Minkina, T., Rawat, S., Mandzhieva, S. *et al.* (2021) Nano-enabled products: challenges and opportunities for sustainable agriculture. *Plants* 10(12), 2727. DOI: 10.3390/plants10122727.

Raliya, R., Saharan, V., Dimkpa, C. and Biswas, P. (2018) Nanofertilizer for precision and sustainable agriculture: current state and future perspectives. *Journal of Agricultural and Food Chemistry* 66(26), 6487–6503. DOI: 10.1021/acs.jafc.7b02178.

Raliya, R., Tarafdar, J.C., Singh, S.K., Gautam, R., Choudhary, K., *et al.* (2014) MgO nanoparticles biosynthesis and its effect on chlorophyll contents in the leaves of clusterbean (*Cyamopsis tetragonoloba* L.). *Advanced Science, Engineering and Medicine* 6, 538–545. DOI: 10.1166/asem.2014.1540.

Rathnaweera, D.N., Pabodha, D., Sandaruwan, C., Priyadarshana, G., Deraniyagala, S.P. *et al.* (2019) Urea modified calcium carbonate nanohybrids as a next-generation fertilizer. In: *International Research Conference, KDU-ERAt:* General Sir John Kotelawala Defence University, Sri Lanka. Available at: http://ir.kdu.ac.lk/handle/345/2353

Rizwan, M., Ali, S., ur Rehman, M.Z., Malik, S., Adrees, M. *et al.* (2019) Effect of foliar applications of silicon and titanium dioxide nanoparticles on growth, oxidative stress, and cadmium accumulation by rice (*Oryza sativa*). *Acta Physiologiae Plantarum* 41(3), 35. DOI: 10.1007/s11738-019-2828-7.

Sabir, S., Arshad, M. and Chaudhari, S.K. (2014) Zinc oxide nanoparticles for revolutionizing agriculture: synthesis and applications. *Scientific World Journal* 2014, 925494. DOI: 10.1155/2014/925494.

Sathish Kumar, R. and Arthanareeswaran, G. (2019) Nano-curcumin incorporated polyethersulfone membranes for enhanced anti-biofouling in treatment of sewage plant effluent. *Materials Science & Engineering C* 94, 258–269. DOI: 10.1016/j.msec.2018.09.010.

Semra, C. and Nadaroglu, H. (2015) The use of nanotechnology in the agriculture. *Advances in Nano Research* 3(4), 207–223. DOI: 10.12989/anr.2015.3.4.207.

Shah, V. and Belozerova, I. (2009) Influence of metal nanoparticles on the soil microbial community and germination of lettuce seeds. *Water, Air, and Soil Pollution* 197(1–4), 143–148. DOI: 10.1007/s11270-008-9797-6.

Shahryari, F., Rabiei, Z. and Sadighian, S. (2020) Antibacterial activity of synthesized silver nanoparticles by sumac aqueous extract and silver-chitosan nanocomposite against *Pseudomonas syringae* pv. *syringae*. *Journal of Plant Pathology* 102(2), 469–475. DOI: 10.1007/s42161-019-00478-1.

Sharma, D., Kanchi, S. and Bisetty, K. (2019) Biogenic synthesis of nanoparticles: a review. *Arabian Journal of Chemistry* 12(8), 3576–3600. DOI: 10.1016/j.arabjc.2015.11.002.

Sharma, P., Sharma, A., Sharma, M., Bhalla, N., Estrela, P. *et al.* (2017) Nanomaterial fungicides: *in vitro* and *in vivo* antimycotic activity of cobalt and nickel nanoferrites on phytopathogenic fungi. *Global Challenges* 1(9), 1700041. DOI: 10.1002/gch2.201700041.

Sharon, M., Choudhary, A.K., and Kumar, R. (2010) Nanotechnology in agricultural diseases and food safety. *Journal of Phytology* 2, 83–92.

Singh, H., Sharma, A., Bhardwaj, S.K., Arya, S.K., Bhardwaj, N. *et al.* (2021) Recent advances in the applications of nano-agrochemicals for sustainable agricultural development. *Environmental Science* 23(2), 213–239. DOI: 10.1039/D0EM00404A.

Siriwardena, T.A.D.P., Priyadarshana, G., Sadaruwan, C., de Silva, M. and Kottegoda, N. (2017) Urea modified silica nanoparticles: next generation slow release plant nutrients. In: *Third Biennial International Symposium on Polymer Science and Technology,* 13–15 July, Colombo, Sri Lanka. Available at: http://dr.lib.sjp.ac.lk/handle/123456789/7596

Song, M.-R., Cui, S.-M., Gao, F., Liu, Y.-R., Fan, C.-L. *et al.* (2012) Dispersible silica nanoparticles as carrier for enhanced bioactivity of chlorfenapyr. *Journal of Pesticide Science* 37(3), 258–260. DOI: 10.1584/jpestics.D12-027.

Sousa, G.F.M., Gomes, D.G., Campos, E.V.R., Oliveira, J.L., Fraceto, L.F. *et al.* (2018) Post-emergence herbicidal activity of nanoatrazine against susceptible weeds. *Frontiers in Environmental Science* 6, 12. DOI: 10.3389/fenvs.2018.00012.

Stadler, T., Buteler, M., Valdez, S.R. and Gitto, J.G. (2018) Particulate nanoinsecticides: a new concept in insect pest management. In: Begum, G. (ed.) *Insecticides: Agriculture and Toxicology*. InTech Open, pp. 83–106. DOI: 10.5772/intechopen.68177.

Sun, D., Hussain, H.I., Yi, Z., Rookes, J.E., Kong, L. *et al.* (2016) Mesoporous silica nanoparticles enhance seedling growth and photosynthesis in wheat and lupin. *Chemosphere* 152, 81–91. DOI: 10.1016/j.chemosphere.2016.02.096.

Taban, A., Saharkhiz, M.J. and Khorram, M. (2020) Formulation and assessment of nano-encapsulated bioherbicides based on biopolymers and essential oil. *Industrial Crops and Products* 149, 112348. DOI: 10.1016/j.indcrop.2020.112348.

Ur Rahim, H., Qaswar, M., Uddin, M., Giannini, C., Herrera, M.L. *et al.* (2021) Nano-enable materials promoting sustainability and resilience in modern agriculture. *Nanomaterials* 11(8), 2068. DOI: 10.3390/nano11082068.

Usman, M., Farooq, M., Wakeel, A., Nawaz, A., Cheema, S.A. *et al.* (2020) Nanotechnology in agriculture: current status, challenges and future opportunities. *Science of The Total Environment* 721, 137778. DOI: 10.1016/j.scitotenv.2020.137778.

Wagner, G., Korenkov, V., Judy, J.D. and Bertsch, P.M. (2016) Nanoparticles composed of Zn and ZnO inhibit peronospora tabacina spore germination *in vitro* and *P. tabacina* infectivity on tobacco leaves. *Nanomaterials* 6(3), 50. DOI: 10.3390/nano6030050.

Wang, X., Han, H., Liu, X., Gu, X., Chen, K. *et al.* (2012) Multi-walled carbon nanotubes can enhance root elongation of wheat (*Triticum aestivum*) plants. *Journal of Nanoparticle Research* 14(6), 841. DOI: 10.1007/s11051-012-0841-5.

Wang, Y.-N., Tsang, Y.F., Wang, H., Sun, Y., Song, Y. *et al.* (2020) Effective stabilization of arsenic in contaminated soils with biogenic manganese oxide (BMO) materials. *Environmental Pollution* 258, 113481. DOI: 10.1016/j.envpol.2019.113481.

Wang, Z., Wei, F., Liu, S.-Y., Xu, Q., Huang, J.-Y. *et al.* (2010) Electrocatalytic oxidation of phytohormone salicylic acid at copper nanoparticles-modified gold electrode and its detection in oilseed rape infected with fungal pathogen *Sclerotinia sclerotiorum*. *Talanta* 80(3), 1277–1281. DOI: 10.1016/j.talanta.2009.09.023.

Worrall, E.A., Hamid, A., Mody, K.T., Mitter, N. and Pappu, H.R. (2018) Nanotechnology for plant disease management. *Agronomy* 8(12), 285. DOI: 10.3390/agronomy8120285.

Wu, J., Xie, Y., Fang, Z., Cheng, W. and Tsang, P.E. (2016) Effects of Ni/Fe bimetallic nanoparticles on phytotoxicity and translocation of polybrominated diphenyl ethers in contaminated soil. *Chemosphere* 162, 235–242. DOI: 10.1016/j.chemosphere.2016.07.101.

Xin, X., Judy, J.D., Sumerlin, B.B. and He, Z. (2020) Nano-enabled agriculture: from nanoparticles to smart nanodelivery systems. *Environmental Chemistry* 17(6), 413. DOI: 10.1071/EN19254.

Zehra, A., Rai, A., Singh, S.K., Aamir, M., Ansari, W.A., *et al.* (2021) An overview of nanotechnology in plant disease management, food safety, and sustainable agriculture. In: *Food Security and Plant Disease Management*. Woodhead Publishing, Sawston, UK, pp. 193–219. DOI: 10.1016/b978-0-12-821843-3.00009-x.

Zhao, P., Cao, L., Ma, D., Zhou, Z., Huang, Q. *et al.* (2018) Translocation, distribution and degradation of prochloraz-loaded mesoporous silica nanoparticles in cucumber plants. *Nanoscale* 10(4), 1798–1806. DOI: 10.1039/C7NR08107C.

Zheng, L., Hong, F., Lu, S. and Liu, C. (2005) Effect of nano-TiO$_2$ on strength of naturally aged seeds and growth of spinach. *Biological Trace Element Research* 104(1), 83–92. DOI: 10.1385/BTER:104:1:083.

6 Nanoformulations of Agrochemicals Used to Apply Pesticides and Fertilizers for Crop Improvement

Yehia A. Osman*, Eladl Eltanahy, Amany Saad Hegazy and Ahmed A. Razak
Botany Department, Faculty of Science, Mansoura University, Egypt

Abstract

Agriculture productivity would not have been possible without the intensive use of traditionally formulated pesticides and fertilizers, as well as other agricultural practices. However, the use of these chemicals is associated with drawbacks that affect their efficacy. Recently, nanoformulations of these agrochemicals (and their inert ingredients) have significantly contributed to the use of lower doses, reducing the amount of active ingredients, and hence minimizing and/or eliminating any associated potential hazards. The result of these nanoformulations of pesticides and fertilizers is improved efficiencies, stabilities, solubility, bioavailability, dispersion, enhanced uptake, reduced soil adsorption and fixation, prolonged effective duration, and reduced environmental loads.

Keywords: nanoformulations, pesticides, fertilizers, crop improvements, nanoagrochemicals

6.1 A Brief History of Agriculture and Crop Improvements

Early human inhabitants of Earth obtained their food from their natural surroundings and later domesticated the desired crops and animals. Since the first domestication of crop plants, our ancestors have worked to improve crops and animals for food and feed. There is no collection of their exact trials, but we perceive them to have been centered on the productivity and quality of these crops. Moreover, these new traits or genetic alterations must have been started with careful screening and selection from among the thousands of plant species and animals grown in the wild (Diamond, 2010). This process of domestication depended on the identification of useful wild species and monitoring the genetic changes that met the selection criteria at the time. Such criteria would have included edible organ size and shape, edible parts free of a bitter taste, separation of seeds in grains, seed dormancy, changes in lifespan in terms of the growth of roots or tubers and in crops grown for fruit or seeds (Frary and Doğanlar, 2003).

Through the years, the ingenuity of humans became evident in the different techniques used to select and develop naturally occurring variants of crops. Although many of these development techniques are classified as traditional, others that depend on changing certain genetic characteristics and on genetic engineering now dominate the development of new crop varieties with countless possibilities. Additional

*Corresponding author: yaolazeik@mans.edu.eg

methods include asexual (vegetative) propagation, grafting, and tissue culture (Maluszynski *et al.*, 1995). All of these techniques opened up opportunities to increase genetic variability from natural intercrosses with wild populations, while modern gene-transfer techniques have reduced the time to construct newly desired crops to satisfy all sorts of life necessities (Majid *et al.*, 2017).

In the 18th and 19th centuries, agricultural scientists kept records of the performance of selected crop lineages from one seed generation to the next. They came up with the idea that the mating process between relevant crops would lead to genetic variations. This led seed producers to develop pedigree selection and progeny testing to evaluate the genetic improvements of the progeny. This was the birth of plant-breeding programs, even without an understanding of the genetic mechanisms involved in the inheritance of traits. This was later understood after Gregor Mendel published his work with garden peas in 1866, in which he established the basic principles of inheritance such as dominance and recessiveness, as well as genotypes and phenotypes (Purugganan, 2019).

DNA and gene-transfer technology were then more recently introduced to overcome the limitations of classical plant breeding in crop improvement. This technology centered on improving a selected trait directly with minimum screening and selection required. Moreover, it is based on a specific gene(s) rather than mixing of the whole organism's genomes and relies on cell and tissue culture. This technology reduced chemical pesticide use, increased crop yield, and was able to free up the agricultural use of fragile environments. However, the greatest problem for the commercialization of foods from transgenic or genetically altered crops is consumer acceptability. Despite this, no convincing evidence has proved that genetically altered crops have produced harmful products. To overcome the worries and resistance of consumers, risk assessments for genetically altered crops must be conducted, and only those shown to be safe for human use should be commercialized. The molecular biology revolution will allow further crop improvement efforts, working side by side with conventional plant-breeding techniques (Ahmar *et al.*, 2021).

6.2 The Role of Agrochemicals in the 'Green Revolution'

The 'Green Revolution' started in the mid-20th century by introducing high-yielding varieties of wheat and rice in developing countries (e.g. India and Mexico) to alleviate poverty and malnutrition (Davies, 2003; Pingali, 2012). This resulted in a large increase in the productivity of these food grains. This then led to changes in agricultural practices, which translated into planting two crops annually instead of one, improved irrigation facilities, crop protection measures, an increase in the farming area, modernized farm equipment, and increased use of fertilizers and pesticides (Singh, 2000; Mariyono, 2015). However, several negative impacts arose from these changes, the most notable being the damage inflicted on the environment, as well as the higher cost of fertilizers and pesticides (Rahman, 2015). Various limitations of traditionally formulated pesticides were determined, such as low dispensability, dust drift, high solvent content, and low persistence in the field; it was found that only 1% of the applied pesticide remained on plant surfaces. All of these factors must be minimized to extend the activity of pesticides on crops and increase their efficacies (Sun *et al.*, 2016; Abdollahdokht *et al.*, 2022).

Much of the progress in agricultural practices in the last 20 years has been accredited to the advancements in nanotechnology research. Nanotechnology has the capability to improve and minimize many aspects pertaining to the use of fertilizers and pesticides and their impacts on the environment and on human health. The use of nanopesticides and nanofertilizers in agriculture has many advantages, such as controlled release toward target organisms, improving nutrient uptake by plants, enhancing the water-holding capacity of the soil, reduced eutrophication, reduced hydrolysis and photodecomposition, increased efficacy and durability of pesticides, a reduced number of active ingredients, and minimization or elimination of any hazards (Cicek and Nadaroglu, 2015; Bratovcic *et al.*, 2021).

It is important to understand that pesticides are toxic not only to pests but also to humans. Exposure of humans to pesticides depends on occupation, starting with those working in the pesticide industry, those applying the pesticides and/or working in greenhouses, and exterminators of house pests. Residues of pesticides on food and in water can constitute a danger to people and/or animals in open fields (Wilson and Tisdell, 2001; Maksymiv, 2015; Abdollahdokht *et al.*, 2022). Consequently, biopesticides have been presented as alternatives to alleviate the limitations of traditionally formulated pesticides due to their lower toxicity to non-target pests, absolute specificity to the target pest, high toxicity in small doses, and rapid degradability (Kumar and Singh, 2015).

6.3   Types of Pesticide

The Environmental Protection Agency defines a pesticide as: (i) any substance or mixture of substances intended to prevent, destroy, repel, or mitigate any pest; (ii) any substance or mixture of substances intended for use as a plant regulator, defoliant, or desiccant (Gheorghe *et al.*, 2017); and (iii) any nitrogen stabilizer. The contribution of pesticides to the agricultural revolution resulted in an increased income for farmers in developed countries. However, abuse and unregulated practice of their use led to a deterioration in the environment and constituted health hazards to humans and animals, which required awareness of their dangers and hence taking the necessary actions to alleviate their use (Kah *et al.*, 2013; Carvalho, 2017; Gheorghe *et al.*, 2017).

The exact definitions of pesticides were followed by a system for their classification. This universal system grouped pesticides according to their: (i) origin (from nature or chemically synthesized); (ii) chemical structure; (iii) target pest; (iv) mode of action; (v) spectrum of activity; (vi) route of entry; and (vii) type of formulation (Gheorghe *et al.*, 2017). The conventional pesticide industry and market have undergone major changes over recent decades. These have entailed greater efficiency of pesticide use than in the past through major improvements to pest management technology and practices in the context of integrated pest management (IPM) programs. These developments have significantly improved pest management practices, and in some cases have reduced pesticide use, and have also reduced the growth in demand for chemical pesticides (Christos, 2009; Baesso *et al.*, 2014).

6.3.1   Classical formulation process

A massive number of pesticides exist in the marketplace worldwide. These may be formulated in various formats including solid powders, wettable powders (WPs), and liquid sprays (Pan *et al.*, 2019). Some formulations target adult insects, while others target the larval stages of the same insect. All pesticides consist of active and inert ingredients. Active ingredients (~900 compounds formulated into 20,000 products in the USA) kill the pest, while inert ingredients are the solvents and carriers of the active ingredients to the target pest. Mixing these two categories of chemicals in a formulation enhances the pesticide's effectiveness and handling, and improves its safety features (Zacharia, 2011). The sources of active ingredients can be natural (from microbes or plants) or compounds that are chemically synthesized in the laboratory (Hazra and Purkait, 2019). The types of formulation are usually determined by the degree of solubility of the active ingredient (regardless of their type or source) in either water or oils, depending on their intended pest and the method of delivery. Ideally, the addition of inactive ingredients should not change the pesticidal capabilities of the active ingredients (Capinera, 2008; Kalyabina *et al.*, 2021).

The characteristics of any formulation process include sorption (adhesion to solid surfaces), solution (completely dissolved active ingredient), suspension (solid particles dispersed in a liquid or WP), and emulsion (solid active ingredient dissolved in an oil-based solvent, which, when mixed with water, forms an emulsion) (Hazra and Purkait, 2019; Stejskal *et al.*, 2021). It is important to use the most appropriate pesticide formula for a particular application, which should fulfil the requirements for pest survival, safety for the applicant and the environment, contact surfaces coloration, available equipment, and the cost to farmers (Kah *et al.*, 2013). Any type of pesticide can contain one or more active ingredient, which can be formulated as a solid or liquid (Rasheed *et al.*, 2017; Hazra and Purkait, 2019; Stejskal *et al.*, 2021).

Solid formulations

Six solid formulations are sold either as ready to use (dust, granules, and pellets) or as concentrates (WPs, dry flowables, and soluble powders) to be mixed with water at the time of application.

Dusts

Dusts are manufactured by the sorption of an active ingredient on to a finely ground, solid inert substance such as talc, clay, or chalk. They are relatively lightweight and simple, and provide excellent coverage, but create an inhalation and drift hazard. They are best used indoors applied into cracks and crevices, behind baseboards and cabinets, etc.

Granules

The active ingredient of the granular formulations is adsorbed on to a large particle (passed through a 4-mesh sieve), while the inert solid may be plant materials, clay, or sand. Granules are applied dry and are usually intended for soil applications. Granules have low inhalation and dermal hazard risks. The bulky size of the granules is their greatest disadvantage, as well as handling and lack of uniformity in application.

Pellets

Pellets look like granules, but the active ingredient is mixed with inert materials to form a slurry (a thick liquid mixture) where the particles are uniform in size and shape. The slurry is then extruded under pressure and cut into pellets that are relatively uniform in size and shape. Pellets are typically used in spot applications and provide a high degree of safety to the applicator.

Wettable powders

WPs are finely divided solids, typically mineral clays, to which an active ingredient is sorbed that is not readily soluble in water. This formulation is diluted with water and a dispersing agent so that it remains suspended in the tank during application as a liquid spray. Agitation is also required to prevent settling out. WPs pose a lower dermal hazard but can present an inhalation hazard to the applicator during mixing.

Dry flowables

Dry flowable formulations are water-dispersible granules and are applied in a spray as for WPs. Inhalation hazards to the applicator are reduced due to their large size, and some dry flowables can be applied in the dry state, as recommended by the manufacturer.

Soluble powders

Soluble powders are as effective as WPs. They are also nonabrasive to application equipment. However, due to their fine particles, they can present an inhalation hazard to applicators during mixing.

Liquid formulations

There are four types of liquid formulation, and their active ingredients are generally mixed with water, crop oil, or light fuel oil as a carrier.

Liquid flowables

Liquid flowables (or suspension concentrates) are like WPs, but differ in the time of mixing the powder, dispersing agents, and wetting agents before packaging. This suspension needs to be diluted in water before use, which makes it safe to the applicator. It does not need constant agitation during application, but may be difficult to rinse from the container after use.

Microencapsulates

The active ingredients of this type of formulation are encapsulated in a starch or plastic coat and sold as a solid or liquid pesticide for enhanced safety of the applicator. Microencapsulates provide timed release of the active ingredient, which is a big advantage for this type of formulation.

Emulsifiable concentrates

This type of formulation is laborious because the active ingredients must be dissolved in oil that has been dissolved in an appropriate oil-based solvent to which an emulsifying agent has been added. The concentrate is then mixed with water and some agitation is required to maintain dispersion of the oil droplets before application as a spray. Emulsifiable concentrates present a dermal hazard to the applicator, have an odor problem, can burn foliage, and can cause the deterioration of rubber and plastic equipment parts.

Solutions

In this type of formulation, the active and inactive ingredients are thoroughly dissolved in water to form a true solution in the spray tank. Solutions are not abrasive to equipment and will not plug strainers and screens. Solutions have few disadvantages; however, some that are produced as dissolved salts can be caustic to human skin.

Miscellaneous liquid formulations

The majority of liquid formulations are designed to be mixed with a carrier before application but some are supplied ready to use. Low- and ultralow-volume concentrates are frequently applied undiluted by specified equipment in specialty places (e.g. space spraying and fogging). While they provide excellent coverage, inhalation problems during application can be quite high due to drift.

Aerosols

This type of pesticide formulation targets indoor insects but is subject to inhalation dangers. Aerosols or mists of very tiny particles will be generated under pressure and delivered to the target pests, which are difficult to confine to the target area.

Fumigants

In contrast to aerosols, fumigants deliver the active ingredient to the target site in the form of a gas with the penetrating power to fill the whole site. Thus, fumigants are used to treat grains in silos or other type of storage and can be applied to treat soil pest insects or objects and structures such as furniture. Fumigants are one of the most hazardous formulations due to their inhalation danger.

6.3.2 Pesticide label information

Labels printed on the pesticide containers must be read carefully at the time of purchase and application. The label will contain information about the trade name, type of formulation, mode of action, concentration of active ingredient (weight basis), precautions, and dose per acre. For example a pesticide with a technical brand name Trigard 75WP means 75% by weight of active ingredient formulated as a WP, while the brand name TIZARA (Lambdacyhalothrin 5% EC) means that it contains $5\%\,l^{-1}$ of the active ingredient formulated as an emulsifiable concentrate (European Food Safety Authority, 2014).

6.4 Emerging Nanotechnology and Nanoformulation of Agrochemicals

The progress of agricultural practices over the last 20 years has been credited to the advancements in nanotechnology research. Nanotechnology has the capability to improve and minimize many aspects pertaining to the uses of fertilizers and pesticides and their impacts on the environment and human health. This technology depends on novel methods of preparing chemical molecules within the nanoscale particle size. The external dimensions of nanomaterials (NMs) ranges from approximately 1 to 100 nm with a large surface area:volume ratio and high mechanical strength. This gives the particles unique physical and chemical properties that enable them to play an important role in the fabrication of many devices (Khan *et al.*, 2019). Further studies of these nanoparticles (NPs) have enhanced the efficacy of pesticides, with a decrease in their toxicity toward humans and the environment (Kookana *et al.*, 2014; Mohd Firdaus *et al.*, 2018). Moreover, the encapsulation of pesticides with NPs reduces their losses in soils due to physical degradation, volatilization, and leaching (Yadav *et al.*, 2020; Chaud *et al.*, 2021). NPs may occur in monomeric, oligomeric, or polymeric particulate systems having a specific scale of size (Chinnamuthu and Boopathi, 2009).

6.4.1 Nanoformulations

Nanoformulations use NPs synthesized by different means. For example, the biologically synthesized silver nitrate NPs required little or no energy input (Benelli, 2016). In this method, leaf extract of *Ocimum sanctum* L. was mixed with silver nitrate at ambient room temperature (Khan *et al.*, 2017). These natural nanofactories under optimized conditions can synthesize stable silver nanoparticles (AgNPs) with a well-defined morphology, size, and composition (Gurunathan *et al.*, 2009). A triad complex can then be synthesized by entrapping the biopesticide with either a protein or a carbohydrate polymer and the AgNPs. These crosslinking nanosilver particles complexed with nanobiopesticide help to release active nanobioinsecticide in a controlled manner and show prolonged pesticidal activity (Ragaei and Sabry, 2014).

6.4.2   Types and synthesis of nanoparticles

There are four major types of NM based on the original chemical used (Jeevanandam *et al.*, 2018):

1. Inorganic-based NMs use a variety of metals (e.g. aluminum (Al), cadmium (Cd), copper (Cu), gold (Au), iron (Fe), lead (Pb), silver (Ag), and zinc (Zn)) and metal oxides (e.g. copper oxide (CuO), cerium oxide (CeO_2), iron oxide (Fe_2O_3), iron oxide (Fe_3O_4), magnesium aluminum oxide ($MgAl_2O_4$), silica (SiO_2), titanium dioxide (TiO_2), and zinc oxide (ZnO)) (Ju-Nam and Lead, 2008).
2. Carbon-based NMs use single-walled carbon nanotubes, multiwalled carbon nanotubes, graphene, fullerene, carbon fibre, carbon black, and activated carbon (Kumar and Kumbhat, 2016).
3. Organic-based NMs or polymer-based are formed from any organic material except carbon materials (e.g. dendrimers, cyclodextrin, liposome, and micelles) (Sahu *et al.*, 2021).
4. Composite-based NMs are prepared from any combination of the previous three types and are characterized as having complicated structures like a metal–organic framework (Jeevanandam *et al.*, 2018).

6.4.3   Toxicity and impact on the environment

Some pesticides mimic or antagonize natural hormones in our body, and it has been reported that immune suppression, reduced intelligence, cancer, and reproductive defects are linked to long-term, low-dose exposure to some pesticides. Specifically, organophosphates, carbamates, pyrethroids, and organochlorine pesticides have been directly linked to neurological disorders such as Parkinson disease and Alzheimer disease (Asghar *et al.*, 2016). The annual effects of pesticide poisoning have been estimated to reach 1 million people causing both death and chronic diseases worldwide.

The safe use of NPs has caused worry among all concerned parties due to the ability of NPs to interact with their surroundings, including humans. The final consequences of these nano–bio–eco interactions with humans are at the heart of the safety debate. Comprehensive studies on the interactions between some metal-based NPs and nuclear DNA at the molecular level have shown that inhalation of vaporized metal-based NPs by humans produces hydroxyl radicals, which can attach to the nuclear DNA bases. These DNA bases may then be damaged by oxidation or expansion or even lost, resulting in a point mutation that can lead to cancer (Shukla *et al.*, 2021). Other studies have shown that the degree of toxicity of NPs varies from high in pure form to very low in natural ecosystems (Wiesner *et al.*, 2009). Many studies have proposed that the NP toxicity mechanism in plant cells is primarily due to the production of reactive oxygen species (Khan *et al.*, 2019a). A seed germination assay was developed to determine the toxicity of NMs in plants (Wang *et al.*, 2001). Transport, fate, and toxicity of emerging and NM contaminants in aquatic ecosystems have also been studied. However, the inherent characteristics of nanopesticides and nanofertilizers, such as controlled release and effective absorption, have led to reduced frequency of application and the concentrations used.

6.5   Nanoformulation of Pesticides

The intensified use of classically formulated pesticides was associated with the agricultural revolution. These chemicals are commonly used to improve crop yield by resisting and/or withstanding damage caused by a wide variety of pests. However, their use caused great damage to the environment and has affected human health negatively (Aktar *et al.*, 2009). Nanopesticides are an emerging research field, and offer new methods to design active ingredients at nanoscale dimensions. The availability of nanopesticide-based formulations has the potential for the development of more effective and safer pesticides/biopesticides. These new classes of nano-sized pesticides, fungicides, herbicides, and fertilizers will increase bioavailability and water solubility, allow efficient control of pathogens, insects, and weeds, have prolonged shelf life, and allow us to monitor and manage plant health and soil improvement (Abdollahdokht *et al.*, 2022).

6.5.1   Nano-encapsulation of pesticides

Pesticides can be encapsulated in a number of nanomolecules via complexation interactions, incorporation, or electrostatic or covalent bonding, and directly delivered to their target. Nanocapsules can be synthesized from lipids, organic materials, plant-derived NPs, and inorganic materials. The encapsulated pesticides will be released in a controlled way and will be protected from premature degradation and/or losses by leaching and volatilization. As an example, lipid-based nano-encapsulation of pesticides encapsulated hydrophobic or hydrophilic pesticides in the absence of organic solvents. The encapsulation of pesticides into different nanocarriers enhanced their activities and toxicities (Abdollahdokht *et al.*, 2022). However, pesticide encapsulation can lead to deteriorative reactions such as physico-chemical instability and formation of acid monomers due to polymer degradation, leading to their decomposition. Moreover, it has been reported that oxygen and sunlight are the most important damaging agents for encapsulated nanopesticides as well as heat and mechanical stress (Gijsman, 2008).

Researchers have produced nanoscale materials that can remedy the prohibitive costs for more efficient nanoformulated fertilizers and pesticides, leading to a second food production

revolution to feed billions around the world. A good example is the preparation of urea–hydroxyapatite nanohybrids for the slow release of nitrogen (Kottegoda *et al.*, 2017). This nanoformulated fertilizer maximized the use of urea, the richest nitrogen source, required by plants, and no multiple applications were required to boost crop production. Moreover, this fertilizer resisted breakdown into ammonia, the cause of eutrophication of water bodies, and prevented its conversion to nitrogen dioxide, one of the main greenhouse gases.

6.6   Nanoformulation of Fertilizers

NPs are one of the modern technologies used effectively in synthesizing active compounds that can be used in multiple applications, including agricultural applications (Fatima *et al.*, 2021). The negatives of using traditional chemical fertilizers on soil, crops, and human health are already known, as well as the occurrence of leakage of these chemicals into the cycles of elements such as the nitrogen cycle in nature (Mosier *et al.*, 2013). These had led to significant imbalances in the ecosystem. Recently, there have been several attempts to create new materials and agricultural growth promoters, pesticides and fertilizers that are less harmful to health and the environment from different sources such as bacteria (Syed-ab-Rahman *et al.*, 2019), cyanobacteria (Alsenani *et al.*, 2020), fungi (Li *et al.*, 2022), microalgae (González-González *et al.*, 2019), and seaweeds (Hussein *et al.*, 2021). The most important of these attempts is the manufacture of NPs from biological sources that are not harmful to the environment. These molecules are prepared by the process of reduction in an aqueous solution of a metal salt as a precursor to the manufacturing process (Salem and Fouda, 2021). This method is uncomplicated, and its cost is relatively low because it requires less energy in production than other methods (Ahmad *et al.*, 2019)

The composition, shape, and size distribution of NPs influence their usefulness, stability, and physico-chemical qualities. Aggregation caused by physical collision/interaction affects the stability of NPs due to particle size changes. In terms of thermodynamics, the van der Waals interaction potential increases as the size of NPs decreases. Changes in the surface potential of NPs in the solution can be quantified as the zeta potential as an indicator of the chemical stability of the NP surface. NPs are deemed stable when their zeta potential is greater than $+30\,mV$ or less than $-30\,mV$, but the type of solvent used and the pH of the solution can significantly impact these values (Fouconnier *et al.*, 2021).

Biological synthesis can create high yields of homogeneous, stable NPs by altering factors such as pH, temperature, mixing velocity, incubation interval, and the ratios of the metal precursor and biological source. Furthermore, biological source materials are natural capping or stabilizing agents, which inhibit aggregation and provide excellent stability to NPs by promoting repulsive interactions between particles (Sardar and Mazumder, 2019). However, a detailed understanding of the mechanism of biological synthesis is still lacking (Salem and Fouda, 2021).

6.6.1 Nanofertilizers biosynthesis by microorganisms

Many microorganisms thrive and grow on many agricultural and industrial wastes, as well as under various harsh conditions of high temperatures, pH, salinity, or even atmospheric pressure and radiation. These conditions may sometimes be one of the requirements for the occurrence of some chemical reactions to produce NPs, so the use of microorganisms has become a global trend in the production of NPs. Microorganisms are divided into groups such as bacteria, fungi, actinomycetes, and algae. Most of these microorganisms, under conditions that are not suitable for growth, and also under normal conditions, can produce reducing metabolic enzymes such as nitrate reductase and α-NADPH-dependent reductases that can reduce metal salts and convert them to the alternative NPs of the metal (Salunke *et al.*, 2016). Additionally, it has been found that there are many active compounds in both dead and live microbial cells (Binupriya *et al.*, 2010) that have the ability to bioform NPs such as sugars and other secondary metabolites, which make these cells nanofactories for the synthesis of NPs. The method of production and extraction

is determined based on the extracellular or intracellular locations of the active substance that produces the NPs (Hulkoti and Taranath, 2014). If the active substances are produced extracellularly, this process is less expensive because it only requires sedimentation of the cells at an appropriate speed using a centrifuge, collection of the supernatant that contains the metabolites, addition of the metal salt to the solution, and incubation in suitable conditions for the formation of NPs.

In contrast, for intracellular compounds, after centrifugation the cell pellet is washed and then transferred to the solution of metal ions and incubated to allow the ions to enter the cell via passive transport. The cells are then treated by ultrasound to break the cell wall and plasma membrane to allow released of the NPs (Singh *et al.*, 2016). Several studies have used microalgae in the production of NPs, both inside and outside the cell (El-Sheekh *et al.*, 2022). Other studies have shown the possibility of controlling the properties of the resulting NPs by changing the type of microalgae and the conditions of production and incubation. For example, at normal temperature, *Anabaena*, *Calothrix*, and *Klebsormidium* spp. generate intracellularly akaganeite–β-FeOOH nanorods with well-controlled size and a unique morphology (Brayner *et al.*, 2009). A straightforward, inexpensive, and previously unknown process for the 'green' manufacture of AgNPs utilizing cell extracts from *Anabaena doliolum* has been demonstrated (Garvita *et al.*, 2014). In addition, using *Anabaena* sp. (strain L31) cell extract is considered a direct approach for manufacturing ZnONPs (Singh *et al.*, 2014).

In green algae, *Chlamydomonas reinhardtii* was used as a model system to investigate the involvement of cellular proteins in the production of AgNPs and how *C. reinhardtii* influences their size, shape, and monodispersity (Barwal *et al.*, 2011). Furthermore, using *Chlorella pyrenoidosa* extract in the production of stable AuNPs, the most influential parameters were pH 8, 100°C, and 100 ppm of aurochlorate salt. The activity of nitrate reductase in the algal extract was determined to be $0.7245\,\mu mol\ min^{-1}\ g^{-1}$, which was reduced to $0.5244\,\mu mol\ min^{-1}\ g^{-1}$ after the production of the AuNPs (Oza *et al.*, 2012).

6.6.2 Major types of nanofertilizer

Nanofertilizers are considered a safer and eco-friendly alternative to the usual chemical fertilizers. Therefore, scientists have devoted much effort to determining the effect of these nanofertilizers on plants at high and low concentrations and their potential negative effects on the environment. Most of these fertilizers fall within the group of solutions of metal salts in which nanometal particles are prepared as described earlier and applied directly. The most important of these fertilizers are discussed.

CuNPs can be synthesized directly using plants and microbes or indirectly using enzymes, proteins, starch, amino acids, etc. CuNPs produced through the enzymatic and nonenzymatic interactions of copper salt with organic molecules are both green (environmentally friendly) and cost-effective (Santhoshkumar *et al.*, 2019). Cu has an important and effective role in plant metabolism as it affects the rate of photosynthesis and respiration and thus changes the metabolic pathways for the formation of proteins and carbohydrates because it is used as a catalyst for some enzymes. CuNPs have many uses that may be more economically important than using it as a nanofertilizer, such as in the electronics industry and for heat sensors, gases, and the battery industry (Kasana *et al.*, 2017).

The extracellular technique for biosynthesis of NPs by algae incorporated the use of extracts from various algal species for synthesis of CuONPs through ultrasound-assisted biosynthesis, for example using the brown alga *Cystoseira trinodis* (Gu *et al.*, 2018). In a recent study, it was found that a concentration of 50 ppm of CuNPs led to plant pathogen inhibition, such as inhibition of spore formation and mycelium in *Alternaria alternata*, and inhibition of the growth of *Phytophthora cinnamomi* by 75% (Banik and Pérez-de-Luque, 2017). In contrast, it was found that high concentrations of CuNPs lead to a decrease in the growth of many crops, although they lead to an increase in seed germination rate at low concentrations (Kasana *et al.*, 2017). CuONPs have been applied as nutrients, insecticides, herbicides, plant growth regulators, and extracts for soil improvement, with research on their accumulation in lettuce and cabbage revealing reduced water retention and crop

productivity (Xiong *et al.*, 2017). Foliar spray treatment of CuNPs at concentrations ranging from 50 to 500 ppm resulted in enhanced yields and elevated bioactive abscisic acid and antioxidants in fruits, and induced fruit rigidity (López-Vargas *et al.*, 2018).

Fe plays an important role in the metabolic processes inside the plant, and it is important to provide it in the soil or in irrigation water using standard or nanofertilizers. It is an essential component in the manufacture of chlorophyll, which in turn affects the rates of photosynthesis and metabolism, in addition its importance for maintaining the structure and function of chloroplasts. It is also involved in the synthesis of many cytochrome enzymes of the electron transport chain (Kroh and Pilon, 2020).

Fe_3O_4NPs can be produced in a single step by reducing ferric chloride solution with a brown seaweed (*Sargassum muticum* (Yendo) Fensholt) water extract comprising sulfated polysaccharides as a key ingredient, which works as a reducing agent and effective stabilizer. The average particle diameter, as measured by transmission electron microscopy, was shown to be 18 ± 4 nm and the particles were in a crystalline cubic form, according to X-ray diffraction (Mahdavi *et al.*, 2013). Unfortunately, compared with plant-based synthesis, NMs produced from microbes are less monodispersed and the rate of synthesis is slower. The easiest, most cost-effective and repeatable way is green production of metallic NPs using various plant components such as the leaf, stem, seed, and root. Plants, as opposed to microbes, create more stable metallic NPs and have shown to be the ideal choice for quick and large-scale synthesis (Iravani, 2011). The biosynthesis of FeNPs has mostly been carried out utilizing green tea (*Camellia sinensis*) extract, which is a cheap and readily available resource. Nano-zerovalent iron has also been synthesized and shown to be stable at room temperature without the addition of any surfactant or polymer (Hoag *et al.*, 2009).

Fe_2O_3NPs have been demonstrated to result in shoot and root elongation, and increased biomass, phytohormones, antioxidants, and enzyme content when used as a nanofertilizer for groundnuts (*Arachis hypogaea*) (Rui *et al.*, 2016), while *Citrus maxima* plants (Hu *et al.*, 2017) treated with both Fe_2O_3NPs, and Fe^{3+} showed good growth and high Fe accumulation compared with controls. A detailed review has described the different methods and studies for the preparation of FeNPs and their environmental implications (Saif *et al.*, 2016).

Zn is one of the essential elements of the plant, as it acts as a cofactor and regulator of enzymatic activity and has a vital role in the physiological processes inside the plant. It is found in nature in a divalent form such as ZnO or zinc sulfate (ZnSO), and is of great importance to human health, but the spread of land desertification has led to Zn deficiency in the soil, and as a result, more than 30% of the world population currently suffer from Zn deficiency due to its deficiency in cereals, which are the leading food in most countries (Guilbert, 2003).

One of the most important difficulties facing the solubility of Zn as a fertilizer is its bivalent nature, which requires its presence with chelated materials. However, this does not solve the problem of solubility completely and leads to the loss of a large amount of Zn because it becomes unavailable to the plant. The strongest competitor to traditional Zn fertilizers is nano-Zn fertilizers such as ZnONPs (Jamdagni *et al.*, 2018). When ZnONPs were mixed with chelated $ZnSO_4$ commercial fertilizer in solutions and sprayed on groundnut crops, they boosted seed germination, seed vigour, early flowering, chlorophyll and carotenoid content, pod yield, and shoot and root development (Prasad *et al.*, 2012). The green manufacture of ZnONPs utilizing aqueous extract after calcination of *S. muticum* produced ZnO at 450°C, and the Fourier-transform infrared spectra indicated pure ZnONPs with average sizes ranging from 30 to 57 nm (Azizi *et al.*, 2014). In an experiment to test algal toxicities of Zn compounds, it was found that the toxicity of bulk ZnO was lower than that of ZnONPs, while that of ZnONPs was less than the Zn^{2+} form when the concentration was lower than 50 mg l^{-1}.

However, ZnONPs had higher algal toxicity than Zn^{2+} ions at concentrations > 50 mg L^{-1} (Ji *et al.*, 2011).

6.7 Conclusions

Food crop quality and a sustainable supply of food for our ever-growing world population are top priorities worldwide. The improvement of crops to satisfy these needs includes a huge list of criteria such as the type of seeds, productivity, quality of the products, pest management approaches pre- and postharvest, use of NPs, quality of the soils used for plantation, and use of fertilizers. This chapter has focused on only two areas concerning crop improvements: controlling pests and replenishing soils with deficient nutrients. Diverse populations of pests (e.g. insects, pathogens, and herbs) can exist in a natural balance in open fields. However, one or more of these pests may specifically attack your preferred crop causing unacceptable economic damage that requires control. Currently, use of the integrated approach to control perennial populations of pests by IPM is the preferred route. Nanopesticides and nanofertilizers are at the heart of such IPM programs. The extensive review of the literature in this chapter has demonstrated that the inclusion of NPs and a diverse number of active ingredients available in IPM programs will ensure success and sustain the use of this strategy under different agroecosystems.

The inclusion of nanopesticides and nanofertilizers has led to sustainable agriculture and has converted centuries of inherited farming practices into smart precision ones, which limit the negative impacts on human health, such as pesticide residues, and can help to maintain eco-balance, soil qualities, and ecosystems.

References

Abdollahdokht, D., Gao, Y., Faramarz, S., Poustforoosh, A., Abbasi, M. *et al.* (2022) Conventional agrochemicals towards nano-biopesticides: an overview on recent advances. *Chemical and Biological Technologies in Agriculture* 9(1), 13. DOI: 10.1186/s40538-021-00281-0.

Ahmad, F., Ashraf, N., Ashraf, T., Zhou, R.-B. and Yin, D.-C. (2019) Biological synthesis of metallic nanoparticles (MNPs) by plants and microbes: their cellular uptake, biocompatibility, and biomedical applications. *Applied Microbiology and Biotechnology* 103(7), 2913–2935. DOI: 10.1007/s00253-019-09675-5.

Ahmar, S., Mahmood, T., Fiaz, S., Mora-Poblete, F., Shafique, M.S. *et al*. (2021) Advantage of nanotechnology-based genome editing system and its application in crop improvement. *Frontiers in Plant Science* 12, 663849. DOI: 10.3389/fpls.2021.663849.

Aktar, W., Sengupta, D. and Chowdhury, A. (2009) Impact of pesticides use in agriculture: their benefits and hazards. *Interdisciplinary Toxicology* 2(1), 1–12. DOI: 10.2478/v10102-009-0001-7.

Alsenani, F., Tupally, K.R., Chua, E.T., Eltanahy, E., Alsufyani, H. *et al*. (2020) Evaluation of microalgae and cyanobacteria as potential sources of antimicrobial compounds. *Saudi Pharmaceutical Journal* 28(12), 1834–1841. DOI: 10.1016/j.jsps.2020.11.010.

Asghar, U., Malik, M. and Javed, A. (2016) Pesticide exposure and human health: a review. *Journal of Ecosystem & Ecography* S5, 005. DOI: 10.4172/2157-7625.S5-005.

Azizi, S., Ahmad, M.B., Namvar, F. and Mohamad, R. (2014) Green biosynthesis and characterization of zinc oxide nanoparticles using brown marine macroalga *Sargassum muticum* aqueous extract. *Materials Letters* 116, 275–277. DOI: 10.1016/j.matlet.2013.11.038.

Baesso, M.M., Teixeira, M.M., Ruas, R.A.A., and Baesso, R.C.E. (2014) Pesticide application technologies. *Revista Ceres* 61, 780–785.

Banik, S. and Pérez-de-Luque, A. (2017) *In vitro* effects of copper nanoparticles on plant pathogens, beneficial microbes and crop plants. *Spanish Journal of Agricultural Research* 15(2), e1005. DOI: 10.5424/sjar/2017152-10305.

Barwal, I., Ranjan, P., Kateriya, S. and Yadav, S.C. (2011) Cellular oxido-reductive proteins of *Chlamydomonas reinhardtii* control the biosynthesis of silver nanoparticles. *Journal of Nanobiotechnology* 9, 56. DOI: 10.1186/1477-3155-9-56.

Benelli, G. (2016) Plant-mediated biosynthesis of nanoparticles as an emerging tool against mosquitoes of medical and veterinary importance: a review. *Parasitology Research* 115(1), 23–34. DOI: 10.1007/s00436-015-4800-9.

Binupriya, A.R., Sathishkumar, M. and Yun, S.-I. (2010) Myco-crystallization of silver ions to nanosized particles by live and dead cell filtrates of *Aspergillus oryzae* var. *viridis* and its bactericidal activity toward *Staphylococcus aureus* KCCM 12256. *Industrial & Engineering Chemistry Research* 49(2), 852–858. DOI: 10.1021/ie9014183.

Bratovcic, A., Hikal, W.M., Said-Al Ahl, H.A.H., Tkachenko, K.G., Baeshen, R.S. *et al*. (2021) Nanopesticides and nanofertilizers and agricultural development: scopes, advances and applications. *Open Journal of Ecology* 11(04), 301–316. DOI: 10.4236/oje.2021.114022.

Brayner, R., Yéprémian, C., Djediat, C., Coradin, T., Herbst, F. *et al*. (2009) Photosynthetic microorganism-mediated synthesis of akaganeite (beta-FeOOH) nanorods. *Langmuir* 25(17), 10062–10067. DOI: 10.1021/la9010345.

Capinera, J.L. (2008) *Encyclopedia of Entomology*, 2nd edn. Springer Science and Business Media, Heidelberg. DOI: 10.1007/978-1-4020-6359-6.

Carvalho, F.P. (2017) Pesticides, environment, and food safety. *Food and Energy Security* 6(2), 48–60. DOI: 10.1002/fes3.108.

Chaud, M., Souto, E.B., Zielinska, A., Severino, P., Batain, F. *et al*. (2021) Nanopesticides in agriculture: benefits and challenge in agricultural productivity, toxicological risks to human health and environment. *Toxics* 9(6), 131. DOI: 10.3390/toxics9060131.

Chinnamuthu, C. and Boopathi, P.M. (2009) Nanotechnology and agroecosystem. Madras Agricultural Journal 96, 17–31.

Christos, A.D. (2009) Understanding benefits and risks of pesticide use. *Scientific Research Essays* 4, 945–949.

Cicek, S. and Nadaroglu, H. (2015) The use of nanotechnology in the agriculture. *Advances in Nano Research* 3(4), 207–223. DOI: 10.12989/anr.2015.3.4.207.

Davies, W.P. (2003) An historical perspective from the Green Revolution to the gene revolution. *Nutrition Reviews* 61(6 Pt 2), S124–S134. DOI: 10.1301/nr.2003.jun.S124-S134.

Diamond, J.M. (2010) *The Worst Mistake in the History of the Human Race*. Oplopanax Publishing.

El-Sheekh, M.M., Morsi, H.H., Hassan, L.H.S. and Ali, S.S. (2022) The efficient role of algae as green factories for nanotechnology and their vital applications. *Microbiological Research* 263, 127111. DOI: 10.1016/j.micres.2022.127111.

European Food Safety Authority (2014) Conclusion on the peer review of the pesticide risk assessment of the active substance lambda-cyhalothrin. *EFSA Journal* 12(5), 3677. DOI: 10.2903/j.efsa.2014.3677.

Fatima, F., Hashim, A. and Anees, S. (2021) Efficacy of nanoparticles as nanofertilizer production: a review. *Environmental Science and Pollution Research International* 28(2), 1292–1303. DOI: 10.1007/s11356-020-11218-9.

Fouconnier, B., Lopez-Serrano, F., Puente Lee, R.I., Terrazas-Rodriguez, J.E., Roman-Guerrero, A. *et al.* (2021) Hybrid microspheres and percolated monoliths synthesized via pickering emulsion co-polymerization stabilized by *in situ* surface-modified silica nanoparticles. *Express Polymer Letters* 15(6), 554–567. DOI: 10.3144/expresspolymlett.2021.47.

Frary, A. and Doğanlar, S. (2003) Comparative genetics of crop plant domestication and evolution. *Turkish Journal of Agriculture Forestry* 27, 59–69.

Garvita, S., Piyoosh, K.B., Shailesh, K.S., Rajeshwar, P.S. and Madhu, B.T. (2014) Green synthesis of silver nanoparticles using cell extracts of *Anabaena doliolum* and screening of its antibacterial and antitumor activity. *Journal of Microbiology and Biotechnology* 24(10), 1354–1367. DOI: 10.4014/jmb.1405.05003.

Gheorghe, I., Popa, M., Marutescu, L., Saviuc, C., Lazar, V., and Chifiriuc, M.C. (2017) Lessons from inter-regn communication for the development of novel, ecofriendly pesticides. In: Grumezescu, A.M. (ed.) *New Pesticides and Soil Sensors*. Academic Press, London, pp. 1–45.

Gijsman, P. (2008) Review on the thermo-oxidative degradation of polymers during processing and in service. *e-Polymers* 8(1), 1–34. DOI: 10.1515/epoly.2008.8.1.727.

González-González, L.M., Eltanahy, E. and Schenk, P.M. (2019) Assessing the fertilizing potential of microalgal digestates using the marine diatom *Chaetoceros muelleri*. *Algal Research* 41, 101534. DOI: 10.1016/j.algal.2019.101534.

Gu, H., Chen, X., Chen, F., Zhou, X. and Parsaee, Z. (2018) Ultrasound-assisted biosynthesis of CuO-NPs using brown alga *Cystoseira trinodis*: characterization, photocatalytic AOP, DPPH scavenging and antibacterial investigations. *Ultrasonics Sonochemistry* 41, 109–119. DOI: 10.1016/j.ultsonch.2017.09.006.

Guilbert, J. (2003) The world health report 2002—reducing risks, promoting healthy life. *Education for Health* 16, 230–230.

Gurunathan, S., Kalishwaralal, K., Vaidyanathan, R., Venkataraman, D., Pandian, S.R.K. *et al.* (2009) Biosynthesis, purification and characterization of silver nanoparticles using *Escherichia coli*. *Colloids and Surfaces B: Biointerfaces* 74(1), 328–335. DOI: 10.1016/j.colsurfb.2009.07.048.

Hazra, D.K. and Purkait, A. (2019) Role of pesticide formulations for sustainable crop protection and environment management: a review. *Journal of Pharmacognosy and Phytochemistry* 8, 686–693.

Hu, J., Zhang, J., Ji, N. and Zhang, C. (2017) A new regularized restricted Boltzmann machine based on class preserving. *Knowledge-Based Systems* 123, 1–12. DOI: 10.1016/j.knosys.2017.02.012.

Hoag, G.E., Collins, J.B., Holcomb, J.L., Hoag, J.R., Nadagouda, M.N., *et al.* (2009) Degradation of bromothymol blue by 'greener' nano-scale zero-valent iron synthesized using tea polyphenols. *Journal of Materials Chemistry* 19(45), 8671–8677. DOI: 10.1039/b909148c.

Hulkoti, N.I. and Taranath, T.C. (2014) Biosynthesis of nanoparticles using microbes – a review. *Colloids and Surfaces B: Biointerfaces* 121, 474–483. DOI: 10.1016/j.colsurfb.2014.05.027.

Hussein, M.H., Eltanahy, E., Al Bakry, A.F., Elsafty, N. and Elshamy, M.M. (2021) Seaweed extracts as prospective plant growth bio-stimulant and salinity stress alleviator for *Vigna sinensis* and *Zea mays*. *Journal of Applied Phycology* 33(2), 1273–1291. DOI: 10.1007/s10811-020-02330-x.

Iravani, S. (2011) Green synthesis of metal nanoparticles using plants. *Green Chemistry* 13(10), 2638–2650. DOI: 10.1039/c1gc15386b.

Jamdagni, P., Rana, J.S. and Khatri, P. (2018) Comparative study of antifungal effect of green and chemically synthesised silver nanoparticles in combination with carbendazim, mancozeb, and thiram. *IET Nanobiotechnology* 12(8), 1102–1107. DOI: 10.1049/iet-nbt.2018.5087.

Jeevanandam, J., Barhoum, A., Chan, Y.S., Dufresne, A. and Danquah, M.K. (2018) Review on nanoparticles and nanostructured materials: history, sources, toxicity and regulations. *Beilstein Journal of Nanotechnology* 9, 1050–1074. DOI: 10.3762/bjnano.9.98.

Ji, J., Long, Z. and Lin, D. (2011) Toxicity of oxide nanoparticles to the green algae *Chlorella* sp. *Chemical Engineering Journal* 170(2–3), 525–530. DOI: 10.1016/j.cej.2010.11.026.

Ju-Nam, Y. and Lead, J.R. (2008) Manufactured nanoparticles: an overview of their chemistry, interactions and potential environmental implications. *The Science of the Total Environment* 400(1–3), 396–414. DOI: 10.1016/j.scitotenv.2008.06.042.

Kah, M., Beulke, S., Tiede, K. and Hofmann, T. (2013) Nanopesticides: state of knowledge, environmental fate, and exposure modeling. *Critical Reviews in Environmental Science and Technology* 43(16), 1823–1867. DOI: 10.1080/10643389.2012.671750.

Kalyabina, V.P., Esimbekova, E.N., Kopylova, K.V. and Kratasyuk, V.A. (2021) Pesticides: formulants, distribution pathways and effects on human health - a review. *Toxicology Reports* 8, 1179–1192. DOI: 10.1016/j.toxrep.2021.06.004.

Kasana, R.C., Panwar, N.R., Kaul, R.K. and Kumar, P. (2017) Biosynthesis and effects of copper nanoparticles on plants. *Environmental Chemistry Letters* 15(2), 233–240. DOI: 10.1007/s10311-017-0615-5.

Khan, M., Tarek, F., Nuzat, M., Momin, M. and Hasan, M. (2017) Rapid biological synthesis of silver nanoparticles from *Ocimum sanctum* and their characterization. *Journal of Nanoscience* 2017, 1–6. DOI: 10.1155/2017/1693416.

Khan, I., Saeed, K. and Khan, I. (2019a) Nanoparticles: properties, applications and toxicities. *Arabian Journal of Chemistry* 12(7), 908–931. DOI: 10.1016/j.arabjc.2017.05.011.

Khan, Z., Shahwar, D., Yunus Ansari, M.K. and Chandel, R. (2019) Toxicity assessment of anatase (TiO_2) nanoparticles: a pilot study on stress response alterations and DNA damage studies in *Lens culinaris* Medik. *Heliyon* 5(7), e02069. DOI: 10.1016/j.heliyon.2019.e02069.

Kookana, R.S., Boxall, A.B.A., Reeves, P.T., Ashauer, R., Beulke, S. *et al.* (2014) Nanopesticides: guiding principles for regulatory evaluation of environmental risks. *Journal of Agricultural and Food Chemistry* 62(19), 4227–4240. DOI: 10.1021/jf500232f.

Kottegoda, N., Sandaruwan, C., Priyadarshana, G., Siriwardhana, A., Rathnayake, U.A. *et al.* (2017) Urea-hydroxyapatite nanohybrids for slow release of nitrogen. *ACS Nano* 11(2), 1214–1221. DOI: 10.1021/acsnano.6b07781.

Kroh, G.E. and Pilon, M. (2020) Regulation of iron homeostasis and use in chloroplasts. *International Journal of Molecular Sciences* 21(9), 3395. DOI: 10.3390/ijms21093395.

Kumar, N. and Kumbhat, S. (2016) *Essentials in Nanoscience and Nanotechnology*. Wiley, Hoboken, NJ. DOI: 10.1002/9781119096122.

Kumar, S. and Singh, A. (2015) Biopesticides: present status and the future prospects. *Journal of Fertilizers & Pesticides* 06(02), 100–129. DOI: 10.4172/2471-2728.1000e129.

Li, Y., Li, H., Han, X., Han, G., Xi, J. *et al.* (2022) Actinobacterial biofertilizer improves the yields of different plants and alters the assembly processes of rhizosphere microbial communities. *Applied Soil Ecology* 171, 104345. DOI: 10.1016/j.apsoil.2021.104345.

López-Vargas, E., Ortega-Ortíz, H., Cadenas-Pliego, G., de Alba Romenus, K., Cabrera de la Fuente, M. *et al.* (2018) Foliar application of copper nanoparticles increases the fruit quality and the content of bioactive compounds in tomatoes. *Applied Sciences* 8(7), 1020. DOI: 10.3390/app8071020.

Mahdavi, M., Namvar, F., Ahmad, M.B. and Mohamad, R. (2013) Green biosynthesis and characterization of magnetic iron oxide (Fe_3O_4) nanoparticles using seaweed (*Sargassum muticum*) aqueous extract. *Molecules* 18(5), 5954–5964. DOI: 10.3390/molecules18055954.

Majid, A., Parray, G.A., Wani, S.H., Kordostami, M., Sofi, N.R. *et al.* (2017) Genome editing and its necessity in agriculture. *International Journal of Current Microbiology and Applied Sciences* 6(11), 5435–5443. DOI: 10.20546/ijcmas.2017.611.520.

Maksymiv, I. (2015) Pesticides: benefits and hazards. *Journal of Vasyl Stefanyk Precarpathian National University* 2(1), 70–76. DOI: 10.15330/jpnu.2.1.70-76.

Maluszynski, M., Ahloowalia, B.S. and Sigurbjörnsson, B. (1995) Application of *in vivo* and *in vitro* mutation techniques for crop improvement. *Euphytica* 85(1–3), 303–315. DOI: 10.1007/BF00023960.

Mariyono, J. (2015) Green revolution- and wetland-linked technological change of rice agriculture in Indonesia. *Management of Environmental Quality* 26(5), 683–700. DOI: 10.1108/MEQ-07-2014-0104.

Mohd Firdaus, M.A., Agatz, A., Hodson, M.E., Al-Khazrajy, O.S.A. and Boxall, A.B.A. (2018) Fate, uptake, and distribution of nanoencapsulated pesticides in soil-earthworm systems and implications for environmental risk assessment. *Environmental Toxicology and Chemistry* 37(5), 1420–1429. DOI: 10.1002/etc.4094.

Mosier, A., Syers, J.K., and Freney, J.R. (2013) *Agriculture and the Nitrogen Cycle: Assessing the Impacts of Fertilizer Use on Food Production and the Environment*. Island Press, Washington, DC.

Oza, G., Pandey, S., Mewada, A., Kalita, G., Sharon, M. *et al.* (2012) Facile biosynthesis of gold nanoparticles exploiting optimum pH and temperature of fresh water algae *Chlorella pyrenoidusa*. *Advances in Applied Science Research* 3, 1405–1412.

Pan, X.L, Dong, F.S, Wu, X.H, Jun, X., Liu, X.G *et al.* (2019) Progress of the discovery, application, and control technologies of chemical pesticides in China. *Journal of Integrative Agriculture* 18(4), 840–853. DOI: 10.1016/S2095-3119(18)61929-X.

Pingali, P.L. (2012) Green revolution: impacts, limits, and the path ahead. *Proceedings of the National Academy of Sciences USA* 109, 12302–12308. DOI: 10.1073/pnas.0912953109.

Prasad, T., Sudhakar, P., Sreenivasulu, Y., Latha, P., Munaswamy, V. *et al.* (2012) Effect of nanoscale zinc oxide particles on the germination, growth and yield of peanut. *Journal of Plant Nutrition* 35(6), 905–927. DOI: 10.1080/01904167.2012.663443.

Purugganan, M.D. (2019) Evolutionary insights into the nature of plant domestication. *Current Biology* 29(14), R705–R714. DOI: 10.1016/j.cub.2019.05.053.

Ragaei, M. and Sabry, A.-K.H. (2014) Nanotechnology for insect pest control. *International Journal of Science, Environment and Technology* 3, 528–545

Rahman, S. (2015) Green revolution in India: environmental degradation and impact on livestock. *Asian Journal of Water, Environment and Pollution* 12, 75–80.

Rasheed, E., Libs, E. and Salim, E. (2017) Formulation of essential oil pesticides technology and their application. *Agricultural Research & Technology* 9(2), 555759. DOI: 10.19080/ARTOAJ.2017.09.555759.

Rui, M., Ma, C., Hao, Y., Guo, J., Rui, Y. *et al.* (2016) Iron oxide nanoparticles as a potential iron fertilizer for peanut (*Arachis hypogaea*). *Frontiers in Plant Science* 7, 815. DOI: 10.3389/fpls.2016.00815.

Sahu, T., Ratre, Y.K., Chauhan, S., Bhaskar, L.V.K.S., Nair, M.P. *et al.* (2021) Nanotechnology based drug delivery system: current strategies and emerging therapeutic potential for medical science. *Journal of Drug Delivery Science and Technology* 63, 102487. DOI: 10.1016/j.jddst.2021.102487.

Saif, S., Tahir, A. and Chen, Y. (2016) Green synthesis of iron nanoparticles and their environmental applications and implications. *Nanomaterials* 6(11), 209. DOI: 10.3390/nano6110209.

Salem, S.S. and Fouda, A. (2021) Green synthesis of metallic nanoparticles and their prospective biotechnological applications: an overview. *Biological Trace Element Research* 199(1), 344–370. DOI: 10.1007/s12011-020-02138-3.

Salunke, B.K., Sawant, S.S., Lee, S.-I. and Kim, B.S. (2016) Microorganisms as efficient biosystem for the synthesis of metal nanoparticles: current scenario and future possibilities. *World Journal of Microbiology & Biotechnology* 32(5), 88. DOI: 10.1007/s11274-016-2044-1.

Santhoshkumar, J., Agarwal, H., Menon, S., Rajeshkumar, S., and Venkat Kumar, S. (2019) A biological synthesis of copper nanoparticles and its potential applications. In: Shukla, A.K. and Iravani, S. (eds.) *Green Synthesis, Characterization and Applications of Nanoparticles*. Elsevier, Amsterdam, pp. 199–221.

Sardar, M. and Mazumder, J.A. (2019) Biomolecules assisted synthesis of metal nanoparticles. In: Dasgupta, N., Ranjan, S. and Lichtfouse, E. (eds) *Environmental Nanotechnology*, Vol. 2. Springer, Cham, pp. 1–23. DOI: 10.1007/978-3-319-98708-8.

Shukla, R.K., Badiye, A., Vajpayee, K. and Kapoor, N. (2021) Genotoxic potential of nanoparticles: structural and functional modifications in DNA. *Frontiers in Genetics* 12, 728250. DOI: 10.3389/fgene.2021.728250.

Singh, R.B. (2000) Environmental consequences of agricultural development: a case study from the Green Revolution state of Haryana, India. *Agriculture, Ecosystems & Environment* 82(1–3), 97–103. DOI: 10.1016/S0167-8809(00)00219-X.

Singh, G., Babele, P.K., Kumar, A., Srivastava, A., Sinha, R.P. *et al.* (2014) Synthesis of ZnO nanoparticles using the cell extract of the cyanobacterium, *Anabaena* strain L31 and its conjugation with UV-B absorbing compound shinorine. *Journal of Photochemistry and Photobiology B: Biology* 138, 55–62. DOI: 10.1016/j.jphotobiol.2014.04.030.

Singh, P., Kim, Y.-J., Zhang, D. and Yang, D.-C. (2016) Biological synthesis of nanoparticles from plants and microorganisms. *Trends in Biotechnology* 34(7), 588–599. DOI: 10.1016/j.tibtech.2016.02.006.

Stejskal, V., Vendl, T., Aulicky, R. and Athanassiou, C. (2021) Synthetic and natural insecticides: gas, liquid, gel and solid formulations for stored-product and food-industry pest control. *Insects* 12(7), 590. DOI: 10.3390/insects12070590.

Sun, C., Cui, H., Wang, Y., Zeng, Z., Zhao, X. *et al.* (2016) Studies on applications of nanomaterial and nanotechnology in agriculture. *Journal of Agricultural Science Technology* 18, 18–25.

Syed-ab-Rahman, S.F., Carvalhais, L.C., Chua, E.T., Chung, F.Y., Moyle, P.M. *et al.* (2019) Soil bacterial diffusible and volatile organic compounds inhibit phytophthora capsici and promote plant growth. *Science of the Total Environment* 692, 267–280. DOI: 10.1016/j.scitotenv.2019.07.061.

Wang, X., Sun, C., Gao, S., Wang, L. and Shuokui, H. (2001) Validation of germination rate and root elongation as indicator to assess phytotoxicity with *Cucumis sativus*. *Chemosphere* 44(8), 1711–1721. DOI: 10.1016/s0045-6535(00)00520-8.

Wiesner, M.R., Lowry, G.V., Jones, K.L., Hochella, Jr., M.F., Di Giulio, R.T. *et al.* (2009) Decreasing uncertainties in assessing environmental exposure, risk, and ecological implications of nanomaterials. *Environmental Science & Technology* 43(17), 6458–6462. DOI: 10.1021/es803621k.

Wilson, C. and Tisdell, C. (2001) Why farmers continue to use pesticides despite environmental, health and sustainability costs. *Ecological Economics* 39(3), 449–462. DOI: 10.1016/S0921-8009(01)00238-5.

Xiong, T., Dumat, C., Dappe, V., Vezin, H., Schreck, E. *et al.* (2017) Copper oxide nanoparticle foliar uptake, phytotoxicity, and consequences for sustainable urban agriculture. *Environmental Science Technology* 51, 5242–5251. DOI: 10.1021/acs.est.6b05546.

Yadav, R.K., Singh, N.B., Singh, A., Yadav, V., Bano, C. *et al.* (2020) Expanding the horizons of nanotechnology in agriculture: recent advances, challenges and future perspectives. *Vegetos* 33(2), 203–221. DOI: 10.1007/s42535-019-00090-9.

Zacharia, J.T. (2011) Identity, physical and chemical properties of pesticides. In: Stoytcheva, M. (ed.) *Pesticides in the Modern World – Trends in Pesticides Analysis*. InTechOpen. DOI: 10.5772/17513.

7 Role and Application of Nanosensors in Crop Protection for Disease Identification

Anda Maria Baroi[1,2], Camelia Ungureanu[3], Mirela Florina Călinescu[4], Diana Vizitiu[5], Ionela-Daniela Sărdărescu[3,5], Alina Ortan[2], Radu Claudiu Fierascu[1,3]* and Irina Fierascu[1,2]

[1]*National Institute for Research & Development in Chemistry and Petrochemistry — ICECHIM Bucharest, Bucharest, Romania;* [2]*University of Agronomic Sciences and Veterinary Medicine of Bucharest, Bucharest, Romania;* [3]*University Politehnica of Bucharest, Bucharest, Romania;* [4]*Research Institute for Fruit Growing Pitesti — Maracineni, Arges, Romania;* [5]*National Research and Development Institute for Biotechnology in Horticulture Stefanesti — Arges, Arges, Romania*

Abstract

Crop diseases can lead to economic loss, at the same time raising health concerns for the final consumers. In this area, the development of materials and, in recent decades, nanomaterials, can provide important contributions to avoid both economic losses and health-related problems. In this chapter, the main types of diseases affecting some major crops (including food, horticultural, and floriculture crops), as well as the modern nanosensors developed for their identification are discussed. Some of the latest developments and future development perspectives are also presented.

Keywords: crop diseases, fruit crops, floriculture crops, nanomaterials, nanosensors

7.1 Introduction

In recent years, the impact of climate change has grown rapidly. It is an important factor in all stages of crop development and forces farmers to change and adapt their actions. Farmers need to look for new management strategies to make their agricultural work both ecologically and economically viable, with an emphasis on a return on investment. Given the complex interactions between the plant, soil, and environment, the prediction of diseases that can affect plants and also the optimal time for the application of

fungicides is essential. With a well-thought-out approach, farmers can increase crop yields with a lower impact on the environment, while maintaining a high quality of the final agricultural products (Ungureanu *et al.*, 2019).

The presence of three factors is decisive in the occurrence of a disease: (i) an unprotected host; (ii) a favourable environment; and (iii) the presence of the pathogen. Thus, the pathogen must come into contact with an unprotected host plant under favourable environmental conditions all at the same time for the disease to occur (Scholthof, 2007). The severity with

*Corresponding author: fierascu.radu@icechim.ro

which a certain disease attacks crops is also influenced by other factors, including: the variety resistance to the pathogens historically found in the area (presence of pathogen in soil, crop rotation), the presence in the area of other cultures that can host and multiply the pathogen, the degree of destruction of plant debris caused by the previous year's climate (mild winters help the more virulent development of diseases), humidity in the air (rain, fog, dew) to help spread disease, and high temperatures help to increase the multiplication (infection) of the pathogen (Yuen and Mila, 2015).

Weather stations equipped with sensors for precipitation, temperature, and relative humidity, as well as leaf humidity, solar radiation, and possibly humidity and soil temperature, can be used in order to develop algorithms to evaluate the risks of disease development. In addition, the sensors, especially nanosensors, can detect exactly what disease the respective culture has. In determining the risk of disease, weather programs use algorithms that consider environmental factors to calculate this risk (Martinelli *et al.*, 2015). For example, the risk for *Fusarium* head blight is influenced by the history of the field (number of spores available) and the climatic situation during flowering. Fusariosis will not occur at all or only with a very low level of infection if the climate is not favourable during flowering. If this disease occurs, nanosensors can detect its occurrence. Thus, phytopathology has become increasingly important in recent decades (Del Ponte *et al.*, 2009; Lalancette *et al.*, 2012).

Phytopathology, in its most comprehensive definition, represents the study of plant diseases caused by pathogens in relation to different environmental conditions (Singh *et al.*, 2021a). Among the organisms listed as causing infectious diseases are fungi, bacteria, and many others. Ectoparasites are not included in this definition (Nazarov *et al.*, 2020). Control of crop diseases is essential for food production. Plants, regardless of their cultivation type, possess different levels of natural resistance to various diseases; however, some diseases can have devastating impacts, such as chestnut disease (FAO, 2017). Generally speaking, plant diseases lead to major economic losses (Asseng *et al.*, 2014). Typically, diseases are considered responsible for an average yield reduction of

10% year^{-1}, with up to 20% reduction in developed countries. According to the estimations of the United Nations Food and Agriculture Organization (FAO), the total losses caused by pests and diseases can reach 25% (FAO, 2021). In most cases, the virulence of the disease is dependent on hydrolases, as well as on other cell-wall-degrading proteins. The vast majority of cell-wall-degrading proteins are produced by pathogens and target pectin (e.g. pectin esterase, pectate lyase, and pectinase). For microbes, the polysaccharides in the plant cell wall can constitute a nutrient source or act simply as a barrier that needs to be passed (Abbott and Boraston, 2008).

This chapter discusses the main types of disease affecting some major crops (including food, horticultural, and floriculture crops), as well as the modern nanosensors developed for their identification. Some of the latest developments are discussed, and future development perspectives are also presented.

7.2   Diseases Affecting Food and Non-Food Crops in Romania

Fruit growing, although considered one of the important branches of Romanian agriculture, has some particularities, among which is the very high share of subsistence farms (mainly producing for self-consumption) as one of the most important features. The total surface area of fruit plantations exceeds 150,000 ha (~1.7% of the Romanian arable area). At a national level, the dominant plantations are apple (as the main cultivated area), plum, cherry and sour cherry, pear, apricot, peach, walnut, shrubs, and strawberry (Sumedrea *et al.*, 2014; Rodríguez *et al.*, 2021).These fruits are also of primary importance for consumers due to their high levels of vitamins, antioxidants, and carbohydrates (Raiola *et al.*, 2014). Currently, the fruit and shrub cultivation sectors are facing several problems that have a negative impact on fruit productivity and quality, including diseases that are becoming more frequent and diverse. This section contains original and unpublished pictures from the authors' experience of these diseases.

Fig. 7.1. Effects of *Venturia inaequalis* on apple: (a) small lesions on the surface of the leaves; (b) large and developed lesions on the leaves; (c) lesions on leaves and fruits; (d) deformation of fruit.

Fig. 7.2. Symptoms of powdery mildew on apple: (a) lesions on leaves; (b) lesions on fruit and leaves.

The most common diseases that can completely destroy an apple crop are apple scab (*Venturia inaequalis* (Cooke) G. Winter (1875)) (Bowen *et al.*, 2011; Doolotkeldieva and Bobusheva, 2017), powdery mildew (*Podosphaera leucotricha*) (Naqvi, 2007; Devasahayam, 2009; Strickland *et al.*, 2021), fire blight of apple (*Erwinia amylovora* (Burrill 1882) Winslow *et al.*, 1920) (Beikzadeh and Varasteh, 2020; Paulin, 2000; Gaucher *et al.*, 2022; Slack *et al.*, 2022), moniliosis (brown rot) caused by *Monilinia* spp. (Lesik, 2013; Wang *et al.*, 2018), bitter rot of apple caused by the fungus *Gloeosporium* spp. (Barkai-Golan, 2001; Blažek *et al.*, 2006; Oliveira Lino *et al.*, 2016), bark diseases, latent viruses, and others (Figs 7.1–7.4).

Bacterial fire blight also attacks other decorative or spontaneous species, such as hawthorn (*Crataegus* spp.), rowan (*Sorbus aucuparia*), Japanese quince (*Chaenomeles* spp.) (Fig. 7.5), and is able to dry out orchards in just a few months (Abd Elahi *et al.*, 2008; Ghahremani *et al.*, 2014).

Another economically important crop in Romania is the plum crop. Plum orchards can also be affected by many diseases affecting fruit production (Fig. 7.6). These include brown rot, bacterial canker caused by *Pseudomonas syringae* pv. *morsprunorum* (Hulin *et al.*, 2020), red spot caused by *Polystigma rubrum*, and silver leaf caused by *Chondrostereum purpureum* (Pers.) Pouzar (1959) (CAB International, 2021a).

Cherries and sour cherries can be affected by several diseases and pests: anthracnose of the leaves caused by *Blumeriella jaapii* (Damianov *et al.*, 2011), leaf sieving caused by *Stigmina carpophila* (Lév.) M.B. Ellis, (1959) (Yanderwalle, 1945), and moniliosis (*Monilinia* spp.), which

Fig. 7.3. (a–c) Apple affected by severe fire blight.

Fig. 7.4. Bitter rot of apple caused by *Gloeosporium* sp.

is particularly damaging to cherries and sour cherries when it appears on flowers (Miessner and Stammler, 2010; Hrustic *et al.*, 2012) (Fig. 7.7).

The main diseases that affect apricot and peach plantations are: leaf sieving caused by *Xanthomonas campestris* (Pammel, 1895) (Dowson, 1939), moniliosis (Yin *et al.*, 2014, 2017), powdery mildew (*Podosphaera tridactyla* (Wallr.) de Bary, (1870)), and bacterial cancer, which untreated can compromise the production of the current and following year. Peach cultures can be affected by: leaf blister (leaf curl) caused by *Taphrina deformans* (Britannica, 2018), shot-hole disease caused by *Stigmina carpophila* (Lév.) M.B. Ellis, 1959 (Ivanová *et al.*, 2012), powdery mildew caused by *Sphaerotheca*

pannosa var. *persicae* (Al-Sadi *et al.*, 2012) and peach scab (peach freckles) caused by *Venturia carpophila* (E.E. Fisher (1961) (Agriculture Victoria, 2022) (Fig. 7.8).

Commercial raspberry and strawberry plantations can be affected by several diseases, some with a low frequency but others more harmful and more common, including: raspberry anthracosis caused by *Elsinoë veneta* (Burkh.) Jenkins (1932), brown-purple spot produced by the fungus *Didymella applanata* (Stevic *et al.*, 2017), raspberry rust caused by *Phragmidium rubi-idaei* (DC.) P. Karst. (1879) (Pârlici *et al.*, 2021), gray rot caused by *Botrytis cinerea* Pers. (1794) (Stevic *et al.*, 2017; Petrasch *et al.*, 2019), gray mold disease (Petrasch *et al.*, 2019), wet rot caused by *Rhizopus nigricans* Ehrenberg (1820) (Teixidó *et al.*, 2011), white spot caused by *Mycosphaerella fragariae* (Tul.) Lindau, red spot caused by *Diplocarpon earliana* (Ellis & Everh.) F.A. Wolf, brown spot caused by *Phomopsis obscurans* (Ellis & Everh.) B. Sutton (1965) (Agrios, 2005; Schilder, 2015), powdery mildew caused by *Podosphaera macularis* (Wallr.) U. Braun & S. Takam. (2000) (Sargent *et al.*, 2019), Lanarkshire disease caused by *Phytophthora fragariae* Hickman (1940) (CAB International, 2021b), and strawberry anthracnose caused by *Colletotrichum acutatum* J.H. Simmonds (1968) (Wang *et al.*, 2019a) (Fig. 7.9).

Fig. 7.5. Fire blight on quince crops.

Another category of fruit shrubs that are cultivated in Romania is red and black currants. Diseases of red and black currants are not uncommon, despite the increased immunity of the plant (Fig. 7.10).

Hazelnut cultivation is one of the most profitable crops in Romania, although many years are required from planting to reaching maximum production. However, the hazelnut can be affected by several diseases, including: gray rot produced by *B. cinerea*, which can have a severe impact on production, even after harvest (Belisario and Santori, 2009), powdery mildew produced by *Phyllactinia guttata* (Wallr.) Lév. (Arzanlou *et al.*, 2018), moniliosis caused by *Monilinia laxa* (Belisario and Santori, 2009),

Fig. 7.6. Plum diseases: (a) *Pseudomonas syringae* pv. *morsprunorum*; (b) red spot caused by *Polystigma rubrum* (Pers.) DC.; (c) silver leaf; (d) *Monilinia* sp. causing mummified plum flowers and fruits.

Fig. 7.7. Cherries moniliosis caused by: (a, b) *Monilinia laxa*; (c–e) *Monilinia fructigena*.

Fig. 7.8. (a) Symptoms of powdery mildew caused by *Sphaerotheca pannosa* var. *persicae* on peach leaves. (b) Symptoms of infection by *Monilia* spp. on apricot. (c) Peach leaf curl and shot-hole disease.

and hazelnut anthracnose caused by *Sphaceloma coryli* Vegh & M. Bourgeois (1976) (Minutolo *et al.*, 2016) (Fig. 7.11).

In Romania, there are good conditions for growth and fruiting walnut throughout the country. However, walnut diseases can be a real threat, especially given the relatively high costs of establishment and maintenance of this crop, including: bacterial blight, caused by *Xanthomonas campestris* pv. *juglandis* and *Xanthomonas arboricola* pv. *juglandis* (Burokienė and Puławska, 2012), walnut anthracnose (Yang *et al.*, 2021), and bacterial canker (Amirsardari *et al.*, 2017) (Fig. 7.12).

In Romania, grapevine plantations are spread across areas that differ from each other both from a climatic and an ecopedological point of view. Diseases of vines can be devastating when the degree of infection is high and fungicide treatments are not carried out in time. These diseases include: downy mildew

produced by *Plasmopara viticola* (Berk. & M.A. Curtis) Berl. & de Toni (1888) (Koledenkova *et al.*, 2022; Scott *et al.*, 2022); powdery mildew caused by *Uncinula necator* (Schwein.) Burrill (Jackson, 2008); grapevine anthracnose caused by *Elsinoë ampelina* Shear (1929) (Louime *et al.*, 2011), and gray mold caused by *B. cinerea* (Williamson *et al.*, 2007; ANF, 2018) (Fig. 7.13).

Tomato crops in Romania are quite widespread. Specific diseases and pathogens can be devastating if there is a high level of infection and fungicide treatments are not carried out promptly. These include: gray rot caused by *B. cinerea* (Sarven *et al.*, 2020), cucumber mosaic virus (Ali *et al.*, 2006; Miozzi *et al.*, 2020; Stanković *et al.*, 2021), *Phytophthora parasitica* (Dastur) G.M. Waterh (Petrov *et al.*, 2021), brown leaf spot caused by *Passalora fulva* (Cooke) U. Braun & Crous, (2003) (Kang and Dobinson, 2004), and bacterial spotting caused

Fig. 7.9. (a) Gray mold of strawberry caused by *Botrytis cinerea*. (b) Brown spot on strawberry caused by *P. obscurans*. (c) Strawberry anthracnose. (d) Symptoms of infection by *Monilia* spp. on raspberries.

by *Pseudomonas syringae* pv. *tomato* (Basim *et al.*, 2004) (Fig. 7.14).

Although not a traditional crop in Romania, there are several varieties of roses grown for jam or for decorative purposes. Among the most well-known diseases of roses are: black leaf spot caused by *Diplocarpon rosae* F.A. Wolf (1912) (Noack, 2003), powdery mildew caused by *Sphaerotheca pannosa* var. *rosae* (Pasini *et al.*, 1997), white mold, rose rust caused by *Phragmidium mucronatum* (Pers.) Schltdl. (1824) (Shattock, 2003), downy mildew, and gray bud rot (Fig. 7.15). Pests affecting roses include lice, fleas, spiders, worms, and caterpillars.

7.3 Application of Nanosensors for Disease Identification

Nanotechnology is currently considered a central point for several areas, including industries such as textiles, energy, the environment, electronics, photonics, food, agriculture, biomedicine, and healthcare. The recent advances in nanotechnology have led to evaluation of nanomaterial applications in (bio)detection systems, representing an important advancement in biosensing technology (Kumar *et al.*, 2020). Their use is closely related to their technological advantages, which in turn are related to the enhanced properties of the materials at the nanoscale (Malik *et al.*, 2021).

Fig. 7.10. Examples of the effects of different diseases on red and black currants.

Fig. 7.11. Hazelnut diseases: (a) gray rot; (b) hazelnut anthracnose.

The world we live in has been rapidly dominated by digital information and the need to investigate matter at ever-smaller scales with seemingly ultra-sophisticated instruments that can 'see' atoms and molecules at work through their interactions and assembly at different levels of organization. Sensors make it possible to obtain real-time information about things we cannot see, touch, smell, or hear, and other things we cannot detect, which may be harmful

Fig. 7.12. (a) Symptoms of bacterial blight on walnut caused by *Xanthomonas arboricola* pv. *Juglandis*. (b) Walnut anthracnose caused by *Gnomonia juglandis* (DC.) Traverso.

or helpful to us. The electronic signal collected from the sensor is passed to a circuit where it is digitized by an analog-to-digital converter. The digital information can then be stored in its memory, reproduced visually on a monitor, or made accessible to the real world through a digital communication port (Ungureanu *et al.*, 2022a; Bonciu *et al.*, 2020).

A sensor is a tool for measuring physiological quantities, transforming them into measurable signals. Over time, this process has evolved from simple transducers and measuring instruments into complex systems that include both analogue and digital electronics (Wan *et al.*, 2020).

A biosensor is a compact analytical device, incorporating a biological or biologically derived sensory element, intimately integrated with a physico-chemical transducer. The purpose of the biosensor is to produce an analogue or digital electronic signal that is proportional to the concentration of a single analyte or group of analytes (Bhalla *et al.*, 2016).

There is a wide variety of measurements that can be collected from biosensor transducers and transformed into physical signals. For the measurement of physical and chemical parameters (e.g. mass, concentration, temperature, refractive index, electromagnetic field, potentials, currents) as a result of A-R transfer (analyte: the unknown element to be determined; receptor: the biorecognition molecule, i.e. the biological/biochemical element that interacts with the analyte through a specific reaction), a specific translator is required. This is the interface for the transmission of physical, chemical, and biological processes, which transforms them into signals for their electronic or optical processing, or storage (Chalklen *et al.*, 2020; Purohit *et al.*, 2020)

The main goal of bionanosensors is to reduce the time for pathogen detection from days to hours or even minutes. Bionanosensors are also used to monitor food quality (Kumar *et al.*, 2017).

In the previous section, numerous bacterial or fungal diseases were presented that attack some of the most widespread crops in Romania. here, we will present the modern nanosensors that have been developed for their identification.

Fig. 7.13. Grapevine diseases: (a) downy mildew caused by *Plasmopara viticola*; (b) powdery mildew; (c) grapevine anthracnose; (d) gray mold.

Fig. 7.14. Diseases affecting tomato: (a) gray rot; (b) brown leaf spot; (c) cucumber mosaic virus.

It is very important for growers to prevent and control crop diseases by detecting pathogens quickly to prevent productivity decline. This can be done sensitively and accurately with the help of bionanosensors. The biosensing systems are innovative nanoparticle (NP)-based biosensors. These could be improved from the point of view of detection sensitivity by the use of surface-enhanced optical properties (e.g. localized surface plasmon resonance (LSPR)) (Kaur, 2011) or electron-conductive nanoscale substrates (e.g. graphene) as transducers.

Microbial nanosensors are biosensors where fermentation reactions and electron transfer play an important role. There are two classes of microbial biosensors that use the same principle: measurement of metabolic activity in

Fig. 7.15. Rose diseases: (a) powdery mildew; (b) rose rust caused by *Phragmidium mucronatum*.

the presence of the analyte. The first class uses immobilized microorganisms and measures the products resulting from metabolism. This class has become commonly defined as microbial biosensors. Microbial biosensors contain immobilized microorganisms and a much more diversified transduction chain. Generally, they are used for a single biochemical process (Lim *et al.*, 2015; Singh *et al.*, 2021b). The second class measures the electrical activity of the metabolism of microorganisms when they consume 'biofuel', for example glucose. This class is known generically as bioelectrochemical cells or biofuel cells (Xu and Ying, 2011; Naresh and Lee, 2021).

One of the most studied biomaterials in terms of the composition of biosensors is graphene. Due to its large specific surface area and pronounced electronic conductivity, the biosensing systems containing it are characterized by a remarkably high sensitivity (Mukherjee *et al.*, 2016). LSPR-based plasmonic nanosensors have advantages as effective detection tools. Nocerino *et al.* (2022) recently published a study summarizing the diversity of molecules of interest that have been successfully detected by LSPR. Surface-enhanced Raman spectroscopy (SERS) is another technique used in the field of bionanosensors (Zong *et al.*, 2018). Flexible and environmentally friendly paper-based substrates have captured attention in SERS applications. Ogundare and van Zyl (2019) provided an overview of their development and implementation

in bioanalysis reaching detection limits as low as 100 fM.

Nanobiodetection of crop pathogens involves quantum dots, metallic/metalloid NPs (e.g. gold NPs (AuNPs)), nanosensors, and nanobarcodes (biobarcoded DNA) (Khiyami *et al.*, 2014).

Quantum dots (inorganic fluorophores) are a type of nanomaterial (NM; semiconductor NPs) that have all three dimensions (length, width, and thickness) in the order of nanometers. Due to their photostability, small size, and the ease with which their surface can be functionalized with various biological molecules, they are increasingly used in the field of nanosensors (Arya *et al.*, 2005; Wang *et al.*, 2006).

The simplest nanotechnology-based detection method is represented by colorimetric detection, based on the interaction/binding between functionalized AuNPs and analytes, subsequently leading to the aggregation of AuNPs and a visual colour change from red to blue (Jazayeri *et al.*, 2018). Biodetection based on the change of the LSPR band is based on the change of intensity/spectral position of the LSPR band, induced by the specific interactions of AuNPs (Byun, 2010). The method is less sensitive than aggregation assays, only being suitable for the detection of known analytes. Another requirement of the method is the functionalization of the AuNPs with molecular recognition elements (i.e. antibodies) (Kaur,

2011). For the evaluation of unknown samples, Raman spectroscopy is the most suitable method, being a highly specific technique able to identify molecular species based on their unique Raman vibrational signature. The signal can be amplified when analytes are adsorbed on to metal surfaces (surface-enhanced Raman scattering). For efficient biodetection applications, the SERS substrate is therefore essential. In solution, aggregated or anisotropic nanospherical AuNPs cause the formation of hot spots between the NPs, amplifying the Raman signal. Other important aspects, besides the shape of the AuNPs, are the functionalization and biocompatibility, which must allow the adsorption of analytes (Zong *et al.*, 2018).

The DNA barcoding technique is a tool for species identification based on DNA sequences (Wang *et al.*, 2019b). This technique was designed to advance the study of molecular diversity as a tool for the recognition and identification of organisms by removing the impediments of taxonomy based solely on morphological characters (Pentinsaari *et al.*, 2014). It is used in the study of biodiversity and allows the objective identification of species, based on DNA sequences, by experts and nonexperts. DNA barcode sequences represent short segments of the genome, used to identify plant species. Recently, extensive DNA barcode sequence libraries have been published for the order Coleoptera (Pentinsaari *et al.*, 2014; Hendrich *et al.*, 2015).

Some beetles from the order Coleoptera cause damage to agricultural crops and fruits. The technology of DNA barcoding has made it possible to solve numerous problems associated with the identification of coleopterans such as species delimitation (Jusoh *et al.*, 2014), identification of sister taxa (Hendrich *et al.*, 2010), and highlighting haplotypes of major importance within agro-ecosystems (Greenstone *et al.*, 2011). DNA barcode regions possess significant potential to be widely applied as a tool for coleopteran identification and also as a complement to the toolkit of classical taxonomy. Even in the few cases where DNA barcode regions fail to provide accurate taxon identification

to the species level, they narrow the selection options to a few related species. DNA barcoding offers promising results, especially for immature development stages. Mapping the diversity, ecology, and evolution of coleopteran species is key to a better understanding of the structure and functioning of ecosystems. The discovery of the diversity of undescribed Coleoptera will be accelerated by the application of this technique (Pentinsaari *et al.*, 2014). In addition to its importance in taxonomic identification of species, the 'DNA barcode' region also demonstrates broad prospects for use in research directions such as analysis of the genetic structure and dynamics of coleopteran populations, analysis of morphological and ecological evolution (Jordal *et al.*, 2002; Evans *et al.*, 2003; Mayer *et al.*, 2014), the foundation of knowledge regarding the conservation status of the species (Molfini *et al.*, 2018), and revealing the adaptive potential of economically important species (Munteanu-Molotievskiy *et al.*, 2016).

NMs may be used for mycotoxin identification and detoxication, enhancing plant resistance, plant disease prediction and nanomolecular identification of plant pathogens (Khiyami *et al.*, 2014), and can be used in different forms, from solid particles to polymer and oil–water-based structures. They are being used as plant protection products (fungicides, insecticides, herbicides, additives, and active constituents) and as fertilizers (Gogos *et al.*, 2012). Several examples of bionanosensor examples for plant pathogen detection are presented in Table 7.1.

According to Dobrucka (2020), the use of metal NPs in nanosensors can also lead to an increase in food quality assurance. Their use is related to the development of colorimetric sensor arrays, electrochemical biosensors, surface plasmon resonance biosensors, and sensor chips, among others (Ungureanu *et al.*, 2022a).

In order to obtain more specific and sensitive biosensors, a series of prerequisites (especially related to biofunctionalization of NMs) are necessary. Depending on the selected analyte, the NM can be decorated with different (bio)

Table 7.1. Use of bionanosensors for plant pathogen detection.

Plant pathogen	Crop	Nanomaterial	Limit of detection/ accuracy	Reference
p-Ethylphenol released by *Phytophthora* spp.	Strawberries	Single-walled carbon nanotubes	0.13% *p*-ethylphenol	Wang *et al.* (2020)
Phakopsora pachyrhizi	Soybean	Fluorescent nanoparticles	2.2 ng ml^{-1}	Miranda *et al.* (2013)
Pseudomonas syringae	Tomato	AuNP–ssDNA	214 pM	Li *et al.* (2021)
VOCs exhaled by *Phytophthora infestans* infection	Tomato	Reduced graphene oxide and AuNPs	>97% accuracy	Li *et al.* (2021)
Phytophthora infestan	Potato	AuNPs	0.1 pg ml^{-1}	Zhan *et al.* (2018)
Plum pox virus	Stone fruit trees	Au and pentacene films	–	Greenshields *et al.* (2016)
Bacillus thuringiensis	Various cultures	PtNP–IgGd	10^2 CFU	Hao and Li (2018)
Xanthomonas axonopodis	Citrus	AuNPs	0.01 nM	Haji-Hashemi *et al.* (2018)
Volatile organic compounds exhaled by *Aspergillus* and *Rhizopus* spp. fungi	Strawberries	Nitrogen- and boron-doped multi-walled carbon nanotubes	–	Greenshields *et al.* (2016)
Cucumber mosaic virus	Cucumber	AuNPs	–	Uda *et al.* (2017)
Phytophthora infestans	Tomato	AuNPs	0.4 ppm	Li *et al.* (2019)
Yellow leaf curl virus	Tomato	AuNPs	5 ng µl^{-1}	Razmi *et al.* (2019)
Xanthomonas campestris	*Brassica* spp.	AuNPs	10^2 CFU ml^{-1}	Peng and Chen, 2019)
Xanthomonas arboricola pv. *pruni*	Stone fruit trees	Carbon nanoparticles	10^4 CFU ml^{-1}	López-Soriano *et al.* (2017)
Fusarium oxysporum	Tomato	CdSe/ZnS–MPA quantum dot	–	Zehra *et al.* (2021)
Ralstonia solanacearum	Tomato	AuNP–ssDNA	7.5 ng	Farber *et al.* (2019)
Aflatoxins	Wheat	AuNPs	0.01 nM for aflatoxin B1	Bhardwaj *et al.* (2009)
Fusarium avenaceum	Cereal grains	Carbon dot	NA	Liu *et al.* (2017)

Continued

Table 7.1. Continued

Plant pathogen	Crop	Nanomaterial	Limit of detection/accuracy	Reference
Arabidopsis thaliana (suitable host for tomato yellow leaf curl virus)	Tomato	CdSe–PEI quantum dot	10 mg ml^{-1}	Gao *et al.* (2018)
Valsa mali	Apple	Carbon dot	Sensitivity: pH range of 2.13–9.34	Koo *et al.* (2015)
Botrytis cinerea	Tomato	AgNPs	85% (at 100 µg ml^{-1})–100% (at 1000 µg ml^{-1})	Malandrakis *et al.* (2019)
Botrytis cinerea, *Plasmopara viticola*	Sweet bell pepper	Green-synthesized ZnONPs from marjoram	87.0% (at 400 µg ml^{-1})	Hassan *et al.* (2019)
		Green-synthesized ZnONPs from olive	79.3% (at 400 µg ml^{-1})	Hassan *et al.* (2019)
		Chemically synthesized ZnONPs	69.8% (at 400 µg ml^{-1})	Hassan *et al.* (2019)
	Grapevine	Amorphous SiNPs	150 ppm	Rashad *et al.* (2021)
Citrus tristeza virus	Citrus	AuNP	1000 fg ml^{-1}	Khater *et al.* (2019)
Phthorimaea operculella (potato tuber moth)	Potato	Culture of *Metarhizium anisopliae* formulated with MCM-41 silica nanomaterial	2×10^5 conidia ml^{-1} for eggs and 1.5×10^4 conidia ml^{-1} for neonate larvae	Khorrami *et al.* (2022)
Brevicoryne brassicae	Cabbage	Entomopathogenic fungus *Lecanicillium lecanii* anchored into MCM-41	2.5×10^4 conidia/mL on adults of *B. brassicae* under laboratory conditions and 2.0×10^5 conidia ml^{-1} on adults of *B. brassicae* under greenhouse conditions	Asmar *et al.* (2019)
Trunk diseases (*Neofusicoccum parvum*, *Diplodia seriata*)	Grapevine	Nanocarriers loaded with extracts from *Rubia tinctorum*	65.8 and 91.0 µg ml^{-1} for nanocarriers loaded with *R. tinctorum* against *N. parvum* and *D. seriata*	Sánchez-Hernández *et al.* (2022)

AgNP, silver nanoparticle; AuNP, gold nanoparticle; CdSe, cadmium selenide; CFU, colony-forming unit; MCM-41, mobil composition of matter No. 41; MPA, 3-mercaptopropionic acid; NA, not applicable; PEI, polyethylenimine; PtNP, platinum nanoparticle; ssDNA, single-stranded DNA; ZnS, zinc sulfide receptors, including peptides, synthetic oligo-nucleotides/aptamers, enzymes, or molecularly imprinted polymers (Ungureanu *et al.*, 2022a).

7.4 Conclusions and Future Perspectives

As discussed in this chapter, different types of crops are affected by different pathogens. Whether we are speaking of food, horticulture, floriculture, or industrial crops, the diseases affecting them cause serious economic losses throughout the production chain, from field to the final consumer. Viable instruments are needed from research, development, and innovation (RDI) specialists to avoid such losses; this is particularly critical in countries such as Romania, where crop production is important to its society. Additionally, in crops designated for human consumption, these pathogens could also affect human health, inducing serious illnesses.

The obvious question arising is what can we do to prevent these diseases. The work of field specialists is important in monitoring the appearance of these diseases; however, it is almost impossible to check on individual plants, and when the disease signs become visible to nonspecialists, it is usually too late. Thus, developments by RDI specialists to provide modern instruments for the monitoring and controlling of crops' pathogens are essential.

The breakthroughs in nanotechnology in the last decades could provide such instruments. However, researchers in this area should remember that the role of science is to go beyond the 'scientific trends' and produce results that provide positive effects in society, scaling up from laboratory experiments to practical, industrial, day-to-day applications. The question remains as to what nanotechnology can bring to the table in terms of combating crop pathogens. The answer is complex and should be considered for each particular case. Our research

group has previously demonstrated (following laboratory tests) that phytosynthesized NPs (NPs obtained using natural extracts) can be viable antimicrobial instruments (Fierascu *et al.*, 2020; Ungureanu *et al.*, 2021). Natural vegetal resources (even vegetal wastes) can be a very important source of bioactive compounds (Fierascu *et al.*, 2021; Baroi *et al.*, 2022; Ungureanu *et al.*, 2022b), and these extracts, in turn, could also act as potential antimicrobial agents against crop pathogens (Călinescu *et al.*, 2020). This represents one important approach, which is applicable as a treatment method but also prophylactically, to avoid the development of pathogens. However, these technologies still have some bottlenecks, such as achieving reproducibility of the phytosynthesized NPs with consistent sizes and morphologies and standardization of the natural extracts, all important aspects for developing instruments with practical application.

Another important aspect in the fight against crop diseases is the early identification of the presence of pathogens. In this area, NPs and NMs may also find important uses, for example by acting as a support for sensing elements (e.g. antibodies, aptamers, peptides) or even as active agents for direct colorimetric assays in modern sensing instruments (at low detection limits), as described in Table 7.1 and elsewhere (Ungureanu *et al.*, 2022a). Phytosynthesized materials could also provide important advantages due to their enhanced antimicrobial properties and lower toxicity (an important aspect when proposing the use of NMs).

In conclusion, the use of NMs with applications in crop protection in general, and of nanosensors in particular is an emerging area of research, but several important instruments have already been proposed that could provide important economic and social benefits. We can only hope that in the near future, rapid and inexpensive assays based on NMs that are easy to use by nonspecialists will become commercially available for the early detection of crop pathogens.

Acknowledgment

This work was supported by a grant from the Ministry of Research, Innovation and Digitization, CCCDI–UEFISCDI, project number PN-III-P2-2.1-PED-2021–0042, within PNCDI III. A.M.B., R.C.F., and I.F. also acknowledge the support of a grant from the Ministry of Research, Innovation and Digitization, CCCDI–UEFISCDI, project number PN-III-P2-2.1-PED-2021–0273, within PNCDI III, and support provided by the Ministry of Research, Innovation and Digitization through Program 1—Development of the National Research and Development System, Subprogram 1.2, Institutional Performance—Projects to Finance Excellence in RDI, Contract no. 15PFE/2021.

References

Abbott, D.W. and Boraston, A.B. (2008) Structural biology of pectin degradation by *Enterobacteriaceae*. *Microbiology and Molecular Biology Reviews* 72(2), 301–316. DOI: 10.1128/MMBR.00038-07.

Abd Elahi, H., Ghasemi, A. and Mehrabipour, S. (2008) Evaluation of fire blight resistance in some quince (*Cydonia oblonga* Mill.) genotypes. II. Resistance of genotypes to the disease. *Seed and Plant* 24, 529–541. Available at: https://doi.org/10.22092/spij.2017.110836

Agriculture Victoria (2022) Stone fruit scab (or freckle). Available at: https://agriculture.vic.gov.au/biosecurity/plant-diseases/fruit-and-nut-diseases/stone-fruits/stone-fruit-scab-or-freckle (accessed 13 March 2023).

Agrios, G.N. (2005) *Plant Pathology*, 5th edn. Elsevier Academic Press, Burlington, MA.

Ali, A., Li, H., Schneider, W.L., Sherman, D.J., Gray, S. *et al.* (2006) Analysis of genetic bottlenecks during horizontal transmission of cucumber mosaic virus. *Journal of Virology* 80(17), 8345–8350. DOI: 10.1128/JVI.00568-06.

Al-Sadi, A.M., Al-Raisi, I.J., Al-Azri, M., Al-Hasani, H., Al-Shukaili, M.S. *et al.* (2012) Population structure and management of *Podosphaera pannosa* associated with peach powdery mildew in Oman. *Journal of Phytopathology* 160(11–12), 647–654. DOI: 10.1111/j.1439-0434.2012.01955.x.

Amirsardari, V., Sepahvand, S. and Madani, M. (2017) Identification of deep bark canker agent of walnut and study of its phenotypic, pathogenic, holotypic and genetic diversity in Iran. *Journal of Plant Interactions* 12(1), 340–347. DOI: 10.1080/17429145.2017.1317848.

ANF (National Phytosanitary Authority) (2018) *Ghid pentru recunoaşterea şi combaterea bolilor şi dăunătorilor la viţa-de-vie [Guide to identifying and combating livestock diseases and pests in grapevine]*. Available at: www.anfdf.ro/sanatate/ghid/Ghid_informativ_vita.pdf (accessed 13 March 2023).

Arya, H., Kaul, Z., Wadhwa, R., Taira, K., Hirano, T. *et al.* (2005) Quantum dots in bio-imaging: revolution by the small. *Biochemical and Biophysical Research Communications* 329(4), 1173–1177. DOI: 10.1016/j.bbrc.2005.02.043.

Arzanlou, M., Torbati, M. and Golmohammadi, H. (2018) Powdery mildew on hazelnut (*Corylus avellana*) caused by *Erysiphe corylacearum* in Iran. *Forest Pathology* 48, e12450. DOI: 10.1111/efp.12450.

Asmar, S., Fereshteh, K., Hana, B., Khadijeh, A. and Youbert, G. (2019) Entomopathogenic fungus, *Lecanicillium lecanii* R. Z are & W. Gams anchored into MCM-41: a new and effective bio-insecticide against *Brevicoryne brassicae* (Linnaeus, 1758) (Hom: Aphididae) to protect cabbages. *Acta Agriculturae Slovenica* 114(1). DOI: 10.14720/aas.2019.114.1.11.

Asseng, S., Zhu, Y., Basso, B., Wilson, T. and Cammarano, D. (2014) Simulation modeling: applications in cropping systems. In: van Alfen, N.K. (ed.) *Encyclopedia of Agriculture and Food Systems*. Academic Press, London, pp. 102–112.

Barkai-Golan, R. (2001) Postharvest disease initiation. In: Barkai-Golan, R. (ed.) *Postharvest Diseases of Fruits and Vegetables*. Elsevier, Amsterdam, pp. 3–24.

Baroi, A.M., Popitiu, M., Fierascu, I., Sărdărescu, I.D. and Fierascu, R.C. (2022) Grapevine wastes: a rich source of antioxidants and other biologically active compounds. *Antioxidants* 11(2), 393. DOI: 10.3390/antiox11020393.

Basim, H., Basim, E., Yilmaz, S., Dickstein, E.R. and Jones, J.B. (2004) An outbreak of bacterial speck caused by *Pseudomonas syringae* pv. tomato on tomato transplants grown in commercial seedling companies located in the Western Mediterranean region of Turkey. *Plant Disease* 88(9), 1050–1050. DOI: 10.1094/PDIS.2004.88.9.1050A.

Beikzadeh, N. and Varasteh, A.R. (2020) Investigation of the effects of Apple trees infection by *Erwinia Amylovora* on the expression of pathogenesis-related proteins homologous to allergens. *Reports of Biochemistry & Molecular Biology* 8, 376–382.

Belisario, A. and Santori, A. (2009) Gray necrosis of hazelnut fruit: a fungal disease causing fruit drop. *Acta Horticulturae* 845(845), 501–506. DOI: 10.17660/ActaHortic.2009.845.77.

Bhalla, N., Jolly, P., Formisano, N. and Estrela, P. (2016) Introduction to biosensors. *Essays in Biochemistry* 60(1), 1–8. DOI: 10.1042/EBC20150001.

Bhardwaj, H., Sumana, G. and Marquette, C.A. (2009) A label-free ultrasensitive microfluidic surface plasmon resonance biosensor for aflatoxin B1 detection using nanoparticles integrated gold chip. *Food Chemistry* 307, 125530. DOI: 10.1016/j.foodchem.2019.125530.

Blažek, J., Kloutvorová, J. and Křelinová, J. (2006) Incidence of storage diseases on apples of selected cultivars and advanced selections grown with and without fungicide treatments. *Horticultural Science* 33(3), 87–94. DOI: 10.17221/3744-HORTSCI.

Bonciu, E., Sarac, I., Pentea, M., Radu, F. and Butnariu, M. (2020) The reliability of nanotechnology for sustainable industries. In: Bhat, R.A., Hakeem, K.R. and Al-Saud, N.B.S. (eds) *Bioremediation and Biotechnology Volume 3: Persistent and Recalcitrant Toxic Substances*. Springer, Cham, Switzerland, pp. 195–226. DOI: 10.1007/978-3-030-46075-4.

Bowen, J.K., Mesarich, C.H., Bus, V.G.M., Beresford, R.M., Plummer, K.M. *et al.* (2011) *Venturia inaequalis*: the causal agent of apple scab. *Molecular Plant Pathology* 12(2), 105–122. DOI: 10.1111/j.1364-3703.2010.00656.x.

Britannica (2018) Leaf blister. In: *Encyclopedia Britannica*. Available at: www.britannica.com/science/leaf -blister (accessed 13 March 2023).

Burokienė, D. and Puławska, J. (2012) Characterization of *Xanthomonas arboricola* pv. Juglandis isolated from walnuts in Lithuania. *Journal of Plant Pathology* 94, S1.23–S1.27.

Byun, K.M. (2010) Development of nanostructured plasmonic substrates for enhanced optical biosensing. *Journal of the Optical Society of Korea* 14(2), 65–76. DOI: 10.3807/JOSK.2010.14.2.065.

CAB International (2021a) *Chondrostereum purpureum* (silver blight of stone fruit trees). Available at: www.cabi.org/isc/datasheet/51554 (accessed 13 March 2023).

CAB International (2021b) *Phytophthora fragariae* (strawberry red stele root rot). Available at: www.cabi.o rg/isc/datasheet/40967 (accessed 13 March 2023).

Călinescu, M., Ungureanu, C., Soare, C., Fierascu, R.C., Fierăscu, I. *et al.* (2020) Green matrix solution for growth inhibition of *Venturia inaequalis* and *Podosphaera leucotricha*. *Acta Horticulturae* 1289, 61–66. DOI: 10.17660/ActaHortic.2020.1289.9.

Chalklen, T., Jing, Q. and Kar-Narayan, S. (2020) Biosensors based on mechanical and electrical detection techniques. *Sensors* 20(19), 5605. DOI: 10.3390/s20195605.

Damianov, S., Simeria, G., Grozea, I., Ștef, R., Virteiu, A. *et al.* (2011) Researches regarding the introduction of anthracnose-resistant varieties in cherry tree (*Coccomyces hiemalis* sin *Blumeria jaapi*) in the orchard centre Lugoj Herindești. *Research Journal of Agricultural Science* 43, 28–33.

Del Ponte, E.M., Fernandes, J.M.C., Pavan, W. and Baethgen, W.E. (2009) A model-based assessment of the impacts of climate variability on fusarium head blight seasonal risk in Southern Brazil. *Journal of Phytopathology* 157(11–12), 675–681. DOI: 10.1111/j.1439-0434.2009.01559.x.

Devasahayam, H.L. (2009) *Illustrated Plant Pathology: Basic Concepts*. New India Publishing, Delhi, India.

Dobrucka, R. (2020) Metal nanoparticles in nanosensors for food quality assurance. *Scientific Journal of Logistics* 16(2), 271–278. DOI: 10.17270/J.LOG.2020.390.

Doolotkeldieva, T. and Bobusheva, S. (2017) Scab disease caused by *Venturia inaequalis* on apple trees in Kyrgyzstan and biological agents to control this disease. *Advances in Microbiology* 07(06), 450–466. DOI: 10.4236/aim.2017.76035.

Evans, J., Pettis, J., Hood, M. and Shimanuki, H. (2003) Tracking an invasive honey bee pest: mitochondrial DNA variation in North American small hive beetles. *Apidologie* 34(2), 103–109. DOI: 10.1051/apido:2003004.

FAO (2017) *The Future of Food and Agriculture: Trends and Challenges*. Food and Agriculture Organization of the United Nations, Rome. Available at: www.fao.org/3/i6583e/i6583e.pdf (accessed 13 March 2023).

FAO (2021) Climate change fans spread of pests and threatens plants and crops, new FAO study. Food and Agriculture Organization of the United Nations, Rome. Available at: www.fao.org/news/story/en/item/1402920/icode/ (accessed 13 March 2023).

Farber, C., Mahnke, M., Sanchez, L. and Kurouski, D. (2019) Advanced spectroscopic techniques for plant disease diagnostics. A review. *TrAC Trends in Analytical Chemistry* 118, 43–49. DOI: 10.1016/j.trac.2019.05.022.

Fierascu, R.C., Fierascu, I., Lungulescu, E.M., Nicula, N., Somoghi, R. *et al.* (2020) Phytosynthesis and radiation-assisted methods for obtaining metal nanoparticles. *Journal of Materials Science* 55(5), 1915–1932. DOI: 10.1007/s10853-019-03713-3.

Fierascu, R.C., Fierascu, I., Baroi, A.M. and Ortan, A. (2021) Selected aspects related to medicinal and aromatic plants as alternative sources of bioactive compounds. *International Journal of Molecular Sciences* 22(4), 1521. DOI: 10.3390/ijms22041521.

Gao, G., Jiang, Y.-W., Sun, W., and Wu, F.-G. (2018) Fluorescent quantum dots for microbial imaging. *Chinese Chemical Letters* 29, 1475–1485.

Gaucher, M., Righetti, L., Aubourg, S., Dugé de Bernonville, T., Brisset, M.-N. *et al.* (2022) An *Erwinia amylovora* inducible promoter for improvement of apple fire blight resistance. *Plant Cell Reports* 41(8), 1499–1513. DOI: 10.1007/s00299-022-02887-6.

Ghahremani, Z., Alipour, M., Ahmadi, S., Abdollahi, H., Mohamadi, M. *et al.* (2014) Selecting effective indices for evaluation of fire blight resistance in quince germplasm under orchard settings. *Acta Horticulturae* 1056(1056), 247–251. DOI: 10.17660/ActaHortic.2014.1056.41.

Gogos, A., Knauer, K. and Bucheli, T.D. (2012) Nanomaterials in plant protection and fertilization: current state, foreseen applications, and research priorities. *Journal of Agricultural and Food Chemistry* 60(39), 9781–9792. DOI: 10.1021/jf302154y.

Greenshields, M.W.C.C., Cunha, B.B., Coville, N.J., Pimentel, I.C., Zawadneak, M.A.C. *et al.* (2016) Fungi active microbial metabolism detection of *Rhizopus* sp. and *Aspergillus* sp. Section Nigri on strawberry using a set of chemical sensors based on carbon nanostructures. *Chemosensors* 4, 19. DOI: 10.3390/chemosensors4030019.

Greenstone, M.H., Vandenberg, N.J. and Hu, J.H. (2011) Barcode haplotype variation in North American agroecosystem lady beetles (Coleoptera: Coccinellidae). *Molecular Ecology Resources* 11(4), 629–637. DOI: 10.1111/j.1755-0998.2011.03007.x.

Haji-Hashemi, H., Norouzi, P., Safarnejad, M.R., Larijani, B., Habibi, M.M. *et al.* (2018) Sensitive electrochemical immunosensor for citrus bacterial canker disease detection using fast Fourier transformation square-wave voltammetry method. *Journal of Electroanalytical Chemistry* 820, 111–117. DOI: 10.1016/j.jelechem.2018.04.062.

Hao, W. and Li, L.L. (2018) *In Vivo Self-Assembly Nanotechnology for Biomedical Applications*. Springer Nature, Singapore.

Hassan, M., Zayton, M.A., and El-Feky, M. (2019) Role of green synthesized ZnO nanoparticles as antifungal against post-harvest gray and black mold of sweet bell pepper. *Journal of Biotechnology and Bioengineering* 3, 8–15.

Hendrich, L., Pons, J., Ribera, I. and Balke, M. (2010) Mitochondrial cox1 sequence data reliably uncover patterns of insect diversity but suffer from high lineage-idiosyncratic error rates. *PloS One* 5(12), e14448. DOI: 10.1371/journal.pone.0014448.

Hendrich, L., Morinière, J., Haszprunar, G., Hebert, P.D.N., Hausmann, A. *et al.* (2015) A comprehensive DNA barcode database for Central European beetles with a focus on Germany: adding more than 3500 identified species to BOLD. *Molecular Ecology Resources* 15(4), 795–818. DOI: 10.1111/1755-0998.12354.

Hrustic, J., Mihajlovic, M., Grahovac, M., Delibasic, G., Bulajic, A. *et al.* (2012) Genus *Monilinia* on pome and stone fruit species. *Pesticidi i Fitomedicina* 27(4), 283–297. DOI: 10.2298/PIF1204283H.

Hulin, M.T., Jackson, R.W., Harrison, R.J. and Mansfield, J.W. (2020) Cherry picking by pseudomonads: after a century of research on canker, genomics provides insights into the evolution of pathogenicity towards stone fruits. *Plant Pathology* 69(6), 962–978. DOI: 10.1111/ppa.13189.

Ivanová, H., Kaločaiová, M., and Bolvanský, M. (2012) Shot-hole disease on *Prunus persica* – the morphology and biology of *Stigmina carpophila*. *Folia Oecologica* 39, 21–27.

Jackson, R.S. (2008) Vineyard practice. In: Jackson, R.S. (ed.) *Wine Science*, 3rd edn. Academic Press, Burlington, MA, pp. 108–238.

Jazayeri, M.H., Aghaie, T., Avan, A., Vatankhah, A. and Ghaffari, M.R.S. (2018) Colorimetric detection based on gold nano particles (GNPs): an easy, fast, inexpensive, low-cost and short time method in detection of analytes (protein, DNA, and ion). *Sensing and Bio-Sensing Research* 20, 1–8. DOI: 10.1016/j.sbsr.2018.05.002.

Jordal, B.H., Normark, B.B., Farrell, B.D. and Kirkendall, L.R. (2002) Extraordinary haplotype diversity in haplodiploid inbreeders: phylogenetics and evolution of the bark beetle genus *Coccotrypes*. *Molecular Phylogenetics and Evolution* 23(2), 171–188. DOI: 10.1016/S1055-7903(02)00013-1.

Jusoh, W.F.A., Hashim, N.R., Sääksjärvi, I.E., Adam, N.A. and Wahlberg, N. (2014) Species delineation of Malaysian mangrove fireflies (Coleoptera: Lampyridae) using DNA barcodes. *The Coleopterists Bulletin* 68(4), 703–711. DOI: 10.1649/0010-065X-68.4.703.

Kang, S. and Dobinson, K.F. (2004) Molecular and genetic basis of plant-fungal pathogen interactions. In: Arora, D.K. and Khachatourians, G.G. (eds) *Applied Mycology and Biotechnology*, Vol. 4. Elsevier, Amsterdam, pp. 59–97.

Kaur, K. (2011) Optical biosensing using localized surface plasmon resonance of gold nanoparticles. PhD thesis, University of Waterloo, Waterloo, Canada. Available at: https://uwspace.uwaterloo.ca/bitstream/handle/10012/5983/Kanwarjeet_Kaur.pdf (accessed 3 March 2023).

Khater, M., de la Escosura-Muñiz, A., Altet, L. and Merkoçi, A. (2019) *In situ* plant virus nucleic acid isothermal amplification detection on gold nanoparticle-modified electrodes. *Analytical Chemistry* 91(7), 4790–4796. DOI: 10.1021/acs.analchem.9b00340.

Khiyami, M.A., Almoammar, H., Awad, Y.M., Alghuthaymi, M.A. and Abd-Elsalam, K.A. (2014) Plant pathogen nanodiagnostic techniques: forthcoming changes? *Biotechnology & Biotechnological Equipment* 28(5), 775–785. DOI: 10.1080/13102818.2014.960739.

Khorrami, F., Soleymanzade, A., Batmani, H. and Ghosta, Y. (2022) Entomopathogenic fungus, *Metarhizium anisopliae*, anchored onto MCM-41 silica nanomaterial: a novel and effective insecticide against potato tuber moth in stored potatoes. *Hellenic Plant Protection Journal* 15(1), 21–29. DOI: 10.2478/hppj-2022-0003.

Koledenkova, K., Esmaeel, Q., Jacquard, C., Nowak, J., Clément, C. *et al.* (2022) *Plasmopara viticola* the causal agent of downy mildew of grapevine: from its taxonomy to disease management. *Frontiers in Microbiology* 13, 889472. DOI: 10.3389/fmicb.2022.889472.

Koo, Y., Wang, J., Zhang, Q., Zhu, H., Chehab, E.W. *et al.* (2015) Fluorescence reports intact quantum dot uptake into roots and translocation to leaves of *Arabidopsis thaliana* and subsequent ingestion by insect herbivores. *Environmental Science & Technology* 49(1), 626–632. DOI: 10.1021/es5050562.

Kumar, V., Guleria, P. and Mehta, S.K. (2017) Nanosensors for food quality and safety assessment. *Environmental Chemistry Letters* 15(2), 165–177. DOI: 10.1007/s10311-017-0616-4.

Kumar, H., Kuča, K., Bhatia, S.K., Saini, K., Kaushal, A. *et al.* (2020) Applications of nanotechnology in sensor-based detection of foodborne pathogens. *Sensors* 20(7), 1966. DOI: 10.3390/s20071966.

Lalancette, N., McFarland, K.A. and Burnett, A.L. (2012) Modeling sporulation of *Fusicladium carpophilum* on nectarine twig lesions: relative humidity and temperature effects. *Phytopathology* 102(4), 421–428. DOI: 10.1094/PHYTO-08-10-0222.

Li, Z., Paul, R., Ba Tis, T., Saville, A.C., Hansel, J.C. *et al.* (2019) Non-invasive plant disease diagnostics enabled by smartphone-based fingerprinting of leaf volatiles. *Nature Plants* 5(8), 856–866. DOI: 10.1038/s41477-019-0476-y.

Lesik, K. (2013) *Monilinia* species causing fruit brown rot, blossom and twig blight in apple orchards in Belarus. *Proceedings of the Latvian Academy of Sciences. Section B* 67(2), 192–194. DOI: 10.2478/prolas-2013-0031.

Li, Z., Liu, Y., Hossain, O., Paul, R., Yao, S. *et al.* (2021) Real-time monitoring of plant stresses via chemiresistive profiling of leaf volatiles by a wearable sensor. *Matter* 4(7), 2553–2570. DOI: 10.1016/j.matt.2021.06.009.

Lim, J.W., Ha, D., Lee, J., Lee, S.K. and Kim, T. (2015) Review of micro/nanotechnologies for microbial biosensors. *Frontiers in Bioengineering and Biotechnology* 3, 61. DOI: 10.3389/fbioe.2015.00061.

Liu, W., Li, C., Sun, X., Pan, W., Yu, G. *et al.* (2017) Highly crystalline carbon dots from fresh tomato: UV emission and quantum confinement. *Nanotechnology* 28(48), 485705. DOI: 10.1088/1361-6528/aa900b.

López-Soriano, P., Noguera, P., Gorris, M.T., Puchades, R., Maquieira, Á. *et al.* (2017) Lateral flow immunoassay for on-site detection of *Xanthomonas arboricola* pv. *pruni* in symptomatic field samples. *PloS One* 12(4), e0176201. DOI: 10.1371/journal.pone.0176201.

Louime, C., Lu, J., Onokpise, O., Vasanthaiah, H.K.N., Kambiranda, D. *et al.* (2011) Resistance to *Elsinoë ampelina* and expression of related resistant genes in *Vitis rotundifolia* Michx. grapes. *International Journal of Molecular Sciences* 12(6), 3473–3488. DOI: 10.3390/ijms12063473.

Malandrakis, A.A., Kavroulakis, N. and Chrysikopoulos, C.V. (2019) Use of copper, silver and zinc nanoparticles against foliar and soil-borne plant pathogens. *Science of The Total Environment* 670, 292–299. DOI: 10.1016/j.scitotenv.2019.03.210.

Malik, P., Gupta, R., Malik, V. and Ameta, R.K. (2021) Emerging nanomaterials for improved biosensing. *Measurement: Sensors* 16, 100050. DOI: 10.1016/j.measen.2021.100050.

Martinelli, F., Scalenghe, R., Davino, S., Panno, S., Scuderi, G. *et al.* (2015) Advanced methods of plant disease detection. A review. *Agronomy for Sustainable Development* 35(1), 1–25. DOI: 10.1007/s13593-014-0246-1.

Mayer, F., Björklund, N., Wallen, J., Långström, B. and Cassel-Lundhagen, A. (2014) Mitochondrial DNA haplotypes indicate two postglacial re-colonization routes of the spruce bark beetle *Ips typographus* through northern Europe to Scandinavia. *Journal of Zoological Systematics and Evolutionary Research* 52, 285–292. DOI: 10.1111/jzs.12063.

Miessner, S. and Stammler, G. (2010) *Monilinialaxa, M. fructigena* and *M. fructicola*: risk estimation of resistance to QoI fungicides and identification of species with cytochrome *b* gene sequences. *Journal of Plant Diseases and Protection* 117(4), 162–167. DOI: 10.1007/BF03356354.

Minutolo, M., Nanni, B., Scala, F. and Alioto, D. (2016) *Sphaceloma coryli*: a reemerging pathogen causing heavy losses on hazelnut in Southern Italy. *Plant Disease* 100(3), 548–554. DOI: 10.1094/PDIS-06-15-0664-RE.

Miozzi, L., Vaira, A.M., Brilli, F., Casarin, V., Berti, M. *et al.* (2020) Arbuscular mycorrhizal symbiosis primes tolerance to cucumber mosaic virus in tomato. *Viruses* 12(6), 675. DOI: 10.3390/v12060675.

Miranda, B.S., Linares, E.M., Thalhammer, S. and Kubota, L.T. (2013) Development of a disposable and highly sensitive paper-based immunosensor for early diagnosis of Asian soybean rust. *Biosensors & Bioelectronics* 45, 123–128. DOI: 10.1016/j.bios.2013.01.048.

Molfini, M., Redolfi de Zan, L., Campanaro, A., Rossi de Gasperis, S., Mosconi, F. *et al.* (2018) A first assessment of genetic variability in the longhorn beetle *Rosalia alpina* (Coleoptera: Cerambycidae) from the Italian Apennines. *The European Zoological Journal* 85(1), 36–45. DOI: 10.1080/24750263.2018.1433243.

Mukherjee, A., Majumdar, S., Servin, A.D., Pagano, L., Dhankher, O.P. *et al.* (2016) Carbon nanomaterials in agriculture: a critical review. *Frontiers in Plant Science* 7, 172. DOI: 10.3389/fpls.2016.00172.

Munteanu-Molotievskiy, N., Moldovan, A., Bacal, S., and Toderas, I. (2016) Beetle population structure at the crossroads of biogeographic regions in Eastern Europe: the case of *Tatiana rhynchites aequatus* (Coleoptera: Rhynchitidae). *North-Western Journal of Zoology* 12, 166–177.

Naqvi, S. (2007) *Diseases of Fruits and Vegetables, Diagnosis and Management*, Vol. I. Springer, Berlin.

Naresh, V. and Lee, N. (2021) A review on biosensors and recent development of nanostructured materials-enabled biosensors. *Sensors* 21(4), 1109. DOI: 10.3390/s21041109.

Nazarov, P.A., Baleev, D.N., Ivanova, M.I., Sokolova, L.M. and Karakozova, M.V. (2020) Infectious plant diseases: etiology, current status, problems and prospects in plant protection. *Acta Naturae* 12(3), 46–59. DOI: 10.32607/actanaturae.11026.

Noack, R. (2003) Selection strategies for disease and pest resistance. In: Roberts, A.V. (ed.) *Encyclopedia of Rose Science*. Elsevier, Amsterdam, pp. 165–169.

Nocerino, V., Miranda, B., Tramontano, C., Chianese, G., Dardano, P. *et al.* (2022) Plasmonic nanosensors: design, fabrication, and applications in biomedicine. *Chemosensors* 10(5), 150. DOI: 10.3390/chemosensors10050150.

Ogundare, S.A. and van Zyl, W.E. (2019) A review of cellulose-based substrates for SERS: fundamentals, design principles, applications. *Cellulose* 26(11), 6489–6528. DOI: 10.1007/s10570-019-02580-0.

Oliveira Lino, L., Pacheco, I., Mercier, V., Faoro, F., Bassi, D. *et al.* (2016) Brown rot strikes *Prunus* fruit: an ancient fight almost always lost. *Journal of Agricultural and Food Chemistry* 64(20), 4029–4047. DOI: 10.1021/acs.jafc.6b00104.

Pârlici, R.-M., Maxim, A., Mang, S.M., Camele, I., Mihalescu, L. *et al.* (2021) Alternative control of *Phragmidium rubi-idaei* infecting two *Rubus* species. *Plants* 10(7), 1452. DOI: 10.3390/plants10071452.

Pasini, C., d'Aquila, F., Curir, P. and Gullino, M.L. (1997) Effectiveness of antifungal compounds against rose powdery mildew (*Sphaerotheca pannosa* var. *rosae*) in glasshouses. *Crop Protection* 16(3), 251–256. DOI: 10.1016/S0261-2194(96)00095-6.

Paulin, J.P. (2000) *Erwinia amylovora*: general characteristics, biochemistry, and serology. In: Vanneste, J.L. (ed.) *Fire Blight: The Disease and Its Causative Agent, Erwinia Amylovora*. CAB International, Wallingford, UK, pp. 87–116. DOI: 10.1079/9780851992945.0000.

Peng, H. and Chen, I.A. (2019) Rapid Colorimetric detection of bacterial species through the capture of gold nanoparticles by chimeric phages. *ACS Nano* 13(2), 1244–1252. DOI: 10.1021/acsnano.8b06395.

Pentinsaari, M., Hebert, P.D.N. and Mutanen, M. (2014) Barcoding beetles: a regional survey of 1872 species reveals high identification success and unusually deep interspecific divergences. *PloS One* 9(9), e108651. DOI: 10.1371/journal.pone.0108651.

Petrasch, S., Knapp, S.J., van Kan, J.A.L. and Blanco-Ulate, B. (2019) Grey mould of strawberry, a devastating disease caused by the ubiquitous necrotrophic fungal pathogen *Botrytis cinerea*. *Molecular Plant Pathology* 20(6), 877–892. DOI: 10.1111/mpp.12794.

Petrov, N., Stoyanova, M., and Gaur, R.K. (2021) Ecological methods to control viral damages in tomatoes. In: Gaur, R.K., Khurana, S.M.P., Sharma, P., and Hohn, T. (eds.) *Plant Virus–Host Interaction*, 2nd edn. Academic Press, London, pp. 469–488.

Purohit, B., Vernekar, P.R., Shetti, N.P. and Chandra, P. (2020) Biosensor nanoengineering: design, operation, and implementation for biomolecular analysis. *Sensors International* 1, 100040. DOI: 10.1016/j.sintl.2020.100040.

Raiola, A., Rigano, M.M., Calafiore, R., Frusciante, L. and Barone, A. (2014) Enhancing the health-promoting effects of tomato fruit for biofortified food. *Mediators of Inflammation* 2014, 139873. DOI: 10.1155/2014/139873.

Rashad, Y.M., El-Sharkawy, H.H.A., Belal, B.E.A., Abdel Razik, E.S. and Galilah, D.A. (2021) Silica nanoparticles as a probable anti-oomycete compound against downy mildew, and yield and quality enhancer in grapevines: field evaluation, molecular, physiological, ultrastructural, and toxicity investigations. *Frontiers in Plant Science* 12, 763365. DOI: 10.3389/fpls.2021.763365.

Razmi, A., Golestanipour, A., Nikkhah, M., Bagheri, A., Shamsbakhsh, M. *et al.* (2019) Localized surface plasmon resonance biosensing of tomato yellow leaf curl virus. *Journal of Virological Methods* 267, 1–7. DOI: 10.1016/j.jviromet.2019.02.004.

Rodríguez, A., Pérez-López, D., Centeno, A. and Ruiz-Ramos, M. (2021) Viability of temperate fruit tree varieties in Spain under climate change according to chilling accumulation. *Agricultural Systems* 186, 102961. DOI: 10.1016/j.agsy.2020.102961.

Sánchez-Hernández, E., Langa-Lomba, N., González-García, V., Casanova-Gascón, J., Martín-Gil, J. *et al.* (2022) Lignin–chitosan nanocarriers for the delivery of bioactive natural products against wood-decay phytopathogens. *Agronomy* 12(2), 461. DOI: 10.3390/agronomy12020461.

Sargent, D.J., Buti, M., Šurbanovski, N., Brurberg, M.B., Alsheikh, M. *et al.* (2019) Identification of QTLs for powdery mildew (*Podosphaera aphanis*; syn. *Sphaerotheca macularis* f. sp. *fragariae*) susceptibility in cultivated strawberry (*Fragaria ×ananassa*). *PloS One* 14(9), e0222829. DOI: 10.1371/journal.pone.0222829.

Sarven, M.S., Hao, Q., Deng, J., Yang, F., Wang, G. *et al.* (2020) Biological control of tomato gray mold caused by *Botrytis Cinerea* with the entomopathogenic fungus *Metarhizium Anisopliae*. *Pathogens* 9(3), 213. DOI: 10.3390/pathogens9030213.

Schilder, A. (2015) Protect strawberries from foliar diseases after renovation. Michigan State University, Michigan. Available at: www.canr.msu.edu/news/protect_strawberries_from_foliar_diseases_after_renovation (accessed 14 March 2023).

Scholthof, K.-B.G. (2007) The disease triangle: pathogens, the environment and society. *Nature Reviews. Microbiology* 5(2), 152–156. DOI: 10.1038/nrmicro1596.

Scott, E.S., Dambergs, R.G., Stummer, B.E., and Petrovic, T. (2022) Fungal contaminants in the vineyard and wine quality and safety. In: Reynolds, A.G. (ed.) *Managing Wine Quality*, 2nd edn, Woodhead Publishing, Duxford, UK, pp. 587–623.

Shattock, R.C. (2003) Rust. In: Roberts, A.V. (ed.) *Encyclopedia of Rose Science*. Elsevier, Amsterdam, pp. 165–169.

Singh, P.P., Kumar, A., Gupta, V., and Prakash, B. (2021a) Recent advancement in plant disease management. In: Kumar, A. and Droby, S. (eds) *Food Security and Plant Disease Management*. Woodhead Publishing, Duxford, UK, pp. 1–18.

Singh, A., Sharma, A., Ahmed, A., Sundramoorthy, A.K., Furukawa, H. *et al.* (2021b) Recent advances in electrochemical biosensors: applications, challenges, and future scope. *Biosensors* 11(9), 336. DOI: 10.3390/bios11090336.

Slack, S.M., Schachterle, J.K., Sweeney, E.M., Kharadi, R.R., Peng, J. *et al.* (2022) In-orchard population dynamics of *Erwinia amylovora* on apple flower stigmas. *Phytopathology* 112(6), 1214–1225. DOI: 10.1094/PHYTO-01-21-0018-R.

Stanković, I., Vučurović, A., Zečević, K., Petrović, B., Nikolić, D. *et al.* (2021) Characterization of cucumber mosaic virus and its satellite RNAs associated with tomato lethal necrosis in Serbia. *European Journal of Plant Pathology* 160(2), 301–313. DOI: 10.1007/s10658-021-02241-8.

Stevic, M., Pavlovic, B. and Tanovic, B. (2017) Efficacy of fungicides with different modes of action in raspberry spur blight *(Didymella applanata)* control. *Pesticidi i Fitomedicina* 32(1), 25–32. DOI: 10.2298/PIF1701025S.

Strickland, D.A., Hodge, K.T. and Cox, K.D. (2021) An examination of apple powdery mildew and the biology of *Podosphaera leucotricha* from past to present. *Plant Health Progress* 22(4), 421–432. DOI: 10.1094/PHP-03-21-0064-RV.

Sumedrea, D., Isac, I., Iancu, M., Olteanu, A., Coman, M. *et al.* (2014) *Pomi, Arbusti Fructiferi, Capsun—Ghidtehnicsi Economic [Trees, Fruit Bushes, Strawberry - Technical and Economic Guide].* Invel Multimedia, Otopeni, Romania.

Teixidó, N., Torres, R., Viñas, I., Abadias, M., and Usall, J. (2011) Biological control of postharvest diseases in fruit and vegetables. In: Lacroix, C. (ed.) *Protective Cultures, Antimicrobial Metabolites and Bacteriophages for Food and Beverage Biopreservation.* Woodhead Publishing, Cambridge, UK, pp. 364–402.

Uda, M.N.A., Hasfalina, C.M., Samsuzana, A.A., Faridah, S., R., R.A, *et al.* (2017) Determination of set potential voltages for cucumber mosaic virus detection using screen printed carbon electrode. *AIP Conference Proceedings* 1808, 020056. DOI: 10.1063/1.4975289.

Ungureanu, C., Calinescu, M., Ferdes, M., Soare, L., Vizitiu, D. *et al.* (2019) Isolation and cultivation of some pathogen fungi from apple and grapevines grown in arges county. *Revista de Chimie* 70(11), 3913–3916. DOI: 10.37358/RC.19.11.7671.

Ungureanu, C., Fierascu, I., Fierascu, R.C., Costea, T., Avramescu, S.M. *et al.* (2021) *In vitro* and *in vivo* evaluation of silver nanoparticles phytosynthesized using *Raphanus sativus* L. waste extracts. *Materials* 14(8), 1845. DOI: 10.3390/ma14081845.

Ungureanu, C., Tihan, G.T., Zgârian, R.G., Fierascu, I., Baroi, A.M. *et al.* (2022a) Metallic and metal oxides nanoparticles for sensing food pathogens-an overview of recent findings and future prospects. *Materials* 15(15), 5374. DOI: 10.3390/ma15155374.

Ungureanu, C., Fierascu, I. and Fierascu, R.C. (2022b) Sustainable use of cruciferous wastes in nanotechnological applications. *Coatings* 12(6), 769. DOI: 10.3390/coatings12060769.

Wan, H., Zhuang, L., Pan, Y., Gao, F., Tu, J, *et al.* (2020) Biomedical sensors. In: Feng, D.D. (ed.) *Biomedical Engineering, Biomedical Information Technology*, 2nd edn. Academic Press, London, pp. 51–79.

Wang, H., Wang, Y., Hou, X. and Xiong, B. (2020) Bioelectronic nose based on single-stranded DNA and single-walled carbon nanotube to identify a major plant volatile organic compound (*p*-ethylphenol) released by *Phytophthora Cactorum* infected strawberries. *Nanomaterials* 10(3), 479. DOI: 10.3390/nano10030479.

Wang, J.R., Guo, L.Y., Xiao, C.L. and Zhu, X.Q. (2018) Detection and identification of six *Monilinia* spp. Causing Brown Rot Using TaqMan real-time PCR from pure cultures and infected apple fruit. *Plant Disease* 102(8), 1527–1533. DOI: 10.1094/PDIS-10-17-1662-RE.

Wang, L., O'Donoghue, M.B., and Tan, W. (2006) Nanoparticles for multiplex diagnostics and imaging, *Nanomedicine* 1, 413–426. https://doi.org/10.2217/17435889.1.4.413

Wang, N.Y., Forcelini, B.B. and Peres, N.A. (2019a) Anthracnose fruit and root necrosis of strawberry are caused by a dominant species within the *Colletotrichum acutatum* species complex in the United States. *Phytopathology* 109(7), 1293–1301. DOI: 10.1094/PHYTO-12-18-0454-R.

Wang, Y., Jin, M., Chen, G., Cui, X., Zhang, Y. *et al.* (2019b) Bio-barcode detection technology and its research applications: a review. *Journal of Advanced Research* 20, 23–32. DOI: 10.1016/j.jare.2019.04.009.

Williamson, B., Tudzynski, B., Tudzynski, P. and van Kan, J.A.L. (2007) *Botrytis cinerea*: the cause of grey mould disease. *Molecular Plant Pathology* 8(5), 561–580. DOI: 10.1111/j.1364-3703.2007.00417.x.

Xu, X. and Ying, Y. (2011) Microbial biosensors for environmental monitoring and food analysis. *Food Reviews International* 27(3), 300–329. DOI: 10.1080/87559129.2011.563393.

Yanderwalle, R. (1945) Observations and researches carried out at the state phytopathological station during the year 1943. *Bulletin de l'Institut Agronomique et des Stations de Recherchesde Gembloux* 14, 65–76.

Yang, H., Cao, G., Jiang, S., Han, S., Yang, C. *et al.* (2021) Identification of the anthracnose fungus of walnut (*Juglans* spp.) and resistance evaluation through physiological responses of resistant vs. susceptible hosts . *Plant Pathology* 70(5), 1219–1229. DOI: 10.1111/ppa.13354.

Yin, L.F., Chen, S.N., Cai, M.L., Li, G.Q. and Luo, C.X. (2014) First report of brown rot of apricot caused by *Monilia mumecola*. *Plant Disease* 98(5), 694. DOI: 10.1094/PDIS-09-13-0995-PDN.

Yin, L., Cai, M., Du, S. and Luo, C. (2017) Identification of two *Monilia* species from apricot in China. *Journal of Integrative Agriculture* 16(11), 2496–2503. DOI: 10.1016/S2095-3119(17)61734-9.

Yuen, J. and Mila, A. (2015) Landscape-scale disease risk quantification and prediction. *Annual Review of Phytopathology* 53, 471–484. DOI: 10.1146/annurev-phyto-080614-120406.

Zehra, A., Rai, A., Singh, S.K., Aamir, M., Ansari, W.A, *et al.* (2021) An overview of nanotechnology in plant disease management, food safety, and sustainable agriculture. In: Kumar, A. and Droby, S. (eds) *Food Security and Plant Disease Management*. Woodhead Publishing, Duxford, UK, pp. 193–219.

Zhan, F., Wang, T., Iradukunda, L. and Zhan, J. (2018) A gold nanoparticle-based lateral flow biosensor for sensitive visual detection of the potato late blight pathogen, *Phytophthora infestans*. *Analytica Chimica Acta* 1036, 153–161. DOI: 10.1016/j.aca.2018.06.083.

Zong, C., Xu, M., Xu, L.-J., Wei, T., Ma, X. *et al.* (2018) Surface-enhanced Raman spectroscopy for bioanalysis: reliability and challenges. *Chemical Reviews* 118(10), 4946–4980. DOI: 10.1021/acs.chemrev.7b00668.

8 Nanomaterials for Postharvest Management and Value Addition

Khaled Sayed-Ahmed[1,2]* and Yasser M. Shabana[3]
[1]Department of Agricultural Biotechnology, Faculty of Agriculture, Damietta University, New Damietta, Egypt; [2]The Center for Excellence in Research of Advanced Agricultural Sciences, Faculty of Agriculture, Damietta University, New Damietta, Egypt; [3]Plant Pathology Department, Faculty of Agriculture, Mansoura University, El-Mansoura, Egypt

Abstract

This chapter highlights the potential postharvest applications of nanotechnology to minimize the losses that represent a challenge to ensuring global food security. Nanotechnology is a new approach that requires mastering, controlling, and understanding the unique properties of matter at the nanoscale. Along the food chain, postharvest diseases decrease the quality, quantity, shelf life, seed viability, and nutritional value of food required for human consumption. Therefore, the use of nanotechnology as an innovative and sustainable way of preventing these losses has attracted the interest of the scientific and agricultural communities. Several nanoparticles(NPs) exhibit antimicrobial activity against various postharvest pathogens. Moreover, NPs can be used alone orin combination with traditional pesticide formulations to develop modes of action and applications that mitigate the harmful impacts of storage pests. Food products can be packaged or coated with nanomaterials to prolong their shelf life and give them outstanding and upgraded properties. This chapter reviews the applicability and sustainability of nanotechnology in postharvest management. The cytotoxicity levels of NPs, based on their type and synthesis techniques, are reported in this chapter. In addition, NP stability and the resistance of postharvest pathogens to NPs are discussed to assess the applicability and sustainability of nanotechnology in different postharvest techniques.

Keywords: Nanotechnology, postharvest management, sustainability, value addition, food security

8.1 Introduction

Nanotechnology, as a multidisciplinary field, has versatile practical applications. It covers several fields from chemistry to physics and biology to materials science (Rastogi *et al.*, 2018).

Nanoparticles (NPs) have sizes ranging from 1 to 100nm and are the building units of this technology. The unique properties of nanotechnology depend mainly on the controlled size and shape (Tiwari *et al.*, 2018).

Based on their outstanding chemical, physical, and biological characteristics, NPs have attracted the interest of plant pathologists and chemists as an alternative to combat phytopathogens instead of conventional chemical products (Bhagyaraj and Oluwafemi, 2018; Il'ina *et al.*, 2022). They also solve the problem

*Corresponding author: drkhaled_1@du.edu.eg

© CAB International 2023. *Nanoformulations for Sustainable Agriculture and Environmental Risk Mitigation* (eds Z. Khan and N.A. Șuțan)
DOI: 10.1079/9781800623095.0008

of resistance, which requires large amounts of chemical compounds to fight resistant phytopathogens. In this respect, NPs can be used singly or mixed with other chemical compounds to fight pyhtopathogens during postharvest and handling processes (Consolo et al., 2020; Akpinar et al., 2021). Nevertheless, more studies are required to estimate the cytotoxicity of NPs and their accumulation and interaction inside the human body to ensure the applicability and sustainability of these NPs. The preparation of NPs using biological and natural sources, as a green method, has attracted more attention to reduce the cytotoxicity and genotoxicity of the prepared NPs as much as possible (Jamuna and Ravishankar, 2014; Magdolenova et al., 2014).

8.2   Introduction to Nanotechnology

8.2.1   The history of nanotechnology

Nanotechnology has an older history than many people realize. For instance, the outstanding colors of dichroic glass fabricated in the ancient Roman period corresponded to the formation of metal NPs (MNPs) during the manufacturing process. The change in colour depended on the size of the MNPs owing to their surface plasmon resonance (SPR). Silver and gold particles on the cup glass ranged from 50 to 100 nm in size. Windows made of stained glass during the sixth to the 15th centuries in Europe represent another example indicating the presence of nanotechnology in this period. The dazzling and different colors corresponded to gold chloride NPs as well as other NPs for various metal oxides. In the Islamic world between the ninth and 17th centuries, the bright, shiny ceramic glazes contained silver NPs (AgNPs), copper NPs (CuNPs) or other MNPs (Tolochko, 2009). Additionally, in the 13th and 18th centuries, the nanowires of cementite as well as carbon nanotubes (CNTs) were used in the manufacturing of Damascus swords to enhance their flexibility, durability, and sharpness (Reibold et al., 2006; Tolochko, 2009).

The greatest development in nanotechnology was reported in 1857 by Michael Faraday, who studied metal colloid properties, specifically gold colloids less than 100 nm. He compared the electrical and optical characteristics of gold in the colloidal form with those of fine leaves of gold and showed that colloidal gold had different characteristics due to its granular structure (Baalousha et al., 2014). In 1925, the Nobel laureate Richard Zsigmondy determined the dimensions of NPs (e.g. colloidal gold) and described the concept of NPs for the first time (Baalousha et al., 2014). It is widely believed that Richard Feynman invented modern nanotechnology. He won the Nobel Prize in 1965 in physics, when he showed that the sizes of elements and compounds can be controlled at the nanoscale by developing unique production methods and measurements (Keiper, 2003).

In 1974, the term nanotechnology was used for the first time by Norio Taniguchi, who described the formation of semiconductors with nanometric precision using a focused ion beam (Keiper, 2003; Hulla et al., 2015). In 1981, Gerd Binnig and Heinrich Rohrer invented the scanning tunneling microscope, which led to a breakthrough in nanotechnology. Scanning tunneling microscopy shows the three-dimensional structure of the surfaces of small objects that cannot be imaged using powerful electron microscopes or conventional microscopes. Based on this invention, in 1986, Binnig and Rohrer received the Nobel Prize in Physics (Filipponi and Sutherland, 2013). In 1986, the first atomic force microscope was developed by Gerd Binnig, Calvin Quate, and Christoph Gerber, and in 1991, researchers of the Japanese company NEC found that CNTs with the shape of a C60 molecule were 100 times stronger than steel and less than six times heavier than steel (Baalousha et al., 2014). At the dawn of the 21st century, the main advances were in nanotechnology applications in a number of fields, including computer technology, biotechnology, medicine, energy, manufacturing, and synthetic materials.

8.2.2   Fundamentals of nanotechnology

Nanotechnology represents an interesting multidisciplinary field with numerous practical applications. In general, this depends on the manufacture of promising and novel materials at a nanoscale (Wan et al., 2016; Rastogi et al., 2018). Basically, the classification of nanomaterials (NMs) involves three main structural categories of three-, two-, or one-dimensional NMs (Wan *et al.*, 2016; Tiwari *et al.*, 2018).

Currently, NMs mostly comprise four types: (i) carbon-based NMs; (ii) composite-based NMs; (iii) inorganic NMs; and (iv) organic NMs. Carbon-based NMs consisting of carbon exhibit various morphological shapes (e.g. ellipsoids, hollow tubes, spheres). In composite-based NMs, NPs are combined with different NPs or bulk materials. Organic NMs are derived from organic matter, while inorganic NMs include metals and metal oxides at the nanoscale and have various morphological shapes, such as spherical, rod, hexagonal, and cuboid (Jeevanandam *et al.*, 2016).

NMs possess outstanding physical and chemical characteristics that rely on their size, shape, aspect ratio, size distribution, surface area, crystallinity, and agglomeration state. NMs are known to have unique optical features owing to their SPR. The differences in their optical properties depend on their particle sizes (Bhagyaraj and Oluwafemi, 2018). The interaction between light and the outer electron bands of MNPs results in this SPR effect. The excitation of outer electrons by light photons causes the outer electrons of NPs to vibrate at a particular wavelength corresponding to this resonance (González *et al.*, 2014). The distinctive magnetic characteristics of NMs are attributed to the interaction between their magnetic spin and electron charges. NMs show magnetic properties due to a high surface area:volume ratio and the magnetic interactions between neighboring atoms (Bhagyaraj and Oluwafemi, 2018). Mechanical characteristics of NMs such as elasticity and hardness contribute to their potential applications and are helpful for estimating their roles and mechanisms of action (Wagner *et al.*, 2011). In addition, several NPs exhibit distinctive catalytic features (i.e. high selectivity and reactivity) compared with their bulk analogs. There is a relationship between NP size and their catalytic activity. Other factors also affect the catalytic selectivity and reactivity of NMs, including their oxidation state, geometry, chemical composition, and the surrounding environment (Mori and Shitara, 1994; Häkkinen *et al.*, 2003).

Numerous techniques have been used to synthesize NMs. They are prepared mainly using two approaches: bottom-up and top-down methods (Fig. 8.1) (Birnbaum and Piqué, 2011). The top-down approach refers to breaking larger particles down to very small nanoscale dimensions and various shapes (Balasooriya *et al.*, 2017) and uses various mechanisms, including laser ablation, milling, nanolithography, attrition, electrochemical explosion technique, and chemical etching. In the bottom-up approach, NPs are produced by smaller units (e.g. atoms and/or molecules). This approach includes several techniques such as the sol–gel process, pyrolysis, spinning, biological methods, plasma, atomic condensation, and chemical vapor deposition. It exhibits more benefits than the top-down approach. NPs prepared via the bottom-up approach usually have a small size, large surface energy, large surface area, reduced imperfections, monodispersion, and spatial confinement.

After the preparation of NPs, they are characterized using techniques such as spectroscopic and morphological techniques, dynamic light scattering and zeta potential, as well as energy dispersive X-ray, X-ray diffraction, thermogravimetric, and nuclear magnetic resonance analyses. The spectroscopic method involves the interaction between the prepared NMs and electromagnetic radiation. The main spectroscopic characterization techniques include ultraviolet/visible, infrared photoelectron, and Raman spectroscopies in addition to mass spectrometry. Morphological characterization techniques of NMs include scanning electron microscopy (SEM) or high-resolution transmission electron microscopy, and field emission (SEM), as well as scanning probe microscopy, scanning tunneling microscopy, near-field scanning optical microscopy, confocal laser scanning microscopy, and atomic force microscopy (Omran, 2020).

8.3   Postharvest Losses

Postharvest losses involve degradation in the quality and quantity of the produce in the period from harvest to table. Quality losses have a large effect on the nutrient/calorific value, edibility, and acceptability of the commodity (Kader, 2003). Quantity losses are those that lead to a loss in the amount of product. In

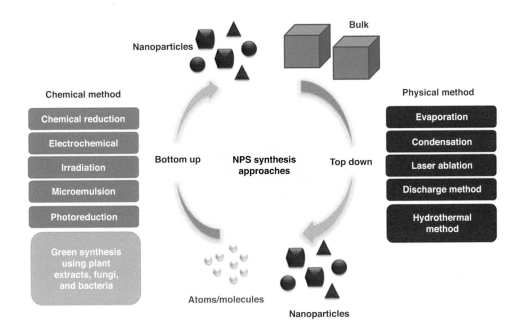

Fig. 8.1. Approaches to nanoparticle synthesis.

developing countries, total postharvest losses are approximately 20–40% of all crops owing to mechanical damage, microbial decay, physiological disorders, and insect damage during harvest, transport, and storage (Kitinoja and Gorny, 1999;). Postharvest losses result from weight loss, quality loss, nutritional loss, seed viability loss and a shorter shelf life.

8.3.1 Weight loss

Weight loss can be determined by measuring the reduction in moisture content of different fruits and vegetables. Fresh vegetables and fruits gradually lose moisture content following the harvest process, leading to weight loss and shrinkage. Weight loss increases with increasing storage period, and the percentage of dry matter increases due to losing more moisture. Most fruits and vegetables are considered unsalable as fresh products if they have lost around 3–10% of their fresh weight (van Hung et al., 2011; Sinha et al., 2019). This weight loss is highly affected by the vapor pressure gradient between the vegetables or fruits and the surrounding environment. During storage time, the transpiration rate is closely related to the relative humidity when the stored products and the surrounding environment are at the same temperature (Sharkey and Peggie, 1984; van Hung et al., 2011). Transpiration is considered the main factor responsible for weight loss in fresh vegetables or fruits, and therefore results in poor quality of produce (Yahia and Carrillo-Lopez, 2018). It is common to store produce under low temperatures and high relative humidity (85–95%) to preserve the quality of the stored fruits and vegetables (Makule et al., 2022). Weight loss can be measured by determining the reduction in the stored produce weight and is expressed as a percentage of change in weight in comparison with the initial fresh weight of the produce (Riaz et al., 2021).

8.3.2 Quality loss

Quality losses are concerned with shape, size, odor, taste, and external features. Texture is felt by mouth when eating the produce. Texture characterizes a mixture of sensations detected by the tongue, lips, and teeth. Texture is an important factor for both eating and cooking qualities. Flavor includes two main issues: aroma induced by volatile compounds, which can be detected by the nose, and taste provoked by the produce components of either sugar and/or acidity. Flavor quality is evaluated based on the perception of taste and aroma (Paltrinieri, 2014). The loss of flavor of the produce through postharvest storage causes dissatisfaction of the consumer. Therefore, the causes of flavor loss must be identified and avoided as much as possible (Forney, 2008). After harvesting fresh produce, there are two main methods, metabolic and diffusional mechanisms, that result in flavor loss. The formation of flavor molecules or those that result in off-flavors causes variations in produce flavor (Voilley and Souchon, 2006). These variations depend on the produce physiology, which is affected by maturity and the surrounding environment. In addition, variation in commodity flavor is also found to be affected by the diffusion of volatile molecules out of the produce. Diffusion relies on the chemical and physical characteristics of the flavor compounds as well as the cutting process, packaging, and produce matrix (Forney, 2008; Pott et al., 2020). The cell walls of plant cells have a crucial effect on texture (Zhang et al., 2019). Degradation of pectin is considered the most common reason for softening during the ripening period (Defilippi et al., 2018). The appealing colour of produce corresponds to two main categories of secondary metabolites, polyphenols and terpenoids, which are also responsible for nutritional and organoleptic features (Pott et al., 2019). Storage has negative impacts on the accumulation of colorants, such as anthocyanin and other compounds corresponding to the colour of the produce. Therefore, storage under low temperatures is an acceptable technique to maintain the content of anthocyanins in produce such as strawberries and blood oranges (Carmona et al., 2017; Li et al., 2019).

8.3.3   Nutritional loss

Nutritional losses are the combination of qualitative and quantitative losses. They refer to the loss of nutritional components such as vitamins.

They may be invisible, but they can increase owing to adverse handling conditions such as high temperatures, a long storage period, and low relative humidity (Yahia et al., 2019). Vegetables and fruits are rich in nutritional value and include health-promoting compounds such as vitamins, proteins, minerals, antioxidant compounds, and dietary fibre (Gogo et al., 2017). The chain level significantly influences the nutritional value of the commodities. For example, 2–13% of carotenoids are lost during the drying process of sweet potatoes (Bechoff et al., 2011). Approximately 53–61% loss of vitamin C results from okra processing (Tekpor, 2011). Sweet potato leaves lose approximately 14%, 40–58%, and 86–95% of pro-vitamin A during the blanching, cooking, and drying processes, respectively (Mosha et al., 1997). Losses of 13%, 58%, 53%, and 16% in the content of protein, ascorbate, vitamin A, and iron, respectively, were seen in cassava leaves during processing after harvesting (Lyimo et al., 1991). Mushrooms possess high nutritional value; however, the preservation methods for mushrooms, including thermal, chemical, and physical processes, also affect their nutritional value and bioactive properties (Marçal et al., 2021). Phytochemical interactions with the surrounding environment during harvesting may alter the concentration of bioactive components in fruits and vegetables. For instance, it was found that low storage temperatures (5 and 12°C) enhanced the antioxidant activity of ripened cluster tomatoes compared with those stored at room temperature (25–27°C). In addition, tomatoes stored at room temperature showed increased lycopene content, while total soluble solids did not change either at room temperature or during low-temperature storage (Javanmardi and Kubota, 2006).

8.3.4   Seed viability loss

Storage of seeds after harvesting in suitable conditions is important for native seed producers and restoration practitioners. This is considered key to maintaining seed viability. Native seeds are stored for specific periods, and then delivered to the site of restoration or propagation. Seed viability and quality decrease if stored in inappropriate conditions (de Vitis *et al.*, 2020). The temperature, moisture, and oxygen proportion have a significant impact on the deterioration of seeds and their viability. Lowering the moisture content of seeds to specific thresholds leads to an increase in seed viability for approximately 90% of species in a predictable manner (Roberts, 1973; de Vitis *et al.*, 2020). Deterioration of seeds occurs largely as a result of oxidative processes, which can deteriorate proteins (e.g. enzymes), lipids, RNA, and DNA (Walters *et al.*, 2010; Fleming *et al.*, 2019; de Vitis *et al.*, 2020). Germination vigour usually decreases with increasing storage period because the metabolic system of seeds begins to break down, leading to slow or poor germination (Roberts, 1973; de Vitis *et al.*, 2020). A high quality of seeds is a sign of good harvesting and postharvest practices. Early-picked cotton has high moisture content due to humid weather in the morning, leading to a loss in material quality, and traditional storage processes do not protect the seeds from the fluctuations of humidity that decrease the cottonseed viability. In contrast, picking the cottonseed during hot and dry weather reduces physical abrasion during ginning and reduces the deterioration of physical quality during processing (Afzal *et al.*, 2020).

8.3.5 Short shelf life

Shelf life is often known as the duration for which the produce is still safe and maintains the required sensory and nutritional qualities (Torres-Sánchez *et al.*, 2020). The storage or handling temperature is the most important factor in the quality of perishable food (de Venuto and Mezzina, 2018). Therefore, a cold chain is required to avoid postharvest losses and short shelf life. By controlling the temperature, a prolonged shelf life for produce can be obtained (Torres-Sánchez *et al.*, 2020).

The majority of produce is perishable and subject to biological damage and physiological disorders. Food quality and safety are the main concerns from harvesting to consumers. Efforts have been made to extend the shelf life of produce to the greatest extent possible (Taghavi *et al.*, 2018). For example, fresh blueberries, blackberries, and raspberries have a pleasant flavor that attracts consumers and maximizes their popularity. However, their potential for market growth and fresh supply is limited due to their seasonality and short shelf life. A high susceptibility to decay, high respiration rates, and sensitive structures are the factors that reduce their storability. The main current practices for an extended shelf life depend mainly on high-humidity storage and cold-chain settings (Huynh *et al.*, 2019).

Another example of the effect of fruit shelf life during storage is guava and banana, which exhibit a short shelf life at high temperatures owing to fast ripening corresponding to alterations such as softening, off-odor development, senescence, anthracnose and crown rot diseases, and injuries that occur during storage (Zhang *et al.*, 2010; Ketsa *et al.*, 2013; Murmu and Mishra, 2018). Furthermore, fungal pathogens are becoming more resistant to the fungicides used (Bazie *et al.*, 2014). The accumulation of enzymatic oxidation of phenolic compounds and reactive oxygen species (ROS) during ripening causes degradation inside plant parts. This deterioration is estimated using physical properties such as colour, the buildup of hydrogen peroxide and superoxide anions, the ion leakage percentage, and the accumulation of thiobarbituric acid-reactive compound. Using modified atmosphere storage and edible coatings extends the shelf life by reducing the respiration, climacteric peak intensity, and ripening time, and the corresponding changes, and avoiding the increase in ROS (Murmu and Mishra, 2018).

8.4 Nanotechnology Applications for Minimizing Postharvest Losses

8.4.1 Antifungal properties of NPs

In vitro studies

NPs of metals or metal oxides possess antifungal activity owing to their distinctive physical and chemical properties. Numerous studies have investigated the antifungal activity of these NPs against the growth of phytopathogens (Il'ina *et al.*, 2022). As previously described, the antifungal activity of NPs is influenced by several factors. Therefore, it is necessary to understand the action mechanisms and interactions between MNPs and fungal phytopathogens. MNPs release ions that can bind to certain protein groups, leading to a disruption in the function of membrane proteins and cell permeability. They also inhibit the germination of conidia and delay their development. MNPs also interfere with the transport of protein-oxidative electrons. They increase gene transcription in response to oxidative stress, which affects the mitochondrial membrane. Moreover, the generation of ROS induced by MNPs causes severe damage to DNA, and interferes with nutrient absorption, thereby causing fungal cell death (Huerta-García *et al.*, 2014; Mikhailova, 2020; Rana *et al.*, 2020; Cruz-Luna *et al.*, 2021).

AgNPs synthesized via the green method using the seed exudate of *Trifolium resupinatum* exhibited antifungal activities against *Rhizoctonia solani* and *Neofusicoccum parvum* (Khatami *et al.*, 2016). AgNPs exhibit antifungal activities against 18 fungal types, including *Fusarium solani*, *Fusarium oxysporum*, *Alternaria solani*, *Botrytis cinerea*, *Pythium spinosum*, and *Pythium aphanidermatum*, which significantly affect potatoes and tomatoes (Kim *et al.*, 2012). Different media such as potato dextrose agar, maize flour agar, and malt extract agar can be utilized to distinguish the antifungal activity induced by NPs. The antifungal activity of AgNPs against several fungal strains can be attributed to the degradation of their membranes.

A comparison of the antifungal action of AgNPs, chitosan, and chitosan NPs (ChNPs) combined with silver against *Aspergillus flavus*, *R. solani*, and *Alternaria alternata* was carried out. ChNPs showed the highest antifungal activity against *A. flavus*. Chitosan–AgNPs represent a potential alternative to traditional fungicides for controlling the growth of phytopathogens (Il'ina *et al.*, 2022). Spherical biosynthesized AgNPs of around 60 nm prepared using the essential oil of *Mentha piperita* exhibited antifungal properties against the fungal pathogen *Aspergillus niger* (Thanighaiarassu *et al.*, 2014). CuNPs possess antimicrobial activities due to their high surface area and interaction with other particles (Karthik and Geetha, 2011). In some cases, surfactants or polymers are used to prevent oxidation of these metals. NPs can be combined with other NMs such as nanocomposites of

chitosan-coupled copper NPs, which have excellent antifungal activities against phytopathogens such as *F. oxysporum*, *F. moniliforme*, *P. aphanidermatum*, *R. solani*, *A. alternata*, *B. cinerea*, *A. flavus*, *Microsporum gypseum*, *Trichoderma viride*, *Penicillium expansum*, and *Curvularia lunata* (Hernández-Díaz *et al.*, 2021). Green-synthesized CuNPs prepared using *Ocimum sanctum* leaf extract exhibited high antifungal activity against fungal phytopathogens such as *Alternaria carthami* and *Aspergillus niger* (Shende *et al.*, 2016). In addition, CuNPs prepared using plant extracts of *Stachys lavandulifolia* and *Citrus medica* exhibited outstanding antifungal activities against *Fusarium* spp. such as *F. culmorum*, *F. graminearum* and *F. oxysporum* (Shende *et al.*, 2015).

Currently, selenium NPs (SeNPs) are utilized in microbial biofilms as antifungal agents due to their role in spore germination inhibition (Hernández-Díaz *et al.*, 2021). Moreover, titanium dioxide (TiO_2) NPs showed high antifungal activity against *P. expansum* and *Alternaria brassicae* (Maneerat and Hayata, 2006). Typically, zinc oxide NPs (ZnONPs) obtained using various techniques via plant tissues and their extracts possess antifungal properties against various plant pathogens, including *P. expansum*, *A. flavus*, *A. niger*, *F. oxysporum*, *B. cinerea*, and *A. alternata* (Rajiv *et al.*, 2013; Jamdagni *et al.*, 2018). Iron oxide NPs (Fe_2O_3NPs) used at 100 μg ml^{-1}, biosynthesized using plant extracts, can be used to control several phytopathogens, including *Mucor piriformis* and *A. niger* (Devi *et al.*, 2019).

In vivo studies

Phytopathogenic fungi cause significant losses in both vegetables and fruits during postharvest, which minimizes shelf life and food quality (González-Estrada *et al.*, 2017). Agrochemicals are usually required to control the growth of these fungi. However, they are associated with environmental risk and harmful impacts on human health. Pests are estimated to destroy up to 40% of the world's food supply and more than US$220 billion year^{-1}. Several factors influence the antifungal activity of NPs such as size distribution, composition, shape, crystallinity, surface chemistry, and agglomeration (Koduru *et al.*, 2018). For instance, small NPs have a high surface area:volume ratio, which enhances their antifungal activity. These factors can be controlled via synthesis routes. Therefore, the synthetic route may greatly influence the antifungal activity as it is sometimes difficult to remove surfactants or metal precursors from the prepared NPs (Cruz-Luna *et al.*, 2021). Nanoencapsulation is considered an outstanding way to reduce the harmful impact of fungal phytopathogens on produce. Numerous investigations have been carried out to estimate the antifungal activity of nanoencapsulations *in vitro* on different types of produce. For example, poly(lactic acid) nanocapsules enclosing essential oil of lemongrass showed higher *in vivo* antifungal activity than essential oil alone against *Colletotrichum gloeosporioides*, which is responsible for bitter rot in apples (Antonioli *et al.*, 2020). ChNPs also exhibit antifungal activity against numerous fungal phytopathogens. For instance, the effect of ChNPs and chitosan gel against fungal phytopathogens of strawberry, including *B. cinerea*, *A. niger*, and *Rhizopus stolonifer* were investigated *in vitro* (Melo *et al.*, 2020). The edible coating–chitosan nanocomposite showed important changes in the morphology of the tested fungi. The best control of fungal growth in artificially infected strawberries was achieved using the ChNPs.

The use of MNPs represents a good alternative to control fungal growth of phytopathogens during the postharvest period. To date, MNPs of various metals (e.g. Cu, Ag, Se, Fe, nickel (Ni), palladium (Pd), and magnesium (Mg)) have been prepared and utilized as fungicides during the postharvest period. AgNPs have been investigated the most owing to their outstanding antifungal activity, followed by CuNPs. The *in vivo* antifungal activity of other MNPs has also been evaluated and has shown outstanding results (Cruz-Luna *et al.*, 2021), and they have been used as fungicides to inhibit the growth of various phytopathogenic fungi in agriculture. For instance, CuNPs, AgNPs, ZnNPs, FeNPs, SeNPs, NiNPs, and PdNPs have displayed outstanding antifungal properties. Chemical control suffers from the development of fungicide-resistant phytopathogens. In contrast, MNPs are eco-compatible and better alternatives to fungicides, and are a tool with high anti-resistance properties (Consolo *et al.*, 2020; Akpinar *et al.*, 2021).

Gray mold caused by the fungus *B. cinerea* is one of the most destructive diseases in terms of reducing the storage period and the marketing quality of strawberries, grapes, and tomatoes. Green-synthesized CuNPs improved the quality of tomato fruits 10 days after artificial inoculation with *B. cinerea*, and therefore represent an acceptable alternative to traditional fungicides to minimize tomato fruit rots (Bikdeloo *et al.*, 2021). In addition, CuNPs and sulfur NPs (SNPs) with sizes of 2–12 and 10–50 nm, respectively, exhibited unique antifungal activities against *B. cinerea* and *Sclerotinia sclerotiorum*. Their antifungal activities were evaluated *in vivo* on cucumber fruits at two temperatures, 10 and 20°C (Sadek *et al.*, 2022). Under these conditions, the SNPs possessed higher antifungal activity than the CuNPs. The pathogen inhibition activity increased with increasing NP concentration, while the growth inhibition of *S. sclerotiorum* and *B. cinerea* was increased at low temperature (Sadek *et al.*, 2022).

The antifungal activities of silicon dioxide NPs (SiO_2NPs), CuNPs, ChNPs, and their combinations were evaluated *in vivo* on table grapes against the gray mold fungus *B. cinerea*. A single application of SiNPs or ChNPs reduced the growth of the gray mold. Using artificial infection, at the end of cold storage (2 ± 1°C for 1 month), ChNPs, SiO_2NPs, CuNPs, SiO_2–CuNPs, and SiO_2–Ch–CuNPs at a concentration of 3g l⁻¹ all significantly reduced the gray mold severity by 50–57%. After 1 week of shelf life at 22 ± 2°C, these NPs decreased the rot by 19–31%. SiO_2–CuNPs, and SiO_2–Ch–CuNPs were the most effective combinations against gray mold. At the end of cold storage, these NPs had reduced the amount of rotted fruit by 5–54% (Hashim *et al.*, 2019).

8.4.2 Postharvest bacterial disease control using NPs

Several bacterial strains, especially Gram-negative bacteria such as *Erwinia*, *Pseudomonas*, and *Pectobacterium* spp., are responsible for most bacterial soft rots, which are destructive diseases of vegetables, fruits, and ornamentals worldwide (Tomlinson, 1988; Abdelghany *et al.*, 2022). *Pectobacterium* (formerly *Erwinia*) *carotovorum* is a phytopathogen that causes soft rot disease in various plants and blackleg disease in potatoes through degradation of the plant cell wall. Bacterial soft rot disease results in more losses in commodities than any other bacterial disease. Postharvest losses due to soft rot disease have been determined to be 15–30% of the total crop yield (Abo-Elyousr *et al.*, 2010; Abdelghany *et al.*, 2022). The antibacterial activities of AgNPs, chromium oxide NPs (Cr_2O_3NPs), and ZnONPs have been studied to control potato soft rot *in vitro* and *in vivo*. *In vitro*, AgNPs were the most effective against the bacterial growth and exhibited the highest inhibition zone at a concentration of $50\,\mu g\ ml^{-1}$, followed by Cr_2O_3NPs and ZnONPs at concentrations of 100 and $150\,\mu g\ ml^{-1}$, respectively. *In vivo*, AgNPs, Cr_2O_3NPs, and ZnONPs controlled the levels of soft rot disease in storage and greenhouse experiments (Abdelghany *et al.*, 2022).

Antibacterial activities of chitosan rely on specific factors: molecular weight, the degree of deacetylation, the degree of substitution, the cell-wall structure of the target bacteria, and physical form (Perinelli *et al.*, 2018). Mohammadi *et al.* (2016) estimated the antibiotic activity of ChNPs of low and medium molecular weights against *Pseudomonas fluorescens*, which causes head rot disease in broccoli, and *P. carotovorum*, which causes soft rot disease in various vegetables. The ChNPs showed good antibacterial activity, whereas no significant differences were observed in the antibacterial activity for ChNPs synthesized with larger molecular weights.

Biosynthesized AgNPs (100 ppm) with sizes ranging from 15 to 45 nm and spherical shapes showed outstanding antibacterial activities against *P. carotovorum* with an inhibition zone of up to 15.3 mm. Thus, AgNPs can be used as a new alternative way to control postharvest bacterial soft rot (Pineda *et al.*, 2022).

The impact of polymeric nanocomposites of SeNPs, AgNPs, and gold NPS (AuNPs) on bacterial plant pathogens from the genera *Pectobacterium*, *Pseudomonas*, *Micrococcus*, and *Xanthomonas* was investigated. The bactericidal activity of those nanoformulations was studied *in vitro* using the agar well diffusion technique. SeNPs and AgNPs combined with triazole-based polymers and fungal biopolymers showed efficient antibacterial activities against these bacterial phytopathogens. SeNPs had a higher bactericidal impact. Green-synthesized SeNPs

with a surface decorated with different carriers and ligands may enhance the efficacy and minimize the toxicity of these promising NPs (Tsivileva *et al.*, 2021). Copper compounds have been widely used to protect crops due to their unique effects on resistant pathogenic bacteria. Copper sulfide NPs possessed excellent *in vitro* antibacterial effects, with a median effective concentration (EC_{50}) of $17\,mg\ l^{-1}$ against *P. carotovorum*. They are highly effective, eco-friendly antibacterial agents, with low-toxicity to non-targets, compared with commercial dicopper chloride trihydroxide (Yan *et al.*, 2022).

8.4.3   NPs as insecticides

The use of nanotechnology to develop novel and alternative insecticides is an important development. Several NPs, such as SiO_2NPs, appear to be able to overcome insect resistance to traditional insecticides (e.g. phosphine and pyrethroids). SiO_2NPs, as novel insecticides, can be used to combat storage insects of maize. El-Naggar *et al.* (2020) examined the insecticidal activity of SiO_2NPs against *Rhyzopertha dominica*, *Sitophilus oryzae*, *Oryzaephilus surinamensis*, and *Tribolium castaneum* after different durations (24, 48, and 72 hours). After 72 hours, *O. surinamenisis* and *R. dominica* exhibited the lowest 50% lethal concentrations (LC_{50}) values ($0.002\,g\ kg^{-1}$ for both), while *T. castaneum* and *S. oryzae* had LC_{50} values of 0.034 and $0.263\,g\ kg^{-1}$, respectively. SiO_2NPs (0.25–$2.0\,g\ kg^{-1}$) led to 100% mortality for *O. surinamenisis* and 93.3% for *Sitophilus oryzae*, which was more resistant to the insecticidal effect of SiO_2NPs.

Casanova *et al.* (2005) prepared a nicotine carboxylate nano-emulsion and studied its insecticidal activity against adult *Drosophila melanogaster* by determining its 50% lethal time (LT_{50}). This nano-emulsion (nicotine carboxylate emulsion droplets) enhanced the insecticidal activity of nicotine. A rise in the concentration of nicotine increased insecticide activity. Wang *et al.* (2007) developed a nano-emulsion (30 nm droplets) using methyl decanoate and polyoxyethylene lauryl ether as surfactants to capture β-cypermethrin. The disintegration of the insecticide was improved and no precipitate was observed in nano-emulsion samples.

Thus, nanotechnology enhances stability and decreases the concentration of insecticides required in traditional sprays, while maintaining their efficiency.

8.4.4 Nanotechnology applications in coatings and packaging

The increasing demand for food safety, high quality, and sustainable development has encouraged researchers to exploit nanotechnology for use in postharvest treatment and the food industry. Packaging is a key step in the postharvest period to ensure food safety and quality; however, packaging materials are not fully resistant to atmospheric gases and water vapor. Food packaging enhances handling, storage, and protection from pollution from the surrounding environment and meets the developing market demands. Some NMs are designed to enhance packaging efficiency, such as mechanical strength of barriers, performance, and thermodynamic stability. In addition, other NMs contain antioxidants, bacteriostatic agents, enzymes, and plant extracts to prolong the shelf life of produce. To date, the majority of packaging materials have been nonbiodegradable, which represents a serious problem in the environment. Nanocomposites can be used as renewable or green materials in packaging to avoid the harmful environmental impact of synthetic packaging. There are two main types of material being used: (i) inorganic and metal NPs; and (ii) plant extracts and mixtures incorporated into biopolymers (e.g. chitosan, cellulose, and starch) (Kuswandi, 2017; Han et al., 2018; Huang et al., 2018).

Overall, NMs used in food packaging can be classified by three major functions: improved, smart, and active packaging. Improved packaging refers to the use of NPs in bionanocomposite substances to improve the mechanistic and blockade characteristics, including acting as a gas barrier, elasticity, and stability at various temperatures and humidity levels. Smart packaging acts as a sensor or for feedback on the real-time quality of the produce and can be considered a tool to detect fake products and fraud. It is used to measure the level of exposure to specific adverse factors, such as high oxygen levels or insufficient temperatures. Active packaging provides preservation and protection based on mechanisms activated using acquired and/or inherent factors, such as antimicrobial and biodegradable activities. It also minimizes the loss of produce owing to the extension of shelf life. Most NMs are still in the stage of demonstration studies, and their application to food packaging is yet to receive approval because of the potential migration of NMs from packaging to the food matrix. Therefore, further studies are required to ensure food safety after packaging using these NMs (Sharma et al., 2017; Wang et al., 2017; Wyrwa and Barska, 2017; Huang et al., 2018).

8.5 Biological Activity of NMs

NPs have garnered considerable attention owing to their chemical, physical, and biological properties and wide applications. Plant extracts, fungi, bacteria, and algae can reduce NP precursors to form stable and small particles at the nanoscale. Different molecules present in plants such as polysaccharides, alkaloids, vitamins, alcoholic compounds, and amino acids contribute to the green synthesis of NPs. Similarly, these bioactive components affect the biological activity of the prepared NPs by combining with them (Shousha et al., 2019).

Nanotechnology is a rapidly growing technology used to create a variety of novel materials with versatile applications in the pharmaceutical, medical, chemical, electronics, mechanical, and environmental industries. Therefore, the biological activity of NPs has attracted the interest of researchers and manufacturers because NPs have become very important in our daily life. Green-synthesized NPs have become the main branch of nanotechnology. Metals (e.g. Au, Ag, aluminum (Al), and Cu) are inert in their bulk form. However, a decrease in their size may increase their inherent toxicity. Prolonged exposure to NPs may have a harmful impact on the environment and on human health (Ravinayagam and Jermy, 2020; Neme et al., 2021). Therefore, detailed safety studies should be done before a recommendation for commercial use of NPs.

The impacts of various NPs on human health via cytotoxicity or genotoxicity have been investigated in several studies. NPs pass into the human body via inhalation, passage across the skin, or injection into the bloodstream. In addition, they are able to pass through the blood–brain barrier owing to their very small size. NPs can react with proteins (e.g. enzymes), influence the biological behaviour of organs or tissues, and change gene expression (Jamuna and Ravishankar, 2014). NP genotoxicity depends on different conditions that impact the genotoxic response. These factors include physico-chemical characteristics and experimental conditions. The types of assays used to determine the genotoxic effects of NPs are usually micronucleus, comet, bacterial reverse mutation, and chromosome aberration tests (Magdolenova *et al.*, 2014).

8.6 Nanosensor Applications in Agriculture

Nanobiosensors have played a pivotal role in the development of agriculture by developing novel diagnostic techniques and tools. They are accurate, cost-effective, and efficient in different food, agricultural, and environmental applications. Nanosensors can be used to identify pollutants, heavy metals, and pathogens. They can also monitor traceability, temperature, and humidity levels. Nanosensors exhibit unique properties such as sensitivity and are relatively inexpensive. These sensors can observe the signaling waves of hydrogen peroxide in plant leaves, where plants use hydrogen peroxide to communicate and activate the plant cells to produce helpful compounds to fend off predators such as insects. Nanosensors include sensitive nanowires for detection, thin films, CNTs, NPs, and other NMs (Johnson *et al.*, 2021).

Nanosensors are used to determine the conditions of produce. This ameliorates the food packaging problems. They inform the consumer about the status of the commodity during storage and handling processes. They also help to provide safe, fresh food to consumers, and reduce the frequency of foodborne illness, which is useful for maintaining food safety. Nanosensors provide a higher sensitivity and specificity than conventional sensors. Nanobiosensors are typically used in food and pharmaceutical industries. For example, multi-enzymatic sensors can be used in the food industry to detect sucrose concentrations. In addition, glutamate sensors are used to determine glutamate concentration. Ethanol sensors are used as microbial sensors to measure ethanol concentration during alcohol fermentation processes (Bhusare and Kadam, 2021).

8.7 Sustainability

8.7.1 Cytotoxicity levels of NPs

The cytotoxicity of NPs has been evaluated by numerous *in vitro* and *in vivo* studies. The *in vitro* assay is not an entirely realistic method, but multiple cell-based assays and animal studies can be performed to check for cytotoxicity. The interaction between NPs and the surrounding environment relies on their size, surface properties, chemical composition, aggregation behaviour, solubility, biokinetics, and biopersistence (Jamuna and Ravishankar, 2014; Ravinayagam and Jermy, 2020).

AgNPs are considered the prominent MNPs in nanotechnology. However, they have harmful effects on the environment owing to their toxic effect, which limits their potential applicability. However, the possibility of preparing such NPs through biosynthesis methods decreases the level of their cytotoxicity and raises their applicability as a clean, eco-friendly technology. Plant extracts, yeasts, bacteria, and filamentous fungi represent the common biological resources utilized for green synthesis of AgNPs (Pineda *et al.*, 2022). For particles with sizes of 50 and 500 nm, the toxic effects of SiO_2NPs on immunocompetent inbred BALB/c mice were estimated. It was shown that mouse cytotoxicity is induced by SiO_2NPs with diameters of around 50 nm. Despite the fact that toxicity decreases with increasing particle size, it increased in the presence of large silica particles (Mohammadpour *et al.*, 2019).

Single-walled and multi-walled CNTs show cytotoxicity owing to their interaction with different organs in the body. Specifically, single-walled CNTs inflict more toxicity due to their dispersive and hydrophobic properties, leading to apoptosis, compared with multi-walled CNTs. Inhalation of CNTs causes higher toxicity than other modes of exposure such as oral administration, injection, and dermal administration. NPs interact with cells inside the body, resulting in the distribution of particles to different organs. Additionally, the functional moieties and particle size have a significant impact on toxicity (Cui *et al.*, 2005; Ravinayagam and Jermy, 2020).

It is possible to synthesize NPs in a variety of sizes and shapes for different applications such as uptake, retention in tissues, and biocompatibility. Several studies have investigated the reasons for nanotoxicity. The main factors affecting cytotoxicity are size and shape. Cytotoxicity usually increases with decreasing size as a result of the larger surface area, resulting in the release of more ions, which leads to oxidative stress in cells. Small NPs are distributed in the kidney and lung, whereas larger NPs accumulate in the spleen and liver. For instance, AuNPs of 40 nm size accumulate in liver more than those of around 20 nm, which localize in the lung. Furthermore, various shapes have been reported to be more toxic than others with the same composition. For example, nanorods possess greater cytotoxicity than nanoflowers. Nanotube cytotoxicity depends on their length and diameter, and they are also more toxic than nanoflakes (Zhao *et al.*, 2019; Jahan *et al.*, 2021; Wang *et al.*, 2021). A low zeta potential leads to aggregation of NPs and results in physical instability owing to the weak forces acting on them. Hence, aggregated and nonaggregated NPs show a high cytotoxicity. However, aggregated NPs interact with the cell membrane more than nonaggregated NPs, and change its structural integrity, leading to fatal damage (Kong *et al.*, 2021).

8.7.2   Stability of NPs

The stability of the NPs is a key factor in their applicability, toxicity, and sustainability. Therefore, several studies have investigated the stability of NPs to enhance their applicability in many fields such as agriculture and the food industry. The chemical composition of NPs significantly affects their stability. For example, metal oxide NPs show high stability against degradation and have lower toxicity compared with other magnetic NPs (Kakakhel *et al.*, 2021).

For NPs to be useful, they must be stable. The large surface areas of MNPs lead to accumulation and oxidation, making them highly reactive. The surface of the NPs oxidizes rapidly at normal temperature and pressure levels, forming a thin layer of oxide that dramatically changes the properties of MNPs (Ansari *et al.*, 2022). Several techniques have been investigated to maintain the stability of MNPs using polymers or surfactants. Most studies maintained MNP stability by coating them with surfactants. Currently, coating with polymers has attracted increasing attention due to the repulsion effects. MNPs can be dispersed by coating with one or two layers of polymer. To prevent oxidation, the coating of MNPs should be dense. In an acidic environment, one or two thin layers can easily be removed from the surface of NPs (Gao *et al.*, 2021; Kianfar and Cao, 2021; Ansari *et al.*, 2022).

8.7.3   Low resistance of postharvest pathogens to NPs

The use of conventional fungicides, pesticides, and herbicides results in harmful impacts such as harm to domestic animals, decreased pollination by insects, tolerance to agrochemicals, and induction of plant pathogen resistance. Phytopathogen resistance to chemical compounds highlights the need for ionic NPs prepared using green chemistry for use as an alternative to conventional strategies for managing diseases and pests (Álvarez *et al.*, 2016). Resistance to phytopathogens is caused primarily by overuse and abuse of agrochemicals without adhering to dosage and toxicity recommendations. This resistance causes economic losses, environmental risks, and harmful impacts on public health because of the need for high doses of these chemical products to combat this resistance. Global pesticide sales (US$ 35 billion year^{-1}) have increased dramatically over the last 50 years. Thus, alternative products such as NPs have been developed for produce protection (Mateo-Sagasta *et al.*, 2018). For example,

CuNPs exhibit antimicrobial activity against several plant pathogens, both fungi and bacteria. Therefore, green-synthesized CuNPs used alone or combined with other chemical compounds can be used effectively to control phytopathogen growth and solve the problem of resistance (Karthik and Geetha, 2011; Hernández-Díaz *et al.*, 2021).

8.8 Conclusions

Nanotechnology provides nanoscale particles with unique physical, biological, and chemical characteristics. The use of nanotechnology in postharvest treatments has shown an outstanding ability to minimize both the quality and quantity of postharvest losses from harvest until consumption. It enhances the nutritional value, edibility, and acceptability of the produce, as well as other benefits such as long shelf life, low phytopathogen resistance, low cost, and sustainability. The antifungal activities of several NPs or NMs against the growth of phytopathogens have been determined using *in vitro* and *in vivo* studies. NPs affect fungal growth by releasing ions, which disrupt the function of membrane proteins and cell permeability and/or generate ROS, resulting in severe damage to DNA. Nanotechnology can also be used to develop novel and alternative insecticides at the nanoscale with high potential efficacy against resistant insects. The efficiency of produce packaging can also be enhanced using novel designed NMs, such as nanocomposites, inorganic NPs, MNPs, and nano-emulsions, as improved, smart, and active food packaging materials. Nanobiosensors improve postharvest treatments using diagnostic techniques and tools. Their accuracy, cost-effectiveness, and efficiency make them ideal for a variety of environmental and agricultural applications. There have been many studies to reduce the toxicity of the NPs as much as possible via green synthesis procedures to enhance their applicability and sustainability.

References

Abdelghany, W.A., Mohamedin, A.H., Abo-Elyousr, K.A.M. and Hussein, M.A.M. (2022) Control of bacterial soft rot disease of potato caused by *Pectobacterium carotovorum* subsp. *carotovorum* using different nanoparticles . *Archives of Phytopathology and Plant Protection* 55(14), 1638–1660. DOI: 10.1080/03235408.2022.2111247.

Abo-Elyousr, K.A., Khan, Z., and Abedel-Moneim, M.F. (2010) Evaluation of plant extracts and *Pseudomonas* spp. for control of root-knot nematode, *Meloidogyne incognita* on tomato. *Nematropica* 40, 289–299.

Afzal, I., Kamran, M., Ahmed Basra, S.M., Khan, S.H.U., Mahmood, A. *et al.* (2020) Harvesting and postharvest management approaches for preserving cottonseed quality. *Industrial Crops and Products* 155, 112842. DOI: 10.1016/j.indcrop.2020.112842.

Akpinar, I., Unal, M. and Sar, T. (2021) Potential antifungal effects of silver nanoparticles (AgNPs) of different sizes against phytopathogenic *Fusarium oxysporum* f. sp. radicis-lycopersici (FORL) strains. *SN Applied Sciences* 3(4). DOI: 10.1007/s42452-021-04524-5.

Álvarez, S.P., López, N.E.L., Lozano, J.M., Negrete, E.A.R. and Cervantes, M.E.S. (2016) Plant fungal disease management using nanobiotechnology as a tool. In: Prasad, R. (ed.) *Advances and Applications through Fungal Nanobiotechnology*. Springer, Cham, Switzerland, pp. 169–192. DOI: 10.1007/978-3-319-42990-8.

Ansari, M.J., Kadhim, M.M., Hussein, B.A., Lafta, H.A. and Kianfar, E. (2022) Synthesis and stability of magnetic nanoparticles. *BioNanoScience* 12(2), 627–638. DOI: 10.1007/s12668-022-00947-5.

Antonioli, G., Fontanella, G., Echeverrigaray, S., Longaray Delamare, A.P., Fernandes Pauletti, G. *et al.* (2020) Poly(lactic acid) nanocapsules containing lemongrass essential oil for postharvest decay control: *in vitro* and *in vivo* evaluation against phytopathogenic fungi. *Food Chemistry* 326, 126997. DOI: 10.1016/j.foodchem.2020.126997.

Baalousha, M., How, W., Valsami-Jones, E., and Lead, J.R. (2014 Overview of environmental nanoscience. In: Lead, J.R., and Valsami-Jones, E. (eds.) *Frontiers of Nanoscience*, Vol. 7. Elsevier, Amsterdam, pp. 1–54.

Balasooriya, E.R., Jayasinghe, C.D., Jayawardena, U.A., Ruwanthika, R.W.D., Mendis de Silva, R. *et al.* (2017) Honey mediated green synthesis of nanoparticles: new era of safe nanotechnology. *Journal of Nanomaterials* 2017, 1–10. DOI: 10.1155/2017/5919836.

Bazie, S., Ayalew, A. and Woldetsadik, K. (2014) Integrated management of postharvest banana anthracnose (*Colletotrichum musae*) through plant extracts and hot water treatment. *Crop Protection* 66, 14–18. DOI: 10.1016/j.cropro.2014.08.011.

Bechoff, A., Tomlins, K., Dhuique-Mayer, C., Dove, R. and Westby, A. (2011) On-farm evaluation of the impact of drying and storage on the carotenoid content of orange-fleshed sweet potato (*Ipomea batata* Lam.). *International Journal of Food Science & Technology* 46(1), 52–60. DOI: 10.1111/j.1365-2621.2010.02450.x.

Bhagyaraj, S.M. and Oluwafemi, O.S. (2018) Nanotechnology: the science of the invisible. In: Bhagyaraj, S.M., Oluwafemi, O.S., Kalarikkal, N., and Thomas, S. (eds) *Synthesis of Inorganic Nanomaterials*. Woodhead Publishing, Sawston, UK, pp. 1–18.

Bhusare, S. and Kadam, S. (2021) Applications of nanotechnology in fruits and vegetables. *Food and Agriculture Spectrum Journal* 2, 231–236.

Bikdeloo, M., Ahsanilrvani, M., Roosta, H.R. and Ghanbari, D. (2021) Green synthesis of copper nanoparticles using rosemary extract to reduce postharvest decays caused by *Botrytis cinerea* in tomato. *Journal of Nanostructures* 11, 834–841. DOI: 10.22052/JNS.2021.04.020.

Birnbaum, A.J. and Piqué, A. (2011) Laser induced extraplanar propulsion for three-dimensional microfabrication. *Applied Physics Letters* 98(13), 134101. DOI: 10.1063/1.3567763.

Carmona, L., Alquézar, B., Marques, V.V. and Peña, L. (2017) Anthocyanin biosynthesis and accumulation in blood oranges during postharvest storage at different low temperatures. *Food Chemistry* 237, 7–14. DOI: 10.1016/j.foodchem.2017.05.076.

Casanova, H., Araque, P. and Ortiz, C. (2005) Nicotine carboxylate insecticide emulsions: effect of the fatty acid chain length. *Journal of Agricultural and Food Chemistry* 53(26), 9949–9953. DOI: 10.1021/jf052153h.

Consolo, V.F., Torres-Nicolini, A. and Alvarez, V.A. (2020) Mycosinthetized Ag, CuO and ZnO nanoparticles from a promising *Trichoderma harzianum* strain and their antifungal potential against important phytopathogens. *Scientific Reports* 10(1), 20499. DOI: 10.1038/s41598-020-77294-6.

Cruz-Luna, A.R., Cruz-Martínez, H., Vásquez-López, A. and Medina, D.I. (2021) Metal nanoparticles as novel antifungal agents for sustainable agriculture: current advances and future directions. *Journal of Fungi* 7(12), 1033. DOI: 10.3390/jof7121033.

Cui, D., Tian, F., Ozkan, C.S., Wang, M. and Gao, H. (2005) Effect of single wall carbon nanotubes on human HEK293 cells. *Toxicology Letters* 155(1), 73–85. DOI: 10.1016/j.toxlet.2004.08.015.

Defilippi, B.G., Ejsmentewicz, T., Covarrubias, M.P., Gudenschwager, O. and Campos-Vargas, R. (2018) Changes in cell wall pectins and their relation to postharvest mesocarp softening of "Hass" avocados (*Persea americana* Mill.). *Plant Physiology and Biochemistry* 128, 142–151. DOI: 10.1016/j.plaphy.2018.05.018.

de Venuto, D. and Mezzina, G. (2018) Spatio-temporal optimization of perishable goods' shelf life by a pro-active WSN-based architecture. *Sensors* 18(7), 2126. DOI: 10.3390/s18072126.

Devi, H.S., Boda, M.A., Shah, M.A., Parveen, S. and Wani, A.H. (2019) Green synthesis of iron oxide nanoparticles using *Platanus orientalis* leaf extract for antifungal activity. *Green Processing and Synthesis* 8(1), 38–45. DOI: 10.1515/gps-2017-0145.

de Vitis, M., Hay, F.R., Dickie, J.B., Trivedi, C., Choi, J. *et al.* (2020) Seed storage: maintaining seed viability and vigor for restoration use. *Restoration Ecology* 28(S3), S249–S255. DOI: 10.1111/rec.13174.

El-Naggar, M.E., Abdelsalam, N.R., Fouda, M.M.G., Mackled, M.I., Al-Jaddadi, M.A.M. *et al.* (2020) Soil application of nano silica on maize yield and its insecticidal activity against some stored insects after the post-harvest. *Nanomaterials* 10(4), 739. DOI: 10.3390/nano10040739.

Filipponi, L. and Sutherland, D. (2013) *Nanotechnologies: Principles, Applications, Implications and Hands-On Activities*. Publications Office of the European Union, Luxembourg.

Fleming, M.B., Hill, L.M. and Walters, C. (2019) The kinetics of ageing in dry-stored seeds: a comparison of viability loss and RNA degradation in unique legacy seed collections. *Annals of Botany* 123(7), 1133–1146. DOI: 10.1093/aob/mcy217.

Forney, C.F. (2008) Flavour loss during postharvest handling and marketing of fresh-cut produce. *Stewart Postharvest Review* 4(3), 1–10. DOI: 10.2212/spr.2008.3.5.

Gao, C., Liao, J., Lu, J., Ma, J. and Kianfar, E. (2021) The effect of nanoparticles on gas permeability with polyimide membranes and network hybrid membranes: a review. *Reviews in Inorganic Chemistry* 41(1), 1–20. DOI: 10.1515/revic-2020-0007.

Gogo, E.O., Opiyo, A.M., Ulrichs, C. and Huyskens-Keil, S. (2017) Nutritional and economic postharvest loss analysis of African indigenous leafy vegetables along the supply chain in Kenya. *Postharvest Biology and Technology* 130, 39–47. DOI: 10.1016/j.postharvbio.2017.04.007.

González, A.L., Noguez, C., Beránek, J. and Barnard, A.S. (2014) Size, shape, stability, and color of plasmonic silver nanoparticles. *The Journal of Physical Chemistry C* 118(17), 9128–9136. DOI: 10.1021/jp5018168.

González-Estrada, R.R., Chalier, P., Ragazzo-Sánchez, J.A., Konuk, D. and Calderón-Santoyo, M. (2017) Antimicrobial soy protein based coatings: application to Persian lime (*Citrus latifolia* Tanaka) for protection and preservation. *Postharvest Biology and Technology* 132, 138–144. DOI: 10.1016/j.postharvbio.2017.06.005.

Häkkinen, H., Abbet, S., Sanchez, A., Heiz, U. and Landman, U. (2003) Structural, electronic, and impurity-doping effects in nanoscale chemistry: supported gold nanoclusters. *Angewandte Chemie International Edition* 42(11), 1297–1300. DOI: 10.1002/anie.200390334.

Han, J.W., Ruiz-Garcia, L., Qian, J.P. and Yang, X.T. (2018) Food packaging: a comprehensive review and future trends. *Comprehensive Reviews in Food Science and Food Safety* 17(4), 860–877. DOI: 10.1111/1541-4337.12343.

Hashim, A.F., Youssef, K. and Abd-Elsalam, K.A. (2019) Ecofriendly nanomaterials for controlling gray mold of table grapes and maintaining postharvest quality. *European Journal of Plant Pathology* 154(2), 377–388. DOI: 10.1007/s10658-018-01662-2.

Hernández-Díaz, J.A., Garza-García, J.J., Zamudio-Ojeda, A., León-Morales, J.M., López-Velázquez, J.C. *et al.* (2021) Plant-mediated synthesis of nanoparticles and their antimicrobial activity against phytopathogens. *Journal of the Science of Food and Agriculture* 101(4), 1270–1287. DOI: 10.1002/jsfa.10767.

Huang, Y., Mei, L., Chen, X. and Wang, Q. (2018) Recent developments in food packaging based on nanomaterials. *Nanomaterials* 8(10), 830. DOI: 10.3390/nano8100830.

Huerta-García, E., Pérez-Arizti, J.A., Márquez-Ramírez, S.G., Delgado-Buenrostro, N.L., Chirino, Y.I. *et al.* (2014) Titanium dioxide nanoparticles induce strong oxidative stress and mito-chondrial damage in glial cells. *Free Radical Biology & Medicine* 73, 84–94. DOI: 10.1016/j.freeradbiomed.2014.04.026.

Hulla, J.E., Sahu, S.C. and Hayes, A.W. (2015) Nanotechnology: history and future. *Human & Experimental Toxicology* 34(12), 1318–1321. DOI: 10.1177/0960327115603588.

Huynh, N.K., Wilson, M.D., Eyles, A. and Stanley, R.A. (2019) Recent advances in postharvest technologies to extend the shelf life of blueberries (*Vaccinium* sp.), raspberries (*Rubus idaeus* L.) and blackberries (*Rubus* sp.). *Journal of Berry Research* 9(4), 687–707. DOI: 10.3233/JBR-190421.

Il'ina, A.V., Shagdarova, B.T. and Varlamov, V.P. (2022) Prospects for the use of metal nanoparticles and chitosan nanomaterials with metals to combat phytopathogens. *Applied Biochemistry and Microbiology* 58(2), 105–109. DOI: 10.1134/S0003683822020090.

Jahan, I., Erci, F. and Isildak, I. (2021) Facile microwave-mediated green synthesis of non-toxic copper nanoparticles using *Citrus sinensis* aqueous fruit extract and their antibacterial potentials. *Journal of Drug Delivery Science and Technology* 61, 102172. DOI: 10.1016/j.jddst.2020.102172.

Jamdagni, P., Khatri, P. and Rana, J.S. (2018) Green synthesis of zinc oxide nanoparticles using flower extract of *Nyctanthes arbor-tristis* and their antifungal activity. *Journal of King Saud University - Science* 30(2), 168–175. DOI: 10.1016/j.jksus.2016.10.002.

Jamuna, B.A. and Ravishankar, R.V. (2014) Environmental risk, human health, and toxic effects of nanoparticles. In: Kharisov, B.I., Kharissova, O.V., and Rasika Dias, H.V. (eds) *Nanomaterials for Environmental Protection*. Wiley, Hoboken, NJ, pp. 523–535.

Javanmardi, J. and Kubota, C. (2006) Variation of lycopene, antioxidant activity, total soluble solids and weight loss of tomato during postharvest storage. *Postharvest Biology and Technology* 41(2), 151–155. DOI: 10.1016/j.postharvbio.2006.03.008.

Jeevanandam, J., Chan, Y.S. and Danquah, M.K. (2016) Biosynthesis of metal and metal oxide nanoparticles. *ChemBioEng Reviews* 3(2), 55–67. DOI: 10.1002/cben.201500018.

Johnson, M.S., Sajeev, S. and Nair, R.S. (2021) Role of Nanosensors in agriculture. In: *2021 International Conference on Computational Intelligence and Knowledge Economy (ICCIKE)*, IEEE, Dubai, United Arab Emirates, pp. 58–63. DOI: 10.1109/ICCIKE51210.2021.9410709.

Kader, A.A. (2003) A perspective on postharvest horticulture (1978-2003). *HortScience* 38(5), 1004–1008. DOI: 10.21273/HORTSCI.38.5.1004.

Kakakhel, M.A., Wu, F., Feng, H., Hassan, Z., Ali, I. *et al.* (2021) Biological synthesis of silver nanoparticles using animal blood, their preventive efficiency of bacterial species, and ecotoxicity in common carp fish. *Microscopy Research and Technique* 84(8), 1765–1774. DOI: 10.1002/jemt.23733.

Karthik, A.D. and Geetha, K. (2011) Synthesis of copper precursor, copper and its oxide nanoparticles by green chemical reduction method and its antimicrobial activity. *Journal of Applied Pharmaceutical Science* 3, 16–21. DOI: 10.7324/JAPS.2013.3504.

Keiper, A. (2003) The nanotechnology revolution. *The New Atlantis* 17–34.

Ketsa, S., Wisutiamonkul, A. and van Doorn, W.G. (2013) Apparent synergism between the positive effects of 1-MCP and modified atmosphere on storage life of banana fruit. *Postharvest Biology and Technology* 85, 173–178. DOI: 10.1016/j.postharvbio.2013.05.009.

Khatami, M., Nejad, M.S., Salari, S. and Almani, P.G.N. (2016) Plant-mediated green synthesis of silver nanoparticles using *Trifolium resupinatum* seed exudate and their antifungal efficacy on *Neofusicoccum parvum* and *Rhizoctonia solani*. *IET Nanobiotechnology* 10(4), 237–243. DOI: 10.1049/iet-nbt.2015.0078.

Kianfar, E. and Cao, V. (2021) Polymeric membranes on base of polymethyl methacrylate for air separation: a review. *Journal of Materials Research and Technology* 10, 1437–1461. DOI: 10.1016/j.jmrt.2020.12.061.

Kim, S.W., Jung, J.H., Lamsal, K., Kim, Y.S., Min, J.S. *et al.* (2012) Antifungal effects of silver nanoparticles (AgNPs) against various plant pathogenic fungi. *Mycobiology* 40(1), 53–58. DOI: 10.5941/MYCO.2012.40.1.053.

Kitinoja, L. and Gorny, J.R. (1999) *Postharvest Technology for Small-Scale Produce Marketers: Economic Opportunities, Quality and Food Safety*. Postharvest Horticultural Series. Department of Pomology, University of California, Davis, CA.

Koduru, J.R., Kailasa, S.K., Bhamore, J.R., Kim, K.-H., Dutta, T. *et al.* (2018) Phytochemical-assisted synthetic approaches for silver nanoparticles antimicrobial applications: a review. *Advances in Colloid and Interface Science* 256, 326–339. DOI: 10.1016/j.cis.2018.03.001.

Kong, L., Wu, Y., Li, C., Liu, J., Jia, J. *et al.* (2021) Nano-cell and nano-pollutant interactions constitute key elements in nanoparticle-pollutant combined cytotoxicity. *Journal of Hazardous Materials* 418, 126259. DOI: 10.1016/j.jhazmat.2021.126259.

Kuswandi, B. (2017) Environmental friendly food nano-packaging. *Environmental Chemistry Letters* 15(2), 205–221. DOI: 10.1007/s10311-017-0613-7.

Li, D., Zhang, X., Li, L., Aghdam, M.S., Wei, X. *et al.* (2019) Elevated CO_2 delayed the chlorophyll degradation and anthocyanin accumulation in postharvest strawberry fruit. *Food Chemistry* 285, 163–170. DOI: 10.1016/j.foodchem.2019.01.150.

Lyimo, M.H., Nyagwegwe, S. and Mnkeni, A.P. (1991) Investigations on the effect of traditional food processing, preservation and storage methods on vegetable nutrients: a case study in Tanzania. *Plant Foods for Human Nutrition* 41, 53–57. DOI: 10.1007/BF02196382.

Magdolenova, Z., Collins, A., Kumar, A., Dhawan, A., Stone, V. *et al.* (2014) Mechanisms of genotoxicity. A review of *in vitro* and *in vivo* studies with engineered nanoparticles. *Nanotoxicology* 8(3), 233–278. DOI: 10.3109/17435390.2013.773464.

Makule, E., Dimoso, N. and Tassou, S.A. (2022) Precooling and cold storage methods for fruits and vegetables in sub-Saharan Africa—a review. *Horticulturae* 8(9), 776. DOI: 10.3390/horticulturae8090776.

Maneerat, C. and Hayata, Y. (2006) Antifungal activity of TiO_2 photocatalysis against *Penicillium expansum in vitro* and in fruit tests. *International Journal of Food Microbiology* 107(2), 99–103. DOI: 10.1016/j.ijfoodmicro.2005.08.018.

Marçal, S., Sousa, A.S., Taofiq, O., Antunes, F., Morais, A. *et al.* (2021) Impact of postharvest preservation methods on nutritional value and bioactive properties of mushrooms. *Trends in Food Science & Technology* 110, 418–431. DOI: 10.1016/j.tifs.2021.02.007.

Mateo-Sagasta, J., Zadeh, S.M. and Turral, H. (eds) (2018) *More People, More Food… Worse Water?— Water Pollution From Agriculture: A Global Review*. UN Food and Agriculture Organization, Rome. Available at: https://www.fao.org/documents/card/en/c/CA0146EN (accessed 30 March 2023).

Melo, N.F.C.B., Pintado, M.M.E., Medeiros, J.A. da C., Galembeck, A., Vasconcelos, M.A. da S. *et al.* (2020) Quality of postharvest strawberries: comparative effect of fungal chitosan gel, nanoparticles and gel enriched with edible nanoparticles coatings. *International Journal of Food Studies* 9(2), 373–393. DOI: 10.7455/ijfs/9.2.2020.a9.

Mikhailova, E.O. (2020) Silver nanoparticles: mechanism of action and probable bio-application. *Journal of Functional Biomaterials* 11(4), 84. DOI: 10.3390/jfb11040084.

Mohammadi, A., Hashemi, M. and Hosseini, S.M. (2016) Effect of chitosan molecular weight as micro and nanoparticles on antibacterial activity against some soft rot pathogenic bacteria. *LWT - Food Science and Technology* 71, 347–355. DOI: 10.1016/j.lwt.2016.04.010.

Mohammadpour, R., Yazdimamaghani, M., Cheney, D.L., Jedrzkiewicz, J. and Ghandehari, H. (2019) Subchronic toxicity of silica nanoparticles as a function of size and porosity. *Journal of Controlled Release* 304, 216–232. DOI: 10.1016/j.jconrel.2019.04.041.

Mori, S. and Shitara, Y. (1994) Tribochemical activation of gold surface by scratching. *Applied Surface Science* 78(3), 269–273. DOI: 10.1016/0169-4332(94)90014-0.

Mosha, T.C., Pace, R.D., Adeyeye, S., Laswai, H.S. and Mtebe, K. (1997) Effect of traditional processing practices on the content of total carotenoid, beta-carotene, alpha-carotene and vitamin A activity of selected Tanzanian vegetables. *Plant Foods for Human Nutrition* 50(3), 189–201. DOI: 10.1007/BF02436056.

Murmu, S.B. and Mishra, H.N. (2018) Post-harvest shelf-life of banana and guava: mechanisms of common degradation problems and emerging counteracting strategies. *Innovative Food Science & Emerging Technologies* 49, 20–30. DOI: 10.1016/j.ifset.2018.07.011.

Neme, K., Nafady, A., Uddin, S. and Tola, Y.B. (2021) Application of nanotechnology in agriculture, post-harvest loss reduction and food processing: food security implication and challenges. *Heliyon* 7(12), e08539. DOI: 10.1016/j.heliyon.2021.e08539.

Omran, B.A. (2020) *Nanobiotechnology: A Multidisciplinary Field of Science*. Springer International, New York. DOI: 10.1007/978-3-030-46071-6.

Paltrinieri, G. (2014) Handling of fresh fruits, vegetables and root crops: a training manual for Grenada. UN Food and Agriculture Organization. Available at: https://www.fao.org/publications/card/en/c/fe7a6d7c-6b42-4abe-b51a-bdebdc606907/ (accessed 30 March 2023).

Perinelli, D.R., Fagioli, L., Campana, R., Lam, J.K.W., Baffone, W. *et al.* (2018) Chitosan-based nanosystems and their exploited antimicrobial activity. *European Journal of Pharmaceutical Sciences* 117, 8–20. DOI: 10.1016/j.ejps.2018.01.046.

Pineda, M.E.B., Lizarazo Forero, L.M. and Sierra Avila, C.A. (2022) Antibacterial activity of biosynthesized silver nanoparticles (AgNps) against *Pectobacterium carotovorum*. *Brazilian Journal of Microbiology* 53(3), 1175–1186. DOI: 10.1007/s42770-022-00757-7.

Pott, D.M., Osorio, S. and Vallarino, J.G. (2019) From central to specialized metabolism: an overview of some secondary compounds derived from the primary metabolism for their role in conferring nutritional and organoleptic characteristics to fruit. *Frontiers in Plant Science* 10, 835. DOI: 10.3389/fpls.2019.00835.

Pott, D.M., Vallarino, J.G. and Osorio, S. (2020) Metabolite changes during postharvest storage: effects on fruit quality traits. *Metabolites* 10(5), 187. DOI: 10.3390/metabo10050187.

Rajiv, P., Rajeshwari, S. and Venckatesh, R. (2013) Bio-Fabrication of zinc oxide nanoparticles using leaf extract of *Parthenium hysterophorus* L. and its size-dependent antifungal activity against plant fungal pathogens. *Spectrochimica Acta Part A: Molecular and Biomolecular Spectroscopy* 112, 384–387. DOI: 10.1016/j.saa.2013.04.072.

Rana, A., Yadav, K. and Jagadevan, S. (2020) A comprehensive review on green synthesis of nature-inspired metal nanoparticles: mechanism, application and toxicity. *Journal of Cleaner Production* 272, 122880. DOI: 10.1016/j.jclepro.2020.122880.

Rastogi, A., Singh, P., Haraz, F.A. and Barhoum, A. (2018) Biological synthesis of nanoparticles: an environmentally benign approach. In: Barhoum, A. (ed.) *Fundamentals of Nanoparticles*. Elsevier, Amsterdam, pp. 571–604. DOI: 10.1016/B978-0-323-51255-8.00023-9.

Ravinayagam, V. and Jermy, B.R. (2020) Nanomaterials and their negative effects on human health. In: Khan, F.A. (ed.) *Applications of Nanomaterials in Human Health*. Springer, New York, pp. 249–273.

Reibold, M., Paufler, P., Levin, A.A., Kochmann, W., Pätzke, N. *et al.* (2006) Materials: carbon nanotubes in an ancient Damascus sabre. *Nature* 444(7117), 286–286. DOI: 10.1038/444286a.

Riaz, A., Aadil, R.M., Amoussa, A.M.O., Bashari, M., Abid, M. *et al.* (2021) Application of chitosan-based apple peel polyphenols edible coating on the preservation of strawberry (*Fragaria ananassa* cv Hongyan) fruit. *Journal of Food Processing and Preservation* 45(1). DOI: 10.1111/jfpp.15018.

Roberts, E.H. (1973) Predicting the storage life of seeds. *Seed Science and Technology* 1, 499–514.

Sadek, M.E., Shabana, Y.M., Sayed-Ahmed, K. and Abou Tabl, A.H. (2022) Antifungal activities of sulfur and copper nanoparticles against cucumber postharvest diseases caused by *Botrytis cinerea* and *Sclerotinia sclerotiorum*. *Journal of Fungi* 8(4), 412. DOI: 10.3390/jof8040412.

Sharkey, P.J. and Peggie, I.D. (1984) Effects of high-humidity storage on quality, decay and storage life of cherry, lemon and peach fruits. *Scientia Horticulturae* 23(2), 181–190. DOI: 10.1016/0304-4238(84)90022-0.

Sharma, C., Dhiman, R., Rokana, N. and Panwar, H. (2017) Nanotechnology: an untapped resource for food packaging. *Frontiers in Microbiology* 8, 1735. DOI: 10.3389/fmicb.2017.01735.

Shende, S., Ingle, A.P., Gade, A. and Rai, M. (2015) Green synthesis of copper nanoparticles by *Citrus medica* Linn. (Idilimbu) juice and its antimicrobial activity. *World Journal of Microbiology & Biotechnology* 31(6), 865–873. DOI: 10.1007/s11274-015-1840-3.

Shende, S., Gaikwad, N. and Bansod, S. (2016) Synthesis and evaluation of antimicrobial potential of copper nanoparticle against agriculturally important phytopathogens. *Synthesis* 1, 41–47.

Shousha, W.G., Aboulthana, W.M., Salama, A.H., Saleh, M.H. and Essawy, E.A. (2019) Evaluation of the biological activity of *Moringa oleifera* leaves extract after incorporating silver nanoparticles, *in vitro* study. *Bulletin of the National Research Centre* 43(1), 212. DOI: 10.1186/s42269-019-0221-8.

Sinha, S.R., Singha, A., Faruquee, M., Jiku, M., Sayem, A. *et al.* (2019) Post-harvest assessment of fruit quality and shelf life of two elite tomato varieties cultivated in Bangladesh. *Bulletin of the National Research Centre* 43(1), 185. DOI: 10.1186/s42269-019-0232-5.

Taghavi, T., Kim, C. and Rahemi, A. (2018) Role of natural volatiles and essential oils in extending shelf life and controlling postharvest microorganisms of small fruits. *Microorganisms* 6(4), 104. DOI: 10.3390/microorganisms6040104.

Tekpor, S.K. (2011) Effect of some processing and storage methods on the quality of okra (*Abelmoschus esculentus*) fruits. PhD thesis, Kwame Nkrumah University of Science and Technology, Kumasi, Ghana.

Thanighaiarassu, R.R., Sivamai, P., Devika, R. and Nambikkairaj, B. (2014) Green synthesis of gold nanoparticles characterization by using plant essential oil *Menthapiperita* and their antifungal activity against human pathogenic fungi. *Journal of Nanomedicine and Nanotechnology* 5, 5. DOI: 10.4172/2157-7439.1000229.

Tiwari, P., Srivastava, M., Mishra, R., Ji, G. and Prakash, R. (2018) Economic use of waste *Musa paradisica* peels for effective control of mild steel loss in aggressive acid solutions. *Journal of Environmental Chemical Engineering* 6(4), 4773–4783. DOI: 10.1016/j.jece.2018.07.016.

Tolochko, N.K. (2009) History of nanotechnology. In: Kharkin, V., Bai, C., Awadelkarim, O.O, and Kapitsa, S. (eds) *Nanoscience and Nanotechnology*. UNESCO, Oxford, UK, pp. 1–18.

Tomlinson, D.L. (1988) A leaf and fruit disease of *Pandanus conoideus* caused by *Erwinia carotovora* subsp. *carotovora* in Papua New Guinea. *Journal of Phytopathology* 121(1), 19–25. DOI: 10.1111/j.1439-0434.1988.tb00948.x.

Torres-Sánchez, R., Martínez-Zafra, M.T., Castillejo, N., Guillamón-Frutos, A. and Artés-Hernández, F. (2020) Real-time monitoring system for shelf life estimation of fruit and vegetables. *Sensors* 20(7), 1860. DOI: 10.3390/s20071860.

Tsivileva, O.M., Perfileva, A.I., Ivanova, A.A., Pozdnyakov, A.S. and Prozorova, G.F. (2021) The effect of selenium- or metal-nanoparticles incorporated nanocomposites of vinyl triazole based polymers on fungal growth and bactericidal properties. *Journal of Polymers and the Environment* 29(4), 1287–1297. DOI: 10.1007/s10924-020-01963-w.

UN Food and Agriculture Organisation (FAO) (2013) *Food Wastage Footprint: Impacts on Natural Resources: Summary Report*. FAO, Rome. Available at: www.fao.org/3/i3347e/i3347e.pdf (accessed 30 March 2013).

van Hung, D.V., Tong, S., Tanaka, F., Yasunaga, E., Hamanaka, D. *et al.* (2011) Controlling the weight loss of fresh produce during postharvest storage under a nano-size mist environment. *Journal of Food Engineering* 106(4), 325–330. DOI: 10.1016/j.jfoodeng.2011.05.027.

Voilley, A. and Souchon, I. (2006) Flavour retention and release from the food matrix: an overview. In: Voilley, A. and Etievent, P. (eds) *Flavour in Food*. Woodhead Publishing, Cambridge, pp. 117–132.

Wagner, R., Moon, R., Pratt, J., Shaw, G. and Raman, A. (2011) Uncertainty quantification in nano mechanical measurements using the atomic force microscope. *Nanotechnology* 22(45), 455703. DOI: 10.1088/0957-4484/22/45/455703.

Walters, C., Ballesteros, D. and Vertucci, V.A. (2010) Structural mechanics of seed deterioration: standing the test of time. *Plant Science* 179(6), 565–573. DOI: 10.1016/j.plantsci.2010.06.016.

Wan, J., Lacey, S.D., Dai, J., Bao, W., Fuhrer, M.S. *et al.* (2016) Tuning two-dimensional nanomaterials by intercalation: materials, properties and applications. *Chemical Society Reviews* 45(24), 6742–6765. DOI: 10.1039/c5cs00758e.

Wang, L., Li, X., Zhang, G., Dong, J. and Eastoe, J. (2007) Oil-in-water nanoemulsions for pesticide formulations. *Journal of Colloid and Interface Science* 314(1), 230–235. DOI: 10.1016/j.jcis.2007.04.079.

Wang, Y.C., Lu, L. and Gunasekaran, S. (2017) Biopolymer/gold nanoparticles composite plasmonic thermal history indicator to monitor quality and safety of perishable bioproducts. *Biosensors & Bioelectronics* 92, 109–116. DOI: 10.1016/j.bios.2017.01.047.

Wang, X.M., Wu, X.W., Zhao, X.Y., Wang, C.W. and Zhou, J.N. (2021) Exposure-time-dependent subcellular staging of gold nanoparticles deposition and vesicle destruction in mice livers. *Nanomedicine: Nanotechnology, Biology and Medicine* 4, 102393. DOI: 10.1016/j.nano.2021.102393.

Wyrwa, J. and Barska, A. (2017) Innovations in the food packaging market: active packaging. *European Food Research and Technology* 243(10), 1681–1692. DOI: 10.1007/s00217-017-2878-2.

Yahia, E.M. and Carrillo-Lopez, A. (2018) *Postharvest Physiology and Biochemistry of Fruits and Vegetables*. Woodhead Publishing, Cambridge.

Yahia, E.M., Fonseca, J.M., and Kitinoja, L. (2019) Postharvest losses and waste. In: Yahia, E.M. (ed.) *Postharvest Technology of Perishable Horticultural Commodities*. Woodhead Publishing, Cambridge, pp. 43–69.

Yan, W., Fu, X., Gao, Y., Shi, L., Liu, Q. *et al.* (2022) Synthesis, antibacterial evaluation, and safety assessment of CuS NPs against *Pectobacterium carotovorum* subsp. *carotovorum*. *Pest Management Science* 78(2), 733–742. DOI: 10.1002/ps.6686.

Zhang, H., Yang, S., Joyce, D.C., Jiang, Y., Qu, H. *et al.* (2010) Physiology and quality response of harvested banana fruit to cold shock. *Postharvest Biology and Technology* 55(3), 154–159. DOI: 10.1016/j.postharvbio.2009.11.006.

Zhang, C., Xiong, Z., Yang, H. and Wu, W. (2019) Changes in pericarp morphology, physiology and cell wall composition account for flesh firmness during the ripening of blackberry (*Rubus* spp.) fruit. *Scientia Horticulturae* 250, 59–68. DOI: 10.1016/j.scienta.2019.02.015.

Zhao, X., Chang, S., Long, J., Li, J., Li, X. *et al.* (2019) The toxicity of multi-walled carbon nanotubes (MWCNTs) to human endothelial cells: the influence of diameters of MWCNTs. *Food and Chemical Toxicology* 126, 169–177. DOI: 10.1016/j.fct.2019.02.026.

9 Nanotechnology for Precision Farming and Smart Delivery Systems

Gayanath Thiranagama[1], Dilki Jayathilaka[1], Chanaka Sandaruwan[1]* and Nilwala Kottegoda[2]

[1]Sri Lanka Institute of Nanotechnology, Mahenwatta, Pitipana, Homagama, Sri Lanka; [2]Department of Chemistry/Center for Advanced Materials Research, Faculty of Applied Sciences, University of Sri Jayewardenepura, Sri Lanka, Sri Soratha Mawatha, Nugegoda

Abstract

Nanotechnology enables humankind to pave the path to the future due to its widespread applicability to different areas. Nanotechnology-based agricultural applications have the potential to sustain the demanding agricultural requirement to feed the rapidly growing population in the world. Most of these nanotechnology-based applications focus on enhancing staple crop yield while reducing agrochemical usage and postharvest losses. Remarkably, most of these nanotechnology interventions are ultimately driven using smart delivery systems. For example, slow- or controlled-release fertilizers are smart delivery systems that can be a top candidate for reducing fertilizer applications. Nanostructures such as nanoparticles, nanotubes, and layered nanomaterials have been used to manipulate the delivery of nutrients intelligently to the soil. Here, nutrients are retained or trapped inside the delivery system using the exceptional properties of these nanomaterials due to the larger surface area:volume ratio at the nanoscale. For example, urea molecules can be incorporated into layered nanostructures to obtain prolonged and smart release of nutrients. These nanotechnology-enabled smart delivery systems will eventually lead to increased precision farming practices worldwide, and will undoubtedly enhance the yield of crops, preserve the environment by reducing the use of agrochemicals, and raise the quality of life.

Keywords: Precision farming, smart delivery systems, nanotechnology, nanostructures

9.1 Introduction

Agricultural communities are dealing with a wide range of issues, including stagnant crop yields, poor nutrient usage efficiency, falling soil organic matter, multi-nutrient deficits, limited arable lands, reduced water availability, and a human resources crisis resulting from a farming exodus. According to the data gathered by the United Nations Food and Agriculture Organization (FAO), depletion of land and water resources will lead to complications in providing enough food and other agricultural products to sustain livelihoods and fulfil the demand of the world's ever-growing population (Godfray *et al.*, 2010). In 2012, the world population was 7.3 billion people, but according to UN estimates, this number will rise by 1.2% per year, reaching 8.5 billion in 2030 and over 10 billion in 2050 (Madre and Devuyst, 2012). Referring

---

*Corresponding author: chanakas@slintec.lk

© CAB International 2023. *Nanoformulations for Sustainable Agriculture and Environmental Risk Mitigation* (eds Z. Khan and N.A. Ṣuṭan)
DOI: 10.1079/9781800623095.0009

to the most recent United Nations predictions published, there will be 7.98 billion people living on Earth as of October 2022. Consequently, there is a significant chance that their prediction will come true, and continuing growth of the human population will drive up food consumption. The widespread usage of fertilizers, pesticides, and other agricultural inputs is a result of the increased worldwide population's impact on the demand for food.

The pesticide industry is clearly expanding; its global market was around US$32 billion in 2007, (Sharma *et al.*, 2017), US$56 billion in 2012 according to the US Environmental Protection Agency and is predicted to approach US$100 billion between 2025 and 2026 according to analysis by the Freedonia Group in 2021. When it comes to the fertilizer industry, the trend is very similar to what we are seeing in the pesticide industry. By 2028, the size of the global fertilizer market is expected to have increased to US$223 billion from its current value of US$186 billion, according to a market report in 2022.

Due to limited resources and low fertilizer-use efficiency, the expenses of farmers are rising by billions of dollars each year. Therefore, if the prevailing farming system, which is currently seen as unsustainable, continues to develop at the same rate, it may eventually become intolerable. According to Madre and Devuyst (2012), to meet the increases in global demand, the supply of food would need to increase by approximately 30% by 2030 and by almost 50% by 2050. Thus, to feed the world's population, significantly larger quantities of basic agricultural products will need to be produced, and our capacity to sustain or raise the existing levels of agricultural output will be crucial in meeting the future food demands. In simple terms, productivity increases occur when fewer agricultural inputs are required to create a given quantity of agricultural output. The agricultural total factor productivity (TFP), which measures the relationship between agricultural output (crops and animals) and input, is a popular measure of agricultural productivity. A rise in TFP indicates that more agricultural products can be generated using the same amount of farming input. The FAO estimates that the world's food supply must increase by 50% by 2050, which implies a yearly TFP increase of 1.73% over the following

27 years, assuming other factors remain constant. However, considering the yearly TFP growth rates over the past few decades, this appears to be a difficult challenge.

Nanotechnology has answers to many of these challenges. In contrast to its bulk counterparts, nanotechnology allows the creation of extremely small particles with unique qualities. These include a higher surface area:volume size ratio and physico-chemical capabilities (National Science and Technology Council Committee on Technology, 2014). By fusing science and engineering, nanotechnology opens up new multidisciplinary opportunities in agricultural and food sciences (Rai *et al.*, 2015).

Nanotechnology allows us to create better systems for distributing nutrients and pesticides when necessary, with the potential to increase yields and nutritional value. This concept is part of the developing field of precision agriculture, which involves using technology by farmers to use inputs such as water, fertilizer, and other factors more effectively (Raliya *et al.*, 2018). However, any attempt to rationalize the use of agricultural inputs must avoid creating new economic or environmental issues. As a result, it is necessary to incorporate or create new compounds that, once used, can quickly biodegrade or be assimilated into the soil (Pereira, 2015).

9.1.1   Aspects of precision farming

Precision farming lowers waste through the focused use of inputs, which lowers either financial costs or environmental implications associated with agrochemical residues. Large farms in industrialized nations are currently the largest beneficiaries of precision farming. With the help of precision farming, agricultural management choices may be adjusted in both time and space. However, the potential environmental advantages might support increased governmental and private sector incentives to promote adoption, even within small-scale agricultural systems in underdeveloped nations.

As a result of the increasing technology, precision farming technologies continue to become more connected, precise, effective, and broadly applicable. They focus on techniques to improve the quality and quantity of crops overall by using certain crops and types of soil. They also focus on preserving soil and environmental quality and the effectiveness of input utilization. There are three concepts governing precision agriculture: (i) economic stability; (ii) profitability; and (iii) fewer negative environmental effects (Allahyari *et al.*, 2016).

To achieve these aspects of precision farming, two main areas need to be considered. Initially, the farmer needs to identify whether plants are under fertilizer stress or any other disease condition, and then has to implement the required solutions. When it comes to the initial stage of problem identification, many precision farming applications depend heavily on georeferencing technology, such as global positioning systems (GPS). With the use of such technologies, farmers can now use drones in fields along with various types of biosensors, and thus have the freedom to easily work with complicated matrices (Yu *et al.*, 2010) without requiring complex treatment methods (Dong *et al.*, 2013), and to conduct quick analyses (Shi *et al.*, 2013), enabling a number of different measurements (Bali *et al.*, 2014). These advancements enable control of traffic and the use of guidance systems during the field application of fertilizer, seeds, and pesticides, helping to shorten the time and laborious work involved in walking or travelling (Weersink *et al.*, 2018).

Smart delivery systems are vital when it comes to field application. They help farmers to use fertilizers and pesticides effectively once the problem has been identified. Therefore, farmers can get the maximum output from their additives and achieve higher profits while having a reduced environmental impact. This chapter discusses how nanotechnology works in smart delivery systems to achieve these aspects of precision farming and the applications of smart delivery systems.

9.2 Zero-, One-, Two-, and Three-Dimensional Nanostructures and Their Applications in Smart Delivery

Nanomaterials (NMs) can be divided into four categories according to their dimensions: zero-dimensional structures, one-dimensional structures, two-dimensional structures, and three-dimensional structures (Kim *et al.*, 2010).

A structure that has all three dimensions at the nanoscale is said to be zero-dimensional. Most zero-dimensional NPs are spherical or nearly spherical NPs, compared with bulk high-dimensional NPs (Facure *et al.*, 2020). One-dimensional NPs can be described as particles that have at least one dimension falling within the size range of 1–100 nm. These are thin films or monolayers (Ranjan *et al.*, 2016). Two-dimensional nanostructures consist of two dimensions outside the nanometric size range, and therefore they differ from bulk material qualities due to their low dimensional characters (Ranjan *et al.*, 2016). Three-dimensional NMs are those whose dimensions are not confined to the nanoscale (Ranjan *et al.*, 2016). Examples of each of these dimensions are listed in Table 9.1.

When it comes to agriculture, each of these dimensions has been used in various types of smart delivery applications, such as improving nitrogen-use efficiency (NUE) and reducing the toxicity of pesticides. As nitrogen (N) is considered the most important component governing the growth of plants, NUE provides a measure of how the yield responds to fertilizer applications and is the utmost important factor when it comes to modern agriculture. There is also a rapidly developing market for agricultural products that have fewer pesticides. We will discuss how these nanostructures used in agricultural applications work to achieve their targets.

9.3 Role of Smart Delivery Systems in Precision Farming

In modern agriculture, farmers use pesticides in high quantities, but due to the bioaccumulative nature of pesticides, global research activities on the use of these compounds in the field usually aims to lower the amounts, without leading to financial losses and lower productivity (Lechenet *et al.*, 2017). In order to ensure a more secure and sustainable agricultural production, the usage of highly efficient pesticides has become the main focus. Because commercial pesticide formulations need to be applied in large quantities and at high frequencies to be effective against pests, this also leads to their environmental accumulation.

Table 9.1. Classification of nanomaterials based on dimensionality.

Dimensions	Examples	Applications	Reference
Zero-dimensional	Quantum dots	Graphene quantum dots have been used as nanofertilizers	Chakravarty *et al.* (2015)
	Nanoclusters	Copper nanoclusters have been reported as copper fertilizers	Wang *et al.* (2021)
	Nanoparticles	Lignin nanocapsules have been used to deliver bioactive compounds to plantlets	Rastogi *et al.* (2019)
One-dimensional	Nanotubes, nanofibers	Carbon nanotubes are considered a slow-release fertilizer and a plant growth booster; poly(butylene adipate-co-terephthalate) nanofibers have been tested as a zinc slow-release system	Natarelli *et al.* (2021); Safdar *et al.* (2022)
Two-dimensional	Nanolayers	Layered characteristics of montmorillonite used for the preparation of slow-release nanocomposites	Azarian *et al.* (2018)
Three-dimensional (monophase)	Dendrimers	Poly(ether hydroxylamine) dendrimers are claimed to improve the active solubility of agrochemicals	Chauhan *et al.* (2020)
	Nanocapsules	Silicon nanoparticles have been used as nanopesticides, nanoherbicides, and nanofertilizers	Falsini *et al.* (2019)
Three-dimensional (nonmonophase)	Core–shells	Core–shell nanoparticles have been used to deliver genetic material and agrochemicals	Dhiman *et al.* (2021)
	Nanocrystalline materials	Cellulose nanocrystals have been used as a waterborne coating of NPK fertilizers to enhance their efficiency	Kassem *et al.* (2021)

NPK, Nitrogen, Phosphorus, Potassium.

However, pesticides used on crops are thought to kill only 0.1% of the intended pests, with the remaining 99.9% building up in the environment and posing major environmental problems (de Oliveira *et al.*, 2020).

When it comes to fertilizers, there are several issues similar to those in the pesticide sector. Fertilizers are usually applied to the soil by broadcasting, subsurface application, or mixed with irrigation water (Nkebiwe *et al.*, 2016). However, plants can only absorb a limited amount of fertilizer (Dawson and Hilton, 2011). This has led to extreme environmental pollution, such as deterioration of downstream water quality, eutrophication of coastal marine ecosystems, development of photochemical smog, and rising global concentrations of greenhouse gases, caused by large amounts of external fertilizer input with decreasing use efficiency. An additional risk is the decreasing availability of some chemical fertilizers within the next 80 years (Tarafdar and Raliya, 2013).

This is where the smart delivery system comes in. Using this system, we can overcome the problems that arise from conventional fertilizers and pesticides. Use of NMs means that the active components of pesticides can be delivered to their respective targets more efficiently and effectively. These pesticide support structures are referred to as nanopesticides, as they serve as nanodelivery systems. Active components encapsulated within the nanoparticle (NP) support move by diffusion from the nucleus of the NM to the shell throughout this process, or sometimes the NP itself can work as a pesticide. Once the NP comes into contact with the environment, it can target local pests by optimizing the targeted distribution (Ni *et al.*, 2011; Irfan *et al.*, 2018). These nanostructure systems have the potential to retain the chemical stability of the pesticide even under the influence of humidity, oxidation, and other environmental factors, along with permitting their release into the outer environment in a regulated, continuous, and protracted way (Irfan *et al.*, 2018).

When it comes to fertilizers, applying nanofertilizers that take advantage of the special qualities of NPs can increase the effectiveness of nutrient usage with respect to conventional fertilizers (Tarafdar *et al.*, 2013). When nanofertilizers are introduced to the soil, aggregation takes place initially, which reduces the area of effect (Louie *et al.*, 2014), and as the aggregates get larger, they become less mobile in porous materials. Hence, the mobility of the NPs can be increased or decreased according to the amount of organic matter present in the soil, as well as the chemical characteristics of the nanofertilizer

(Kah *et al.*, 2014). Additionally, nanofertilizers have a variety of effects on the activity of soil microorganisms (Avila-Quezada *et al.*, 2022).

Nanofertilizers can improve the efficiency of nutrient usage, which is advantageous for nutrition control (Chhipa, 2016). These types of nanofertilizers release nutrients at a slower rate than traditional fertilizers because they are tied to nano-absorbents that are used alone or in combination. This strategy reduces fertilizer loss into groundwater and improves the efficiency of nutrient usage (Fig. 9.1).

Therefore, these smart delivery systems play an important part in achieving precision farming. However, the characteristics of these NPs must be altered depending on whether they are being applied to foliage or soil.

9.4 Factors to Consider in a Smart Delivery System

9.4.1 Adhesion of nanostructures to foliar tissues in smart delivery systems

This technique involves directly spraying nanostructures on to leaves. Typically, it is intended to

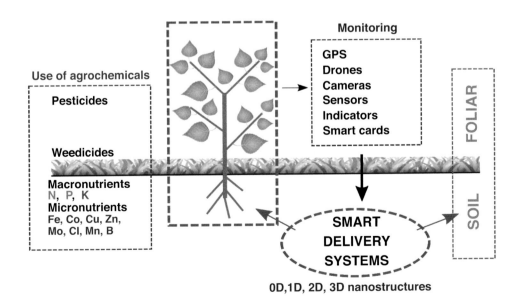

Fig. 9.1. Use of smart delivery systems for soil or foliar applications in different agrochemical applications.

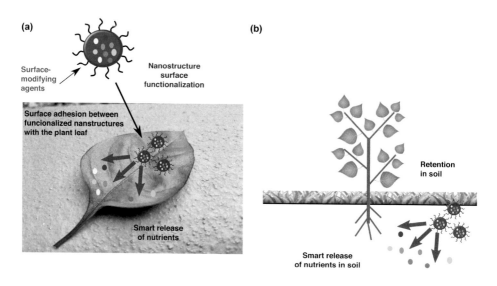

Fig. 9.2. Use of nanostructures (a) with surface-modifying agents in foliar applications, and (b) in soil applications.

deliver insecticides and trace element (Fig. 9.2a). When a plant is growing rapidly, a foliar spray can shorten the time between treatment and plant absorption. It also can get around the issue of a nutrient's restricted absorption from the soil. With this approach, manganese, copper, and iron may be absorbed more effectively than with soil application, where they become trapped in soil particles and are less accessible to the root system (Rai *et al.*, 2015). Therefore, in terms of the effective application of particle delivery platforms intended to protect as well as change plant function, foliar adhesion is of the utmost significance. Carriers and foliage interact strongly, minimizing biocidal leaching and preventing bioaccumulation in unintended organisms. Controlled adherence also enables the localization of the bioactive molecules in the target tissue.

Foliar tissues, particularly their outer layers (cuticle and epidermis), exhibit diverse morphological and chemical characteristics that change depending on the variety of plant and the surroundings (Barthlott *et al.*, 2017). For instance, temperature and humidity may be influenced by differentiating stomata among plant groups. The often-observed multiscale hierarchical structures together with the leaf cuticle restrict adherence. Additionally, the wax layers that

are normally present in the leaf cuticle provide a high water contact angle and low surface energy, and protect it from getting wet. (e.g. via spraying). However, the nanocarrier's form, size, and content could be used to alter how well a plant's surface adheres to it. As plant populations and planting circumstances can cause differences in form and content, the target plant and its surface (e.g. leaf, root, or stem) should be considered when designing carriers. Rod- and platelet-like nanocarriers could be used for foliar treatments due to their larger contact area compared with their spherical counterparts from a straightforward morphological standpoint. However, most carriers are created and utilized as spherical objects for practical reasons. Although the majority of nanocarrier delivery systems only take the methods for controlling the release kinetics into account at present, plant adhesion issues are receiving increasing attention. For example, the surface of nanocarriers (~450 nm diameter) made with poly(lactic acid) loaded with a pesticide was modified to impart positive surface groups, hydrophobicity, and other qualities to improve the adherence to cucumber (*Cucumis sativus*) leaves (Yu *et al.*, 2017). The particle's adherence to cucumber leaves could be modified by grafting acetyl, carboxylic acid, or amine groups on to the particle's surface. Even after many washing processes, those amine-functionalized poly(lactic acid) carriers remained firmly attached to the leaves.

The biocidal performance can be improved by modulating adhesion by adjusting the type of surface contacts. Tannic acid's high binding ability has been used to increase carrier adherence to vegetable tissues and the transport of abamectin or azoxystrobin (Guo *et al.*, 2016; Yu *et al.*, 2019). When compared with unmodified systems, the retention rate of transporters coated with tannic acid on cucumber leaves was significantly increased, rising by more than 50%. The leaf surface and numerous hydrogen bonds, which involved phenolic hydroxyl groups, encouraged high persistence while promoting adhesion. Additionally, the pesticide's photostability increased due to the inclusion of polyphenolic molecules on this carrier, with no adverse impacts on the release profiles (Yu *et al.*, 2019). Other carrier functionalization techniques include coating styrene–methacrylic acid/avermectin carriers with polycatechol (Liang *et al.*, 2017), coating hymexazol-containing graphene oxide nanosheets with polydopamine (Tong *et al.*, 2018), and coating carboxymethylcellulose, encapsulating fipronil (an insecticide), with ethanediamine (Xiao *et al.*, 2019).

Further studies on the impact of carrier shape and surface chemistry are required to better understand interfacial interactions and adhesion function.

9.4.2 Retention of nanostructures in soil using smart delivery systems

Applying pesticides and nutrients to the soil is another method for protecting the plant (Fig. 9.2b). To enhance their effectiveness and efficiency, these additives have been developed with NMs. However, compared with foliar application, the heterogeneous nature of the soil necessitates greater caution.

Numerous changes in biological and environmental media can impact the characteristics and chemical makeup of both the inorganic core and the shell of NPs. These include aggregation, dissolution, adsorption, and ligation of macromolecules, as well as oxidation and reduction (redox) processes. These changes significantly influence the mobility, bioaccumulation, and persistence of the NPs. NPs are also highly susceptible to colliding with other surfaces and sticking to them. These collisions could be with the same type or with a different type of particle in the environment, such as iron oxide, clay, or bacteria. These collisions can have a significant impact on the environmental behaviour and reactivity of NPs. To overcome these issues, capping agents and different binders are used with NMs (Kirschling *et al.*, 2011). Selection of these agents and binders must take into account the necessary rate and extent of the requirement.

To make nanofertilizers more stable in the soil environment, when creating NPs using physical and chemical techniques, anionic nutrients are loaded after surface modification, whereas cationic nutrients are loaded with the desired nutrients exactly as they are (Badgar *et al.*, 2022). One of the four methods listed below can be used to encapsulate fertilizers into nanostructures: (i) the nutrient can be supplied as an emulsion or NP; (ii) the nutrient can be coated with a thin polymer layer; (iii) the nutrient can be encapsulated using nanoporous materials; or (iv) the nutrient can be coated *in situ* on inorganic matrices (Nongbet *et al.*, 2022). Urea-modified hydroxyapatite (HA) pellets provide an example of *in situ* coating, which provides an anchored structure to extend the presence of the delivery system in soil (Kottegoda and Munaweera, 2011).

When it comes to nanopesticides, biodegradable polymers are used to create nanodelivery supports, such as chitosan, cyclodextrins, xanthan-like polysaccharides, poly(V-caprolactone), poly(lactic acid), and polyethylene glycol. However, a number of variables can potentially significantly impact the controlled-release mechanism of these nanostructures, including the mechanical characteristics of the coating material, degree of biodegradability, thickness, number of active ingredients, soil physiology, and water content (Irfan *et al.*, 2018).

It is crucial to consider the stability of the soil additives and the salinity, texture, and salt sensitivity of the plant when choosing this sort of application approach. Other essential factors to consider are pH, cation exchange capacity, and soil moisture.

9.5   Smart Delivery Systems for Different Agrochemicals

9.5.1   Smart delivery systems in pest and disease control

As described earlier, the notion of precision agriculture was developed as a result of the necessity to identify more environmentally friendly options for the use of agrochemicals in the environment. Nanotechnology is one of the tools that can enhance the controlled distribution of agrochemicals in farming (Allahyari *et al.*, 2016). When it comes to the general processes for releasing active chemicals from their matrix, there are various application methods for these nanostructures to increase the efficiency of *in situ* pest control.

Different types of NM and a wide range of agricultural inputs have been converted to nanopesticides, for example as a foliar application in vegetable cultivation. Sharma *et al.* (2017) created copper selenide NPs adorned with graphene NPs loaded with the insecticide chlorpyrifos (Sharma *et al.*, 2017) and observed superior results against some foliage insects compared with conventional pesticides.

Cyclodextrins were used by Campos *et al.* (2018) with the aim of improving the use of chitosan through a nano-complexation process. The obtained mixture was combined with carvacrol and linalool, two phenolic monoterpenes derived from plants with insecticidal and repellant characteristics. Due to better control over the volatilization and release of the encapsulated essential chemicals, the nanocomplex formation of the biopolymers enabled extension of the product's lifespan. The same team used zein, a protein extracted from maize kernel endosperm, to encapsulate natural insecticides such as eugenol, citronella, cinnamaldehyde, and geraniol (de Oliveira *et al.*, 2019). When these active compounds were used without the complexation process, volatilization reduced their effect toward the target, emphasizing the importance of a smart delivery system.

Guo *et al.* (2015) used silica–epichlorohydrin–carboxymethylcellulose to encapsulate emamectin benzoate through emulsion polymerization. The results indicated that the created microcapsules had an exceptional capacity to load emamectin benzoate ($\sim 35\%$ weight/weight) and could shield emamectin benzoate from thermal and photodegradation. They observed a sustained pesticide efficacy from these emamectin benzoate microcapsules against *Myzus persicae*, providing another example of how synthesized microcapsules make the system smarter than the conventional delivery system.

Kumar *et al.* (2015) developed nanocapsules containing the insecticide acetamiprid. Chitosan and alginate polyelectrolytes were complexed to create the nanocapsules, which were then examined for regulated *in vitro* release under three different pH settings. The findings indicated that the NPs displayed regulated release of the acetamiprid for up to 36 hours, with the most significant release occurring at pH 10 and the lowest release occurring at pH 4. This is another example of microcapsules enhancing the efficiency of a delivery system.

Two fungal species that attack crops and legumes, *Fusarium oxysporum* and *Aspergillus niger*, were treated using silica NPs (20–40 nm) (Suriyaprabha *et al.*, 2014). When the authors used nanosilica to treat maize, they found that leaf extracts from the treated plants had higher concentrations of phenolic compounds (2056 mg ml^{-1} and 743 mg ml^{-1} in leaves treated with *F. oxysporum* and *A. niger*, respectively) and lower concentrations of stress-responsive enzymes, compared with treatment with bulk silica. Although the treated crops displayed increased resistance to *A. niger*, only treatments performed with nanosilica produced positive results that were statistically different from those produced with bulk silica in terms of total phenol, disease index, peroxidases, phenylalanine ammonia lyase, and polyphenol oxidase content.

Due to its low price and strong selectivity, 2,4-dichlorophenoxy acetic acid (2,4-D) is among the herbicides that are used the most frequently to suppress broad-leaf weeds after they have emerged. 2,4-D resides primarily in anionic form and is only marginally retained by soil particles due to its comparatively high water solubility. As a result, it has a significant leaching potential and can have an impact on surface water and groundwater, especially if heavy rains follow herbicide application. Cao *et al.* (2018) used positively charged mesoporous silica NPs to create a stimuli-responsive controlled anionic

2,4-D release system. These porous structures provide another example of the use of such structures in smart delivery systems. The bioactivity of the nanoformulation was examined using cucumber (*C. sativus* L.) as the target and wheat (*Triticum aestivum* L.) as the non-target. Their study observed high 2,4-D loading and potent plant bioactivity without any adverse effects on other plants due to the effect of nanomeric carriers.

In order to test the insecticidal effectiveness of garlic (*Allium sativum*) essential oil on adult *Tribolium castaneum* (brown beetles), Yang *et al.* (2009) created NPs using polyethylene glycol to entrap the oil. Using the fusion dispersion approach, they produced spherical NPs with an excellent size distribution and an average size of 240 nm. They tested the activity of the encapsulated oil against *T. castaneum*, and found that its activity remained above 80% even 5 months after administration of the NPs, whereas the use of free garlic oil only achieved 11% of efficacy at the same concentration. Thus, this study demonstrates the use of a support polymer to make the system more efficient and smart.

In the fight against *Aspergillus parasiticus* and *F. oxysporum*, Kumar *et al.* (2016) created chitosan–pectin NPs coated with carbendazim. They created NPs of 70–90 nm that completely inhibited these fungi at doses of 0.5 and 1 ppm compared with pure carbendazim, and which demonstrated 80 and 97% inhibition at the same concentrations, respectively. Based on extensive testing, they concluded that nanoformulated carbendazim is more efficient and safe for the germination and root development of *C. sativus* seeds than direct administration of carbendazim. In this example, the authors used the active compound as the coating around the NP and used the characteristic features of the NPs to make the system more effective in delivering the active substance.

The examples given above show that nanopesticides used with smart delivery mechanisms have the potential to upgrade agricultural technology toward precision farming, producing safe, environmentally friendly, and highly effective solutions.

9.5.2 Smart delivery systems in fertilizers

Fertilizers supply the nutrients that plants require for maximum productivity. Nanofertilizers can be produced by adding nutrients alone or combining adsorbents with nanodimensions. Numerous NPs, including carbon nanotubes, and NPs of copper (Cu), silver (Ag), molybdenum (Mo), manganese (Mn), zinc (Zn), silicon (Si), iron (Fe), and titanium (Ti), as well as their oxides, and nanoformulations of standard agricultural inputs such as urea, phosphorus (P), and sulfur (S), have been converted into nanofertilizers. This nanoform has produced promising outcomes for seed germination production and plant growth when used at the recommended dose.

Kottegoda *et al.* (2014) studied the release characteristics of a nanohybrid composite synthesized by stabilizing urea in layered double hydroxides (LDHs) and montmorillonite (MMT) supporting matrices (Kottegoda *et al.*, 2014). Within the first 50 hours, pure urea was rapidly leached. In contrast, the N release from the urea–MMT and urea–LDH nanohybrids was both regulated and long lasting. The urea curve displayed a gradient of 0.042 after 40 hours, whereas urea–MMT and urea–LDH displayed gradients of 0.024 and 0.030, respectively. At this point, urea–MMT had released 42% of its N, urea–LDH had released 53%, and pure urea had released 82% (Kottegoda *et al.*, 2014). In comparison with pure urea, both nanohybrid compounds exhibit significant slow-release behaviour, allowing their use in smart delivery systems.

Fertilizers consisting of hydroxyapatite NPs (HANPs) are a nano-enabled delivery system capable of providing plants with calcium (Ca) and P. Kottegoda *et al.* (2017) created HANPs treated with urea for a slow release of N to support crop growth. This system was designed to obtain an initial quick release of urea followed by a prolonged release. This dual-mode release mechanism provides an initial high level of N required for the vegetative growth of the plant followed by continuous release of N. In this system, urea molecules are retained on the surface of the HANPs due to interactions between the urea molecules and HANP.

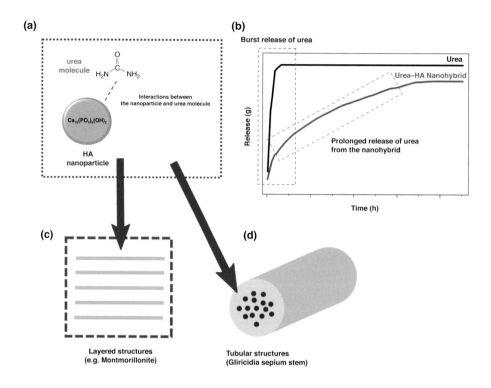

Fig. 9.3. Use of modified nanoparticles (a) with prolonged nutrient release properties (b), used with layered clay structures of montmorillonite (MMT) or layered double hydroxides (c), or with *Gliricidia sepium* plant stem tubular structures (d) in smart delivery systems. HA, hydroxyapatite.

Urea molecules are released from the urea–HA nanohybrid system into the soil in the presence of water. The rate of release was found to be 12 times lower than that of pure urea. Field trials confirmed the controlled-release properties: 48% NUE was observed with 50% less fertilizer application, plus a 10% yield increment. In this example, the interaction between the active compound and the NP was used to optimize the release of the active compound.

Two further major smart fertilizer delivery systems have been developed using the same technology. In both approaches, the synthesized urea–HA nanohybrid system was intercalated into hollow structures (plant vascular tissues) or galleries of layered structures (MMT) (Fig. 9.3) (Gunaratne *et al.*, 2016). In the first approach, the urea–HA nanohybrid system was pressurized into the vascular tissues of *Gliricidia sepium* wood chips. This delivery system (urea–HA–*Gliricidia*) utilized the release mechanism described by Kottegoda *et al.* (2017). The decay of wood chips provided additional controlled-release properties. In contrast to commercial fertilizer, which only displays release for the first 30 days of plant growth, these nanofertilizers demonstrated an initial burst and then a slow release of N for up to 60 days of plant growth. The significant amount of urea attachment on the HA surface is made possible by the material's immense surface area. The delayed and controlled release of urea is facilitated by a robust interaction between the HANPs and urea and the protection from extreme conditions (Kottegoda *et al.*, 2011). Similarly, MMT layered clays have been used to protect the urea–HA nanohybrid system (Madusanka *et al.*, 2017). The release of N from commercial fertilizer compositions using pure urea as the N source ended after 30 days, and about 80% of the N was released after 10 days. On day 60, the commercial fertilizer had released nearly 90% of its N content, whereas the urea–HA–MMT smart delivery system had only released about 65%.

All of these fertilizer systems have been studied extensively using pot trials, field trials, and farmer's field trials to confirm their smart action in the field. For example, the urea–HA–*Gliricidia* and urea–HA–MMT systems performed well in a pot experiment conducted using *Festuca arundinacea* for 60 weeks (Gunaratne *et al.*, 2016). Urea–HA fertilizer has also shown promising field results for both tea and paddy (Kottegoda *et al.*, 2017; Raguraj *et al.*, 2020).

Naseem *et al.* (2020) examined how *Oryza sativa* responded to commercial urea and mesoporous silica NPs ($ZnAl_2Si_{10}O_{24}$; 64 nm) complexed with urea administered to the soil in the same quantities in a pot experiment. Urea release from the nanocomposite began and then accelerated within the first 3 days and continued for up to 336 hour (14 days). Furthermore, when this was used instead of traditional fertilizer, the resulting yield was considerably higher. Similar results have been recorded by de Silva *et al.* (2020) in their study related to urea–silica nanohybrids (de Silva *et al.*, 2020). These studies provide examples of the use of porous structures to enhance the effectiveness of fertilizers.

Some research has also been done on amorphous calcium phosphate (ACP) NPs. The initial phase is crystallized from a calcium cation and phosphate anion supersaturated solution. Although the substance does have an apatitic short-range architecture, X-ray diffraction tests showed that its crystals are so minute that they appear amorphous. These substances have a larger capacity to adsorb smaller molecules (e.g. urea) on their surface than HA, allowing higher loading of macronutrients. They are also more soluble and reactive than HA. Fertilization studies performed on *C. sativus* grown hydroponically showed that nano-urea–ACP (69%) had a greater NUE than urea (49%) (Fellet *et al.*, 2021).

Samavini *et al.* (2018) synthesized citric acid-modified HANPs (CMHANPs) for controlled release of P to overcome issues related to commercial P fertilizers. They carried out P-release studies for HANP, Eppawala Rock phosphate (ERP), and CMHANPs using the standard dialysis membrane method (d'Souza, 2014) and showed that CMHANP has optimum P-releasing ability compared with HANPs and ERP. To confirm these results, they also carried out a bioavailability study for P nutrients using HANPs, CMHANPs, ERP, and triple superphosphates as treatments, with maize (*Zea mays*) as the target plant, while keeping other parameters constant, which have supported and confirmed the results of the P-release study. These results provide an example of a different approach compared with the other smart delivery approaches where the surface of the NP was modified to make it more available for the plants.

The distribution of macronutrients using chitosan-based NMs has been used as a creative method to increase the effectiveness of using so-called NPK (nitrogen, phosphorus, potassium) fertilizers by more precisely delivering the nutrients and delaying their release (Abdel-Aziz *et al.*, 2016). Using nano-chitosan in fertilizer guarantees an extra benefit pertinent to the supply of macronutrients. The additional 9–10% N included in the raw material should be considered (Yen *et al.*, 2009), in addition to the 24% P content of the triphenylphosphine catalyst, which is utilized in ionotropic gelation to create polymers (Bhagat *et al.*, 2019). Dhlamini *et al.* (2020) devised a new methodological approach by using a nanoformulation of chitosan that is characterized by its amorphous nature and filled with NPK and S fertilizers. Compared with standard fertilization methods, greenhouse tests on maize show that chitosan nanofertilizer containing these four macronutrients increased stem diameter, plant height, leaf number, and chlorophyll content. These two studies thus give examples of the benefit of nanostructure support in smart delivery systems.

Nanocomposite fertilizers have proven favourable effects on rhizosphere bacteria by encouraging the production of secondary metabolites, which enhance plant development by encouraging colonization of the root surface (Panichikkal *et al.*, 2019). Additionally, studies have indicated that applying 'controlled/smart release fertilizer' increased wheat production and soil residual mineral N by 6 and 10%, respectively, while reducing N leaching and runoff loss by 25 and 22%, respectively, in comparison with traditional fertilizers (Liu *et al.*, 2016; Jiang *et al.*, 2021).

9.6 Conclusions

With the advancements in nanotechnology, nanotechnology-enabled smart delivery systems will rapidly become more effective and efficient. Although there are currently many studies on these smart delivery systems, there remain some gaps to be filled. For example, almost all fertilizer studies and pesticide studies focus on encapsulating only one ingredient at a time. Hence, farmers still have to use fertilizers and pesticides from time to time, based on the requirements of the plant. This consumes a lot of time, energy, and manpower. However, as nanotechnology develops, a point may eventually be reached where, when a farmer applies fertilizer or pesticide to a field, these added inputs could themselves evaluate the environmental circumstances and release the necessary components accordingly. To achieve this, these inputs will need to be complexed with sensors and different active components. The conclusions that can be reached from the results to date discussed in this chapter are that nanotechnology has immense potential to improve smart delivery systems, with the ultimate goal of achieving the aims of precision farming.

References

Abdel-Aziz, H.M.M., Hasaneen, M.N.A. and Omer, A.M. (2016) Nano chitosan-NPK fertilizer enhances the growth and productivity of wheat plants grown in sandy soil. *Spanish Journal of Agricultural Research* 14(1), e0902. DOI: 10.5424/sjar/2016141-8205.

Allahyari, M.S., Mohammadzadeh, M. and Nastis, S.A. (2016) Agricultural experts' attitude towards precision agriculture: evidence from Guilan Agricultural Organization, Northern Iran. *Information Processing in Agriculture* 3(3), 183–189. DOI: 10.1016/j.inpa.2016.07.001.

Avila-Quezada, G.D., Ingle, A.P., Golińska, P. and Rai, M. (2022) Strategic applications of nano-fertilizers for sustainable agriculture: benefits and bottlenecks. *Nanotechnology Reviews* 11(1), 2123–2140. DOI: 10.1515/ntrev-2022-0126.

Azarian, M.H., Kamil Mahmood, W.A., Kwok, E., Bt Wan Fathilah, W.F. and Binti Ibrahim, N.F. (2018) Nanoencapsulation of intercalated montmorillonite-urea within PVA nanofibers: hydrogel fertilizer nanocomposite. *Journal of Applied Polymer Science* 135(10), 45957. DOI: 10.1002/app.45957.

Badgar, K., Abdalla, N., El-Ramady, H. and Prokisch, J. (2022) Sustainable applications of nanofibers in agriculture and water treatment: a review. *Sustainability* 14(1), 464. DOI: 10.3390/su14010464.

Bali, B., Jauhari, D. and Tiwari, M.P. (2014) Biosensors and bioelectronics doubly imprinted polymer nanofilm-modified electrochemical sensor for ultra-trace simultaneous analysis of glyphosate and glufosinate. *Biosensors and Bioelectronics* 59, 81–88. DOI: 10.1016/j.bios.2014.03.019.

Barthlott, W., Mail, M., Bhushan, B. and Koch, K. (2017) Plant surfaces: structures and functions for biomimetic innovations. *Nano-Micro Letters* 9(2), 23. DOI: 10.1007/s40820-016-0125-1.

Bhagat, D., Pal, A. and Raliya, R. (2019) Chitosan nanofertilizer to foster source activity in maize. *International Journal of Biological Macromolecules* 145, 226–234. DOI: 10.1016/j.ijbiomac.2019.12.155.

Campos, E.V.R., Proença, P.L.F., Oliveira, J.L., Melville, C.C., Della Vechia, J.F. *et al.* (2018) Chitosan nanoparticles functionalized with β-cyclodextrin: a promising carrier for botanical pesticides. *Scientific Reports* 8(1), 2067. DOI: 10.1038/s41598-018-20602-y.

Cao, L., Zhou, Z., Niu, S., Cao, C., Li, X. *et al.* (2018) Positive-charge functionalized mesoporous silica nanoparticles as nanocarriers for controlled 2,4-dichlorophenoxy acetic acid sodium salt release. *Journal of Agricultural and Food Chemistry* 66(26), 6594–6603. DOI: 10.1021/acs.jafc.7b01957.

Chakravarty, D., Erande, M.B. and Late, D.J. (2015) Graphene quantum dots as enhanced plant growth regulators: effects on coriander and garlic plants. *Journal of the Science of Food and Agriculture* 95(13), 2772–2778. DOI: 10.1002/jsfa.7106.

Chauhan, A., Patil, C., Jain, P. and Kulhari, H. (2020) Dendrimer-based marketed formulations and miscellaneous applications in cosmetics, veterinary, and agriculture. In: *Pharmaceutical Applications of Dendrimers: Micro and Nano Technologies*. Elsevier, Amsterdam, pp. 325–334.

Chhipa, H. (2016) Nanofertilizers and nanopesticides for agriculture. *Environmental Chemistry Letters* 15(1), 15–22. DOI: 10.1007/s10311-016-0600-4.

Dawson, C.J. and Hilton, J. (2011) Fertiliser availability in a resource-limited world: production and recycling of nitrogen and phosphorus. *Food Policy* 36(Suppl. 1), S14–S22. DOI: 10.1016/j.foodpol.2010.11.012.

de Oliveira, J.L., Campos, E.V.R., Germano-Costa, T., Lima, R., Vechia, J.F.D. *et al.* (2019) Association of zein nanoparticles with botanical compounds for effective pest control systems. *Pest Management Science* 75(7), 1855–1865. DOI: 10.1002/ps.5338.

de Oliveira, C.R.S., Mulinari, J., Reichert, F.W.J., da Silver, A.H.J., *et al.* (2020) Nano-delivery systems of pesticides active agents for agriculture applications - an overview. DOI: 10.31692/ICIAGRO.2020.0051.

de Silva, M., Siriwardena, D.P., Sandaruwan, C., Priyadarshana, G., Karunaratne, V. *et al.* (2020) Ureasilica nanohybrids with potential applications for slow and precise release of nitrogen. *Materials Letters* 272, 127839. DOI: 10.1016/j.matlet.2020.127839.

Dhiman, S., Yadav, A., Debnath, N. and Das, S. (2021) Application of core/shell nanoparticles in smart farming: a paradigm shift for making the agriculture sector more sustainable. *Journal of Agricultural and Food Chemistry* 69(11), 3267–3283. DOI: 10.1021/acs.jafc.0c05403.

Dhlamini, B., Paumo, H.K., Katata-Seru, L. and Kutu, F.R. (2020) Sulphate-supplemented NPK nanofertilizer and its effect on maize growth. *Materials Research Express* 7(9), 095011. DOI: 10.1088/2053-1591/abb69d.

Dong, J., Fan, X., Qiao, F., Ai, S. and Xin, H. (2013) A novel protocol for ultra-trace detection of pesticides: combined electrochemical reduction of Ellman's reagent with acetylcholinesterase inhibition. *Analytica Chimica Acta* 761, 78–83. DOI: 10.1016/j.aca.2012.11.042.

d'Souza, S. (2014) A review of *in vitro* drug release test methods for nano-sized dosage forms. *Advances in Pharmaceutics* 2014, 1–12. DOI: 10.1155/2014/304757.

Facure, M.H.M., Schneider, R., Mercante, L.A. and Correa, D.S. (2020) A review on graphene quantum dots and their nanocomposites: from laboratory synthesis towards agricultural and environmental applications. *Environmental Science: Nano* 7(12), 3710–3734. DOI: 10.1039/D0EN00787K.

Falsini, S., Clemente, I., Papini, A., Tani, C., Schiff, S. *et al.* (2019) When sustainable nanochemistry meets agriculture: lignin nanocapsules for bioactive compound delivery to plantlets. *ACS Sustainable Chemistry & Engineering* 7(24), 19935–19942. DOI: 10.1021/acssuschemeng.9b05462.

Fellet, G., Pilotto, L., Marchiol, L. and Braidot, E. (2021) Tools for nano-enabled agriculture: fertilizers based on calcium phosphate, silicon, and chitosan nanostructures. *Agronomy* 11(6), 1239. DOI: 10.3390/agronomy11061239.

Godfray, H.C.J., Beddington, J.R., Crute, I.R., Haddad, L., Lawrence, D. *et al.* (2010) Food security: the challenge of feeding 9 billion people. *Science* 327(5967), 812–818. DOI: 10.1126/science.1185383.

Irfan, S.A., Razali, R., KuShaari, K., Mansor, N., Azeem, B. *et al.* (2018) A review of mathematical modeling and simulation of controlled-release fertilizers. *Journal of Controlled Release* 271, 45–54. DOI: 10.1016/j.jconrel.2017.12.017.

Guo, J., Tardy, B.L., Christofferson, A.J., Dai, Y., Richardson, J.J. *et al.* (2016) Modular assembly of superstructures from polyphenol-functionalized building blocks. *Nature Nanotechnology* 11(12), 1105–1111. DOI: 10.1038/nnano.2016.172.

Gunaratne, G.P., Kottegoda, N., Madusanka, N., Munaweera, I., Sandaruwan, C. *et al.* (2016) Two new plant nutrient nanocomposites based on urea coated hydroxyapatite: efficacy and plant uptake. *Indian Journal of Agricultural Sciences* 86, 494–499.

Guo, M., Zhang, W., Ding, G., Guo, D., Zhu, J. *et al.* (2015) Preparation and characterization of enzyme-responsive emamectin benzoate microcapsules based on a copolymer matrix of silica-epichlorohydrin-carboxymethylcellulose. *RSC Advances* 5(113), 93170–93179. DOI: 10.1039/C5RA17901G.

Jiang, M., Song, Y., Kanwar, M.K., Ahammed, G.J., Shao, S. *et al.* (2021) Phytonanotechnology applications in modern agriculture. *Journal of Nanobiotechnology* 19(1), 430. DOI: 10.1186/s12951-021-01176-w.

Kah, M., Machinski, P., Koerner, P., Tiede, K., Grillo, R. *et al.* (2014) Analysing the fate of nanopesticides in soil and the applicability of regulatory protocols using a polymer-based nanoformulation of atrazine. *Environmental Science and Pollution Research International* 21(20), 11699–11707. DOI: 10.1007/s11356-014-2523-6.

Kassem, I., Ablouh, E.-H., El Bouchtaoui, F.-Z., Kassab, Z., Khouloud, M. *et al.* (2021) Cellulose nanocrystals-filled poly (vinyl alcohol) nanocomposites as waterborne coating materials of NPK fertilizer with slow release and water retention properties. *International Journal of Biological Macromolecules* 189, 1029–1042. DOI: 10.1016/j.ijbiomac.2021.08.093.

Kim, Y.T., Han, J.H., Hong, B.H. and Kwon, Y.U. (2010) Electrochemical synthesis of CdSe quantum-dot arrays on a graphene basal plane using mesoporous silica thin-film templates. *Advanced Materials* 22(4), 515–518. DOI: 10.1002/adma.200902736.

Kirschling, T.L., Golas, P.L., Unrine, J.M., Matyjaszewski, K., Gregory, K.B. *et al.* (2011) Microbial bioavailability of covalently bound polymer coatings on model engineered nanomaterials. *Environmental Science & Technology* 45(12), 5253–5259. DOI: 10.1021/es200770z.

Kottegoda, N. and Munaweera, I. (2011) A green slow-release fertilizer composition based on urea-modified hydroxyapatite nanoparticles encapsulated wood, 73–78.

Kottegoda, N., Munaweera, I., Madusanka, N., and Karunaratne, V. (2011) A green slow-release fertilizer composition based on urea-modified hydroxyapatite nanoparticles encapsulated wood. *Current Science* 101, 73–78.

Kottegoda, N., Sandaruwan, C., Perera, P., Madusanka, N. and Karunaratne, V. (2014) Modified layered nanohybrid structures for the slow release of urea. *Nanoscience & Nanotechnology-Asia* 4, 94–102. DOI: 10.2174/2210681204021505211247729.

Kottegoda, N., Sandaruwan, C., Priyadarshana, G., Siriwardhana, A., Rathnayake, U.A. *et al.* (2017) Urea-hydroxyapatite nanohybrids for slow release of nitrogen. *ACS Nano* 11(2), 1214–1221. DOI: 10.1021/acsnano.6b07781.

Kumar, S., Chauhan, N., Gopal, M., Kumar, R. and Dilbaghi, N. (2015) Development and evaluation of alginate-chitosan nanocapsules for controlled release of acetamiprid. *International Journal of Biological Macromolecules* 81, 631–637. DOI: 10.1016/j.ijbiomac.2015.08.062.

Kumar, S., Kumar, D. and Dilbaghi, N. (2016) Preparation, characterization, and bio-efficacy evaluation of controlled release carbendazim-loaded polymeric nanoparticles. *Environmental Science and Pollution Research* 24, 926–937. DOI: 10.1007/s11356-016-7774-y.

Lechenet, M., Dessaint, F., Py, G., Makowski, D. and Munier-Jolain, N. (2017) Reducing pesticide use while preserving crop productivity and profitability on arable farms. *Nature Plants* 3, 17008. DOI: 10.1038/nplants.2017.8.

Liang, J., Yu, M., Guo, L., Cui, B., Zhao, X. *et al.* (2017) Bioinspired development of P(St-MAA)-avermectin nanoparticles with high affinity for foliage to enhance folia retention. *Journal of Agricultural and Food Chemistry* 66(26), 6578–6584. DOI: 10.1021/acs.jafc.7b01998.

Liu, R., Kang, Y., Pei, L., Wan, S., Liu, S. *et al.* (2016) Use of a new controlled-loss-fertilizer to reduce nitrogen losses during winter wheat cultivation in the Danjiangkou Reservoir area of China. *Communications in Soil Science and Plant Analysis* 47(9), 1137–1147. DOI: 10.1080/00103624.2016.1166245.

Louie, S.M., Ma, R. and Lowry, G.V. (2014) Transformations of nanomaterials in the environment. *Frontiers of Nanoscience* 7, 55–87. DOI: 10.1016/B978-0-08-099408-6.00002-5.

Madre, Y. and Devuyst, P. (2012) How will we feed the world in the next decades? An analysis of the demand and supply factors for food. Available at: www.farm-europe.eu/travaux/how-will-we-feed-the-world-in-the-next-decades-an-analysis-of-the-demand-and-supply-factors-for-food/ (accessed 21 March 2023).

Madusanka, N., Sandaruwan, C., Kottegoda, N., Sirisena, D., Munaweera, I. *et al.* (2017) Urea-hydroxyapatite-montmorillonite nanohybrid composites as slow release nitrogen compositions. *Applied Clay Science* 150, 303–308. DOI: 10.1016/j.clay.2017.09.039.

Naseem, F., Zhi, Y., Farrukh, M.A., Hussain, F. and Yin, Z. (2020) Mesoporous ZnAl2Si10O24 nanofertilizers enable high yield of *Oryza sativa* L. *Scientific Reports* 10(1), 10841. DOI: 10.1038/s41598-020-67611-4.

Natarelli, C.V.L., Lopes, C.M.S., Carneiro, J.S.S., Melo, L.C.A., Oliveira, J.E. *et al.* (2021) Zinc slow-release systems for maize using biodegradable PBAT nanofibers obtained by solution blow spinning. *Journal of Materials Science* 56(7), 4896–4908. DOI: 10.1007/s10853-020-05545-y.

National Science and Technology Council Committee on Technology (2014) *National Nanotechnology Initiative Strategic Plan*. National Science and Technology Council Committee on Technology, Washington, DC. Available at: www.nano.gov/sites/default/files/pub_resource/2014_nni_strategic_plan.pdf (accessed 16 March 2023).

Ni, B., Liu, M., Lü, S., Xie, L. and Wang, Y. (2011) Environmentally friendly slow-release nitrogen fertilizer. *Journal of Agricultural and Food Chemistry* 59(18), 10169–10175. DOI: 10.1021/jf202131z.

Nkebiwe, P.M., Weinmann, M., Bar-Tal, A. and Müller, T. (2016) Fertilizer placement to improve crop nutrient acquisition and yield: a review and meta-analysis. *Field Crops Research* 196, 389–401. DOI: 10.1016/j.fcr.2016.07.018.

Nongbet, A., Mishra, A.K., Mohanta, Y.K., Mahanta, S., Ray, M.K. *et al.* (2022) Nanofertilizers: a smart and sustainable attribute to modern agriculture. *Plants* 11(19), 2587. DOI: 10.3390/plants11192587.

Panichikkal, J., Thomas, R., John, J.C. and Radhakrishnan, E.K. (2019) Biogenic gold nanoparticle supplementation to plant beneficial *Pseudomonas monteilii* was found to enhance its plant probiotic effect. *Current Microbiology* 76(4), 503–509. DOI: 10.1007/s00284-019-01649-0.

Pereira, E.I.*et al.* (2015) Perspectives in nanocomposites for the slow and controlled release of agrochemicals: fertilizers and pesticides. In: Rai, M., Ribeiro, C., Mattoso, L. and Duran, N. (eds) *Nanotechnologies in Food and Agriculture*. Springer, Cham, Switzerland, pp. 241–265. DOI: 10.1007/978-3-319-14024-7.

Raguraj, S., Wijayathunga, W.M.S., Gunaratne, G.P., Amali, R.K.A., Priyadarshana, G. *et al.* (2020) Urea–hydroxyapatite nanohybrid as an efficient nutrient source in *Camellia sinensis* (L.) Kuntze (tea) . *Journal of Plant Nutrition* 43(15), 2383–2394. DOI: 10.1080/01904167.2020.1771576.

Rai, M., Rebeiro, C., Mattoso, L. and Duran, N. (eds) (2015) *Nanotechnologies in Food and Agriculture*. Springer, Cham, Switzerland.

Raliya, R., Saharan, V., Dimkpa, C. and Biswas, P. (2018) Nanofertilizer for precision and sustainable agriculture: current state and future perspectives. *Journal of Agricultural and Food Chemistry* 66(26), 6487–6503. DOI: 10.1021/acs.jafc.7b02178.

Ranjan, S., Dasgupta, N. and Lichtfouse, E. (eds) (2016) *Nanoscience in Food and Agriculture*, Vol. 3. Springer, Cham, Switzerland.

Rastogi, A., Tripathi, D.K., Yadav, S., Chauhan, D.K., Živčák, M. *et al.* (2019) Application of silicon nanoparticles in agriculture. *3 Biotech* 9(3), 90. DOI: 10.1007/s13205-019-1626-7.

Safdar, M., Kim, W., Park, S., Gwon, Y., Kim, Y.O. *et al.* (2022) Engineering plants with carbon nanotubes: a sustainable agriculture approach. *Journal of Nanobiotechnology* 20(1), 275. DOI: 10.1186/s12951-022-01483-w.

Samavini, R., Sandaruwan, C., De Silva, M., Priyadarshana, G., Kottegoda, N. *et al.* (2018) Effect of citric acid surface modification on solubility of hydroxyapatite nanoparticles. *Journal of Agricultural and Food Chemistry* 66(13), 3330–3337. DOI: 10.1021/acs.jafc.7b05544.

Sharma, S., Singh, S., Ganguli, A.K. and Shanmugam, V. (2017) Anti-drift nano-stickers made of graphene oxide for targeted pesticide delivery and crop pest control. *Carbon* 115, 781–790. DOI: 10.1016/j.carbon.2017.01.075.

Shi, H., Zhao, G., Liu, M., Fan, L. and Cao, T. (2013) Aptamer-based colorimetric sensing of acetamiprid in soil samples: sensitivity, selectivity and mechanism. *Journal of Hazardous Materials* 260, 754–761. DOI: 10.1016/j.jhazmat.2013.06.031.

Suriyaprabha, R., Karunakaran, G., Kavitha, K., Yuvakkumar, R., Rajendran, V. *et al.* (2014) Application of silica nanoparticles in maize to enhance fungal resistance. *IET Nanobiotechnology* 8(3), 133–137. DOI: 10.1049/iet-nbt.2013.0004.

Tarafdar, J.C. and Raliya, R. (2013) Rapid, low-cost, and ecofriendly approach for iron nanoparticle synthesis using *Aspergillus oryzae* TFR9 . *Journal of Nanoparticles* 2013, 1–4. DOI: 10.1155/2013/141274.

Tarafdar, J.C., Sharma, S. and Raliya, R. (2013) Nanotechnology: interdisciplinary science of applications. *African Journal of Biotechnology* 12(3), 219–226. DOI: 10.5897/AJB12.2481.

Tong, Y., Shao, L., Li, X., Lu, J., Sun, H. *et al.* (2018) Adhesive and stimulus-responsive polydopamine-coated graphene oxide system for pesticide-loss control. *Journal of Agricultural and Food Chemistry* 66(11), 2616–2622. DOI: 10.1021/acs.jafc.7b05500.

Wang, C., Liu, X., Li, J., Yue, L., Yang, H. *et al.* (2021) Copper nanoclusters promote tomato (*Solanum lycopersicum* L.) yield and quality through improving photosynthesis and roots growth. *Environmental Pollution* 289, 117912. DOI: 10.1016/j.envpol.2021.117912.

Weersink, A., Fraser, E., Pannell, D., Duncan, E. and Rotz, S. (2018) Opportunities and challenges for big data in agricultural and environmental analysis. *Annual Review of Resource Economics* 10(1), 19–37. DOI: 10.1146/annurev-resource-100516-053654.

Xiao, D., Liang, W., Li, Z., Cheng, J., Du, Y. *et al.* (2019) High foliar affinity cellulose for the preparation of efficient and safe fipronil formulation. *Journal of Hazardous Materials* 384, 121408. DOI: 10.1016/j.jhazmat.2019.121408.

Yang, F.L., Li, X.G., Zhu, F. and Lei, C.L. (2009) Structural characterization of nanoparticles loaded with garlic essential oil and their insecticidal activity against *Tribolium castaneum* (Herbst) (Coleoptera: Tenebrionidae). *Journal of Agricultural and Food Chemistry* 57(21), 10156–10162. DOI: 10.1021/jf9023118.

Yen, M.-T., Yang, J.-H. and Mau, J.-L. (2009) Physicochemical characterization of chitin and chitosan from crab shells. *Carbohydrate Polymers* 75(1), 15–21. DOI: 10.1016/j.carbpol.2008.06.006.

Yu, Z., Zhao, G., Liu, M., Lei, Y. and Li, M. (2010) Fabrication of a novel atrazine biosensor and its subpart-per-trillion levels sensitive performance. *Environmental Science & Technology* 44(20), 7878–7883. DOI: 10.1021/es101573s.

Yu, M., Yao, J., Liang, J., Zeng, Z., Cui, B. *et al.* (2017) Development of functionalized abamectin poly(lactic acid) nanoparticles with regulatable adhesion to enhance foliar retention. *RSC Advances* 7(19), 11271–11280. DOI: 10.1039/C6RA27345A.

Yu, M., Sun, C., Xue, Y., Liu, C., Qiu, D. *et al.* (2019) Tannic acid-based nanopesticides coating with highly improved foliage adhesion to enhance foliar retention. *RSC Advances* 9(46), 27096–27104. DOI: 10.1039/C9RA05843E.

10 Health, Safety and Environmental Management and Risk Mitigation of Nanomaterials

Codruța Mihaela Dobrescu, Leonard Magdalin Dorobăț* and Monica Angela Neblea

Department of Natural Sciences, University of Pitesti, Pitesti, Romania

Abstract

Smart agriculture is a method of prioritizing issues related to securing the short- and long-term food needs of humanity in the context of the challenges produced by climate and demographic changes. The use of nanomaterials in sustainable agricultural development and plant protection, although an underexplored field, can have a major impact by reducing pollution and increasing agricultural productivity. However, the dangers of using nanomaterials are not well known, and therefore it is necessary to develop risk-management models associated with their use. The interactions between nanoparticles and plants in field conditions must be understood down to the molecular level in order to minimize any possible phytotoxic effects, and to maintain the health of the soil and, implicitly, the productivity and quality of the crops. This chapter reviews the literature on the potential benefits and risks of nanomaterials for plants, the environment, and human health. Issues related to creating a risk governance framework for the sustainable development of nanotechnology are also discussed.

Keywords: nanotechnology, nanomaterials, sustainable agriculture, nanoformulations, risk management

10.1 Management of Nanomaterial Safety in Agriculture

Nanotechnology in agriculture is a leading approach to significantly improve agricultural production in environmentally friendly systems after the uncontrolled use of chemical fertilizers and pesticides has severely affected the soil, groundwater, and ultimately the entire environment (Danish and Hussain, 2019). Using nanotechnology, as many different types of nanomaterials (NMs) as there are elements and compounds can be synthesized, with properties and activities so diverse that a comprehensive assessment of health and environmental risks is impossible. However, much research on nanoparticles (NPs) is oriented toward their effectiveness in the medicine and agricultural sectors. Nano-innovations for agriculture can revolutionize precision agriculture and the development of sustainable agriculture, certain types of environmental challenges, and sustainable resource uses. The use of nanotechnology in agriculture and its acceptance at the level of society raises questions and requires answers regarding the biosecurity of NMs, the exposure levels throughout the life cycle of the products, the adverse effects, the biological reactivity

*Corresponding author: coltanabe@yahoo.com

© CAB International 2023. *Nanoformulations for Sustainable Agriculture and Environmental Risk Mitigation* (eds Z. Khan and N.A. Șuțan)
DOI: 10.1079/9781800623095.0010

acquired once they have been dispersed in environment, their route, and their possible interactions with agrosystem coformulants and stressors.

Most of the references that provide details on the uses of different nanoformulations (NFs) in agriculture contain only preliminary data that seem very favourable, but the safety management of NMs in agriculture should consider several aspects: (i) sometimes the characteristics of NPs are not clearly specified and thus the data presented are illustrative, rather than practical and reproducible; (ii) in many cases, the mechanisms of action of NPs are barely investigated; (iii) many beneficial effects of NPs may be due to the hormesis effect (Kolbert *et al.*, 2022), which consists of a stimulatory action of toxic substances at low doses; and, last but not least, (iv) the difficulty in scalability of the synthetic techniques described (Pestovsky and Martínez-Antonio, 2017).The maximum biological efficiency in agriculture can be obtained using nanofertilizers (nanosized nutrients, nanocoated fertilizers, or NMs based on metal or carbon oxide), nanopesticides (NFs of traditional active ingredients or inorganic NMs), or nanoencapsulated carriers of conventional pesticides that can release targeted and controlled agrochemicals without creating overdose phenomena (Iavicoli *et al.*, 2017). NFs of essential metals and their compounds have a number of favourable properties compared with traditional bulk compounds, that stimulate plant growth and yield, induce phytochemical synthesis, and can trigger plant defense mechanisms against abiotic and biotic stressors (Kolbert *et al.*, 2022).

10.1.1 Role of NFs in crop productivity and quality

Agricultural NMs are multifunctional tools for plant growth, development, and yield, with nanotechnology having multiple applications in the production, processing, storage, packaging, and transportation of agricultural products (Ali *et al.*, 2014). Early and recent research has focused on the potential of NFs to regulate plant growth parameters according to their characteristics (size, shape, concentration, and surface composition), from germination and the

seedling stage, but also, more importantly, under culture conditions in the field. The uptake, translocation, and accumulation at the cell and organ level within the plant have also been addressed, but the need to determine their exact action and specific mechanisms is still a necessity (Malejko *et al.*, 2021). Biofertilizers stimulate plant growth and increase nutrient bioavailability using various mechanisms such as atmospheric nitrogen fixation (Kumari and Singh, 2020), siderophore production (Phillips *et al.*, 2015; Yasmin *et al.*, 2021), phosphorous solubilization (Balogun *et al.*, 2022), and the synthesis of phytohormones (auxins, gibberellins, cytokinins, abscisic acid) in soils, enhancing plant growth (Mehnaz, 2016).

Nanofertilizers and nanopesticides can be tailored to specific crop needs because they allow slow and sustained release, and their application in small quantities, in addition to increasing the efficiency of nutrient use, reduces associated costs, with positive effects on the economy (Arora *et al.*, 2022). Nanotechnology facilitates the expansion of agricultural production through relevant reductions in losses and improved input efficiency (Basit *et al.*, 2022). Nutrients, whether applied individually or in various combinations, bind to nanometer-sized adsorbents, and nutrient release is very slow compared with conventional fertilizers, resulting in more efficient use of nutrients and minimizing runoff. A significant number of NMs behave as growth regulators and therefore show a remarkable improvement in plant biomass production, as well as increases in nutritional quality (fortification) (Chaurasia *et al.*, 2021). NPs have a great potential to improve agricultural productivity, but to manage these aspects, the adverse effects of each individual type of NF must be studied before they are introduced into agriculture for actual use (Pestovsky and Martínez-Antonio, 2017).

A brief review of the recent literature shows that studies are now designed under conditions closer to real life, in contrast to initial laboratory experiments, which tended to apply high doses of NPs, which were toxic to plants. Exposure to lower concentrations has been found to be more appropriate, demonstrating that many NPs positively affect plant growth, health, and quality (Table 10.1). Field studies have been conducted for some agricultural plants: sunflower

Table 10.1. Effects of metal nanoparticles (NPs) on cultivated higher plants.

NP	Experiment	Effects	Reference
Silver NPs (AgNPs)	*Solanum lycopersicum* L., *Raphanus sativus* L. var. *sativus*, *Brassica oleracea* var. *sabellica*, *in vitro* germination Particle size: 23 ± 4 nm Concentrations: 50, 100 mg·l⁻¹ Growth parameters: seed germination, shoot and root length, root and shoot fresh/dry weight Physiological parameters: chlorophyll (total, *a*, *b*, *a/b*), carotenoids, anthocyanins, phenolic compounds, total protein content, and SOD and GPOX activity	*S. lycopersicum:* Increased anthocyanin content and GPOX activity Decreased length and weight of shoots in fresh and dry state, fresh root weight; chlorophyll and carotenoid content *R. sativus* var. *sativus:* Increased fresh and dry weight of shoots, root length Decreased shoot length, fresh weight of roots, anthocyanins and polyphenols and GPOX at 100 mg l⁻¹ *Brassica oleracea* var. *sabellica:* Increased GPOX activity, weight of shoots in fresh and dry state, biometric parameters of the root Decreased chlorophyll, carotenoid, and anthocyanin content	Tymoszuk (2021)
	Phaseolus vulgaris L., seeds, aqueous solutions AgNPs + Nitragina Particle size: ~10 nm Concentrations of AgNPs: 0.25, 1.25, 2.5 mg l⁻¹ Parameters: seed germination, field emergence, physiological parameters	Germination: fast and uniform in laboratory and field Increased average seedling height, fresh and dry weight, and net photosynthesis AgNPs at 0.25 and 1.25 l⁻¹ positively influenced seedling development Strong antimicrobial effect (2.5 mg l⁻¹) beneficial during germination, but unfavorable in the later stage of plant development	Prażak *et al.* (2020)

Continued

Table 10.1. Continued

NP	Experiment	Effects	Reference
	Triticum aestivum L., three-leaf stage Particle size: 8–28 nm Concentrations of AgNPs: 25, 50, 75, 100 mg l^{-1} Growth parameters: seed germination Physiological parameters: photosynthetic pigments, relative water contents, membrane stability index Biochemical parameters: nonenzymatic antioxidants, proline, sugar, protein, glutathione, TASC, TPC, TFC	Increased stem length (22%), roots (5%), leaf area (34%), growth parameters, TPC (2.4%), TFC (2.5%), TASC (2.5%), SOD (1.3%), POD (1.5%), CAT (1.8%), APX (1.2%) and GPOX (1.4%) Significant increase (˜10%) in photosynthetic activity, relative water contents, membrane stability index Decreased proline contents (4%; 50 mg l^{-1}), sugar (5.8%)	Iqbal *et al.* (2019)
Titanium dioxide NPs (TiO$_2$NPs)	*Beta vulgaris* var. *bengalensis*, seeds TiO$_2$NPs, PNPs, ZnONPs, aerated solutions Concentrations: 10, 20, 30, 40, 50 ppm TiO$_2$; 0, 5, 10, 15, 20, 25 ppm ZnO; 10, 20, 30, 40, 50 ppm P Growth parameters: germination percentage, vigour index, root and shoot length, fresh seedling weight Antioxidant activity (CAT, POD, SOD, GR)	Increased plant height, leaf length and width, number of leaves, and stem thickness for TiO$_2$; germination percentage for all NPs (maximum at 30 ppm of TiO$_2$); fresh seedling weight (maximum at 50 ppm TiO$_2$); vigour index (maximum at 30 ppm and 40 ppm TiO$_2$) Significant differences in the activity of antioxidant enzymes in seedling stage (maximum for different concentrations of TiO$_2$)	Das *et al.* (2022)
	Spinacia oleracea var. *viroflay* seeds, inoculated with Ag–TiO$_2$NPs Particle sizes: 7.3, 8.3, 10.9, 26 nm Concentrations: 0.25, 2, 4, 6% Growth parameters: plant height, and number, width and length leaves 64 applied treatments, each with ten repetitions	Increased photosynthetic activity (21.78% for 8.3 nm NP at 0.25% in suspension), germination rate (256.27% for 33 nm NP in suspension) Decreased average germination time by 10.92% (0.25% NP solution and 9 nm NP) Maximum results for Ag–TiO$_2$ NP treatment with 7–8 nm Nps at 2% concentration	Gordillo-Delgado *et al.* (2020)

Continued

Table 10.1. Continued

NP	Experiment	Effects	Reference
	Triticum aestivum, soil applied Particle sizes: <20 nm Concentrations: 0, 20, 40, 60, 80, 100 mg kg^{-1}. Parameters: chlorophyll content, H_2O_2 production, root and shoot length, biomass, phytoavailability of phosphorus Genotoxicity: micronucleus test (MN)	Stimulated plant growth at <60 mg kg^{-1} Micronuclei formation (53.6%), decreased chlorophyll content, increased H_2O_2 (84.4%) at >60 mg kg^{-1} Dose-dependent results: wheat could not tolerate TiO_2NP concentrations >60 mg kg^{-1}	Rafique *et al.* (2018)
Zinc oxide NPs (ZnONPs)	*Zea mays* L., soil or foliar applied Particle sizes: 30–70 nm (width), up to 160 nm (length) Foliar concentration: 2% Zn solution Soil concentration: 8 kg Zn ha^{-1} Parameters: plant height, number of branches, fruit yield, physiological parameters	Increased grain content, plant yield, chlorophyll contents, transpiration rate, photosynthetic rate	Umar *et al.* (2021)
	Linum usitatissimum L., foliar applications Particle sizes: <40 nm Concentrations: 20, 40, 60 mg ZnO l^{-1} Parameters: plant height, number of branches, fruit yield Physiological parameters: oil content	Increased shoot length, root length, number of fruiting branches, capsules, seed and straw yield, weight per 1000 seeds, fresh and dry weight, biological yield per plant, oil content	Sadak and Bakry (2020)
	Helianthus annuus L., foliar application with adjuvant SILWET STAR® Particle size: 20 nm Concentrations: 2.6 mg Zn l^{-1} Parameters: plant height, number of branches, fruit yield, total sugars, amino acids, and phenol, chlorophyll and carotenoid content	Increased anthodium diameter, physiological responses and quantitative and nutritional effects, oil content	Kolenčík *et al.* (2020)

Continued

Table 10.1. Continued

NP	Experiment	Effects	Reference
	Setaria italica L., foliar applied, with adjuvant SILWET STAR®Particle size: 20 nm Concentrations: 2.6 mg Zn l⁻¹ Parameters: plant height, number of branches, fruit yield, physiological parameters	Increased dry mass, oil content, total nitrogen Decreased starch content	Kolenčík *et al.* (2019)
Silica dioxide NPs (SiO₂NPs)	*Solanum tuberosum* L. Particle sizes: NA Concentration: 50 mg l⁻¹ Growth parameters: plant height, number of aerial stems, leaf area, tuber weight, total yield of tubers Physiological parameters: chlorophyll content in leaves/SPAD units, dry matter in leaves, proline content in leaves	Significant differences for all measured indicators Increased vegetative growth, and chlorophyll and proline content under abiotic stress conditions	Ali *et al.* (2021a)
	Saccharum officinarum L., foliar sprays application Particles: SiO₂NPs (5–15 nm), ZnONPs (<100 nm), SeNPs (100 mesh), and graphene nanoribbons (2–15 µm × 40–250 nm) Physiological parameters: assimilator pigment content	Increased chlorophyll and carotenoid content, higher carotenoid accumulation in leaves compared with seedlings, amelioration effects Decreased effects of freezing by upgrading the chlorophyll fluorescence yield and photosynthesis	Elsheery *et al.* (2020)
	Fragaria xananassa Duch., foliar application Particle size: NA Concentrations: 125 mg l⁻¹ Growth parameters: fresh and dry weight of root and shoot, fruit firmness and yield Physiological parameters: photosynthetic pigments, chlorophyll fluorescence, fruit anthocyanins, phenolic compounds, antioxidant activity, relative water content, membrane stability index	Increased number and area of leaves, number of flowers and fruits, fruit size, fruit weight and yield, fresh and dry weight of roots and shoots, concentration of photosynthetic pigments, chlorophyll fluorescence, fruit anthocyanins, total phenolic compounds, ascorbic acid content and antioxidant activity, content of vitamin C and DPPH, carbohydrate, proline	Zahedi *et al.* (2020)

Continued

Table 10.1. Continued

NP	Experiment	Effects	Reference
	Hordeum vulgare L. Particle size: NA Concentrations: 125–250 ppm Physiological parameters: enzymatic and metabolic activities	Increased plant morphophysiological parameters, chlorophyll content (17%), enzymatic and metabolic activities	Ghorbanpour *et al.* (2020)
	Avena sativa L. Particle size: 20 nm Concentrations: 5, 10 mM Parameters: ultrastructure of leaf and root cells, antioxidant enzyme activities, expression of silicon transporter, chemical content, protein content	Significant plant growth No toxic effect	Asgari *et al.* (2018)
Coper oxide NPs (CuONPs)	*Triticum aestivum* L., foliar spray application and soaking of seeds before sowing Particle sizes: 14.0–47.4 nm Concentrations: 50, 100 ppm Growth parameters: root and shoot length, number of leaves, root fresh/dry weight, shoot fresh/dry weight Physiological parameters: photosynthetic pigments, total soluble carbohydrates, POD and SOD	Increased total chlorophylls and carotenoids, carbohydrate and protein contents, antioxidant enzymes Efficacy against *Sitophilus granarius* and *Rhyzopertha dominica* insects (cause damage in stored grain)	Badawy *et al.* (2021)
	Oryza sativa L. Particle sizes: <50 nm Concentrations: 62.5, 125, 250 mg l⁻¹; 250 mg CuONP l⁻¹ supernatant from suspension Growth parameters: root length, shoot length Physiological parameters: antioxidant enzyme activity of the seedlings, chlorophyll and carotenoid contents, lipid peroxidation level, ion leakage in roots	Decreased root and shoot lengths (even at a 62.5 mg l⁻¹ concentration), root weight (31.1%, 67.2%, and 73.5%), leaf weight (38.4%, 62.7%, 72.8%), chlorophyll content The supernatant did not induce observable phenotypic changes and changes in the fresh weight of seedlings Short exposure to high concentrations produced toxicity effects in seedlings	Yang *et al.*, 2020

Continued

Table 10.1. Continued

NP	Experiment	Effects	Reference
	Lens culinaris L., soaked seeds Particle size: NA Concentrations: 0.01, 0.025, 0.05 mg ml⁻¹ Growth parameters: root length, germination percentage, vigour index Physiological parameters: ROS, detection of H_2O_2, POD, enzymatic activity, total protein, total phenol, total proline, flavonoid content, lipid peroxidation rate	Significantly increased percentage of germination, length of the root, and vigour index of the seedling Moderately increased level for all antioxidant enzymes Antioxidant enzymes, defense enzymes, phenol and flavonoid levels, and proline content varied with concentration Dose-dependent results: induced innate immunity and plant vigour (0.025 mg ml⁻¹), delay of all parameters, induction of ROS and H_2O_2 production (0.05 mg ml⁻¹)	Sarkar *et al.* (2020)
	Glycine max L. Particle size: NA Concentrations: 50–500 ppm Physiological parameters: enzymatic and metabolic activities	The size and range of NP treatment determined the severity of the toxicity Anti-oxidative biomarkers changed with NP application	Yusefi-Tanha *et al.* (2020)
Iron oxide NPs (Fe_2O_3NPs)	*Mentha piperita* L. Particle size: NA Concentrations: 10–30 ppm Physiological parameters: enzymatic and metabolic activities	Increased anti-oxidative enzymatic activities Decreased proline and malondialdehyde content An appropriate dose of NPs could be used for stress tolerance strategies	Askary *et al.* (2017)

APX, ascorbate peroxidase; CAT, catalase; DPPH, 2,2-diphenyl-1-picrylhydrazyl; GPOX, guaiacol peroxidase; GR, glutathione reductase; H_2O_2, hydrogen peroxide; NA, not available; PNPs, phosphorus NPs; POD, peroxidase; ROS, reactive oxygen species; SENP, selenium NP; SOD, superoxide dismutase; TASC, total ascorbate content; TFC, total flavonoid content; TPC, total phenolic content.

(Kolenčík et al., 2020), barley (Moaveni et al., 2011), wheat (Du et al., 2015; Hussain et al., 2021; Irshad et al., 2021), millet (Kolenčík et al., 2019), lentil (Khan and Ansari, 2018), carrot (Elizabath et al., 2017), grapevine (Teszlák et al., 2018), and groundnut (Prasad et al., 2012). The results of these studies can provide information that will bring multiple benefits to producers, farmers, and consumers, and, implicitly, will provide economic benefits on a large scale.

10.1.2 Role of NPs in pest control

Nanotechnology can provide new tools for rapid disease detection, molecular treatment of diseases, and other uses in pest and pathogen management (Danish and Hussain, 2019). In modern sustainable agriculture, integrated pest management (against microbial pathogens, fungi, bacteria, and insects) is based on various NFs. NPs can be used successfully as individual insecticides, or as carriers of insecticides, pheromones, or other active substances, or are produced from various natural resources, which makes them 'green' alternatives with great potential to use compared with the traditional control of pests (Athanassiou et al., 2018). Nanocarriers (nanocapsules, nanospheres, micelles, nanogels, and nano-emulsions) modify the properties of active ingredients, and provide controlled release with better dissolution, low dose, improved stability, and long-lasting effectiveness in pest control (Sushil et al., 2021).

Metal NPs (e.g. silver (Ag), gold (Au), coper oxide (CuO), zinc oxide (ZnO), cerium oxide (CeO$_2$) are among the most promising broad-spectrum antibiotic agents, with antibacterial and antifungal action (Aziz et al., 2016). The antibacterial action has bene demonstrated against both Gram-positive bacteria such as *Staphylococcus aureus* (Guan et al., 2021) and Gram-negative bacteria including *Escherichia coli, Klebsiella pneumoniae, Aeromonas hydrophila* (Aziz et al., 2014), *Pseudomonas aeruginosa* (Dorobantu et al., 2015), and *Bacillus subtilis* (Rafińska et al., 2019; Alsamhary, 2020). The mechanism of antimicrobial action of NPs can be one of the following prototypes: oxidative stress and biomolecule damage, ATP depletion and membrane interactions, cation release and intracellular damage, or non-oxidative

mechanisms (Aziz et al., 2014; Slavin et al., 2017; Wang et al., 2017). These mechanisms can even occur simultaneously.

Comparative research on several types of NPs showed greater toxicity on the soil microbial population for AgNPs, CuONPs, and ZnONPs, while titanium dioxide NPs (TiO$_2$NPs) proved to be nontoxic (Asadishad et al., 2018).

ZnONPs, copper NPs (CuNPs), magnesium oxide NPs, calcium oxide NPs, TiO$_2$NPs, manganese oxide NPs, silicon dioxide NPs, and AgNPs are used for early detection of pathogens by applying nanobiosensors (Rai and Kratosova, 2015; Elmer and White, 2016; Kumar et al., 2017) and for the development of ecological pesticides in weed management (Servin et al., 2015; Elmer et al., 2018). Panpatte et al. (2016) believe that nanobiosensors should be used for sustainable agricultural management practices, having excellent potential in detecting insect attacks (Afsharinejad et al., 2015) or pathogens, so that pesticides, insecticides, or fertilizers can be properly applied to ensure an increased crop yield. Some of the most effective NMs used in the fight against insects are aluminosilicate nanotubes, which, when spread on the surface of the plant, are taken up by the insects, who consume them, and die.

Nanoencapsulated NFs transport the pesticides to the target position and release the active ingredient close to the root. In this way, they do not cause damage to other organisms or to the plant itself, nor do they lead to the development of resistance against them (Patil and Solanki, 2016). Other advantages of nanoformulated pesticides are accuracy, controlled solubility, permeability, rigidity, thermal stability, crystallinity, and the fact that they are sustainable in the agroenvironmental system (Khan and Rizvi, 2014; Haq and Ijaz, 2019). However, it should be noted that the intrinsic antifungal and antimicrobial activities of NPs can threaten nitrogen-fixing bacteria and disrupt natural symbiotic relationships (Pestovsky and Martínez-Antonio, 2017).

10.2 Ecotoxicological Implications of Nanoparticles

Although nanotechnology has evolved so that it has become a beneficial and even indispensable strategic tool for the agricultural sector, revealing the antagonistic effects of NFs calls for thorough ecotoxicological research to limit their possible threats to organisms. Rana and Kalaichelvan (2013) considered it important to understand the ecotoxicity of NPs through the lens of the direct link between their adverse effects and the microorganisms, plants, animals, and humans they affect, from the perspective of the trophic levels through which they are related. Research on the ecotoxicological implications of NPs is absolutely legitimate and necessary to regulate the use of NPs in agriculture and the food industry, given the multitude of factors that can induce NP phytotoxicity and that make it difficult to evaluate them from this point of view.

Toxicological research can effectively guide the evaluation and risk management of the impact of NMs on agrosystems and in general on the environment, but also on human health. The phytotoxicity of NPs depends on their physical nature, the size and structure of the NPs, the method of synthesis and production, the structure and surface charge, the coating materials, the degree of aggregation (clumping) or disaggregation of the particles, the plant species, the method of application, and plant growing techniques. Individual modifications of the physico-chemical characteristics of NPs can cause different results in the same system (Paramo *et al.*, 2020). The purpose of these toxicity evaluations is to optimize the application conditions, the effective doses to be administered, and the physiological impact on the plants without producing adverse effects in the food chain. Very recent reviews (Arora *et al.*, 2022; Kolbert *et al.*, 2022) reaffirm the two-way dependence between the plant species and the effects of NFs, depending on the metal accumulation capacity and tolerance of the plant species. NP toxicity can manifest itself at the level of morphological, physiological, and biochemical indices of plants.

Regarding methodological aspects, the duration of exposure, route, and method of administration of NFs influence their effects on plants. The developmental stage of the plant also influences the response shown to the presence of nanometals. Interactions between NPs and plants under field culture conditions require maximum attention in terms of phytotoxic and ecotoxic effects. Maintaining soil health with the help of NFs that limit the use of conventional fertilizers favors the expansion of ecological agroecosystems of crops with good quality and productivity. Cox *et al.* (2016) referred to the need for continuous research in order to be able to provide answers regarding the delayed impact of environmental exposure to NPs, their bioaccumulation in the food chain, the interaction of NPs with other pollutants in the environment, and possible adaptation mechanisms.

10.2.1 Environmental implications of NFs

A topic of widely debated interest is that, due to the almost unlimited synthesis of artificial NPs, their potential release into the environment and implicitly the exposure of different professional categories and the general public to NPs still shows a dramatic upward trend (Patel *et al.*, 2022). Fabricated NFs represent real challenges for environmental systems. The existence in the literature of an impressive number of bibliographic sources on this subject demonstrates that nanoecotoxicology research, over several decades, has shown that the interactions between NMs, cells, plants and animals, the environment, and humans are remarkably complex (Hegde *et al.*, 2015). To assess the impact that NPs have as pollutants that are transmitted to the environment, several of their properties such as mobility, bioavailability, and biodegradability need to be taken into account. Once released into the environment, given the mobility of NMs, they either spread far and wide, with measurements indicating relatively low average concentrations, or are not transferred from the point of release, with accumulation phenomena occurring locally in sediments, in the absence of NP degradation processes. Transport of environmental pollutants takes place largely through water.

The mobility of NPs in soil and sediments depends on the prevailing physico-chemical conditions, especially porosity, but also on the reactivity of each constituent of the matrix, while the bioavailability of dispersed NPs depends on the filtering properties resulting from these circumstances. Lyon *et al.* (2007) considered that the modification of NPs after they come into contact with the constituents of the environmental matrix affects their mobility, aggregation, and toxicity characteristics. Although these aspects have been explained by numerous experimental studies, the structural heterogeneity and organic and mineral composition of natural environments must be taken into account to understand the fate of NPs under natural conditions. NPs with negative effects that persist in the environment can continue to have these effects for a long time. The degradability or persistence of NPs depends on the type of NF. Carbon-based NPs and fullerenes, which theoretically can be degraded and re-enter the carbon biogeochemical cycle, in practice are more likely to accumulate in the environment than to disappear, the explanation being that they are extremely persistent at high temperatures (Cataldo, 2002), and resistant to acids and ozone (Robichaud et al., 2007). Mineral NPs tend to more or less dissolve, and dendrimers and organically coated NPs degrade rapidly in the environment and inside organisms, as reviewed by Hegde et al. (2015). In addition to the toxicity of NPs themselves, the chemicals generated during their synthesis are also toxic and harmful to the environment. There are opinions that, in order to reduce the harmful environmental effects of some NPs, their toxicity should be reduced or even eliminated by their redesign (Dasgupta et al., 2016; Jain et al., 2018). Oomen et al. (2015) suggested the use of agricultural waste for NP synthesis as another possible approach to mitigate their toxicity.

Nanosensors and nanoremediation methods can detect and remove environmental contaminants (Iavicoli et al., 2017). The use of fertilizers, pesticides, and insecticides with NPs or encapsulated in NMs has the effect of reducing environmental pollution by obtaining the expected results using a smaller amount of chemicals, which are released slowly, thus minimizing their loss to the environment. Although concerns regarding the knowledge of environmental risks of NFs are numerous, it is essential to continue research in achieving/ maintaining a 'nontoxic environment'.

NPs have also been tested as a mitigation strategy against various abiotic stresses, including drought (Dimkpa et al., 2020a; Zahedi et al., 2020), flooding (Hashimoto et al., 2020),

high temperatures (Hassan et al., 2018), frost (Jampílek and Kráľová, 2021), salinity (Torabian et al., 2016), metals and metalloids (Rizwan et al., 2019a, 2019b), high and low light intensities and ultraviolet radiation (Tripathi et al., 2017), and nutritional imbalance (Rajput et al., 2021). In nature, plants are sometimes exposed to these stress factors individually but more often to different combinations, and NFs give them greater tolerance to abiotic stress and a considerable added value in the era of climate change, when water scarcity, drought and thermal stress are common in all regions of the world and potentially affect molecular, morphological, and physiological changes, which reduce almost all biological processes and crop plant yields (Gupta and Senthil-Kumar, 2017; Shalaby et al., 2022). Nanofertilizers can be considered true antistressors that act to improve the plant defense system and crop plant performance, and, when used in combination with nanobiofertilizers, offer considerable additional benefits (Zulfiqar et al., 2019).

Significant results for mitigating drought stress have been demonstrated in wheat by applying chitosan (Behboudi et al., 2019), iron NPs (Adrees et al., 2020), ZnONPs (Dimkpa et al., 2019), nanosilicon (Zahedi et al., 2020), and others. As reviewed by Rajput et al. (2021), Shalaby et al. (2022), and Verma et al. (2022) considering different case studies, the application of different NFs in agriculture has resulted in increased resistance of plants to stress and improvement of their productivity when subjected to multiple individual or combined situations of abiotic stress. Numerous studies are constantly adding to the different types of plant responses to abiotic stress, describing a wide variety of mechanisms (Sarraf et al., 2022). However, crop production under stress and integrated management for environmental health and safety still require more studies on adaptive responses.

10.2.2 Implications of NFs for soil and water

Soil is the richest environmental matrix for both primary and agglomerated/aggregated natural

NPs, while water can be the transport pathway and a temporary reservoir for NPs, and the final receptors for any compounds or nonvolatile particles that spread in the environment will be sediments and soil. Soil pollution through the deposition of NPs in sediments therefore has repercussions on groundwater (Rajput *et al.*, 2018a), and the growth of agricultural plants in such contaminated soils also involves the introduction of NPs into the food chain with serious consequences for human health (Rico *et al.*, 2011). The presence of NPs in the soil can have a direct impact on the microorganisms in the biota. The increased reactivity of NPs and their interaction with other soil pollutants can lead either to the formation of new compounds or to modification of the toxicity of existing ones.

NPs behave differently in soil depending on their size: small NPs break down directly and produce ions that can be incorporated into plant tissues, while larger NPs break down in soil into smaller NPs that will later be incorporated into plants (Rajput *et al.*, 2019). To maintain soil conditions favourable to agricultural crops, nanosensors are used to monitor soil conditions (e.g. pH, temperature, humidity, and nutrient availability), as well as microbial pathogens. The use of nanosensors provides useful information about the health of the soil, allows adequate treatment, and facilitates the choice of the right crop for that soil, while the timely detection of microbial pathogens helps to treat them and minimize crop productivity losses (Chaurasia *et al.*, 2021). In fresh water, NPs are influenced by the particular type of organic matter or other natural particles, while in hard water and seawater, NPs tend to aggregate. Modification of the toxicity of NPs in water is influenced by pH, salinity, the presence of organic matter, and the state of dispersion; however, systematic investigations of these phenomena are needed in ecotoxicological studies (Handy *et al.*, 2008). For example, TiO_2 NPs can be used effectively for wastewater treatment. Johnson *et al.* (2011) suggested that up to 99% of TiO_2 NPs remain in wastewater treatment plant sludge, and the presence of microorganisms in this sludge could lead to substances such as selenate and selenite exhibiting even greater toxicity (Jain *et al.*, 2016).

10.2.3   Implications of NFs for plants

Plants are closely related in the 'soil–plant system,' and the influence exerted by NFs on plants depends on soil and environmental characteristics, NP release mechanisms, and plant species. Relevant to consider in the design of research on metal NP effects on plants is the dual impact they can have as a function of concentration, namely, low concentrations positively influence them, while high concentrations threaten plant growth and development (Jha and Pudake, 2016). The accumulation of NPs in plants depends on the absorption modes (through the aerial surface, roots, and seeds), the atmospheric variables with which they interact, the rigidity of the cell wall, and the anatomical, physiological, and biochemical peculiarities of the plant species or cultivar (Mittal *et al.*, 2020). The high efficiency of the foliar entry method is due to the easy absorption of nutrients through the nanopores of plasmodesmata in leaves (Iqbal, 2019; Rajput *et al.*, 2021). Once inside the plant, the NPs move into the tissues through the apoplast and the symplastic pathway, and are translocated through the cytoplasm into the photosynthetic apparatus of the plant (Pérez-de-Luque, 2017).

Plants exposed to NFs show various morphological and physiological changes. as well as ultrastructural changes in cells. NPs can influence germination frequency, shoot and root length, biomass, yield, and photosynthetic efficiency. Physiological changes produced by NPs on plants are particularly important, because they can modify the uptake and translocation of NPs, the response of the plant's primary and secondary metabolism, genetic expression, and, last but not least, their response to biotic and abiotic stress. In a large number of cultivated plants, considerable changes in physiological indices have been revealed, including chlorophyll content, the ability of leaves to capture sunlight, plant performance, and nitrogen and soluble protein metabolism, and antioxidant activity. Changes in biochemical indices and plant enzymatic responses, such as activity of peroxidase, catalase, ascorbate peroxidase, and polyphenol oxidase, activity of antioxidant enzymes (e.g. phosphatase, alkaline phosphatase, phytase, dehydrogenase), and of soluble proteins or proline, have also been documented (Table 10.1).

Among the possible nanotoxicological responses of plants at the anatomical and ultrastructural level are the damage/destruction of some cellular components (e.g. cell wall, cell membrane, chloroplasts, peroxisomes, mitochondria, and nucleus), changes in the shape and size of starch granules, deposition of electron-dense materials near cell walls, and altered appearance of epidermal, mesophyll, cortex, or central cylinder cells (Rajput *et al.*, 2018b). The most destructive impact of NPs on plants is DNA damage, which can occur either through direct pathways or indirectly through the stimulation of reactive oxygen species (ROS) production (Khan *et al.*, 2019). NPs can interact directly with DNA, or with proteins involved in DNA replication or repair, and also with proteins related to cellular homeostasis. The degree of DNA damage and the amount of ROS produced are determined by the chemical composition of the NPs (Zhu *et al.*, 2013; Adisa *et al.*, 2019). In conclusion, it seems an almost impossible mission to quantify and explain the multitude of positive and negative effects of different NFs in order to achieve the desired development of safety protocols for the use and application of NPs in agriculture and on plants intended for human consumption.

10.2.4 Implications of NFs for human health

Due to the lack of knowledge about the human health effects associated with exposure to NPs, whether natural or artificial, the ethical duty is to elucidate their toxicological effect in order to take precautions regarding their occupational exposure or use. Exposure to NPs from the agricultural sector can have a direct or indirect impact on human health. Assessing the risk posed by NPs to humans began with the identification of hazards and professional exposure pathways within the manufacturing process, as this is probably the most immediate situation where safety issues need to be addressed to allow other areas of nanotechnology to advance (Lam *et al.*, 2006; Spruit, 2015). Knowing that, once introduced into an organism, NPs produce a wide variety of effects, an interdisciplinary

approach is needed to identify knowledge gaps and describe the links between NFs, the environment, and human health, so an approach in the form of causal diagrams was proposed by Smita *et al.* (2012) as a tool to assess areas of agreement and disagreement among scientists, policy makers, and the community. Nanotoxicity has varied effects on human health depending on the ease with which NFs enter the human body by different routes, which are mainly by respiratory, dermal or gastrointestinal routes in the case of exposure to natural particles, or by injection in medical applications. It is important to reiterate the possibility of serious problems for human health generated by the increased toxicity of by-products generated during NP synthesis, a situation that could limit the use of NMs as therapeutic and diagnostic tools. Some insoluble NPs can accumulate significantly in the lungs (Wiemann *et al.*, 2017), brain (Fernández-Bertólez *et al.*, 2019), liver (Zhang *et al.*, 2016), spleen (Drozdov *et al.*, 2021), and bones (Gao *et al.*, 2021), as demonstrated using various bioassays with experimental organisms.

According to Forest (2021), in recent years numerous experimental studies have been carried out using mammals as test organisms, which have confirmed various phenomena of toxicity in the cardiac system (Gonzalez *et al.*, 2016; Feng *et al.*, 2017, 2018; Yang *et al.*, 2018; Guo *et al.*, 2021), digestive system (Ahamed *et al.*, 2020; Dussert *et al.*, 2020; Cao *et al.*, 2021), and respiratory system (Nho, 2020; Chaud *et al.*, 2021), but also in the reproductive and renal systems and at the skin level. Research in this field must continue to contribute to the knowledge of the interactions between the multitude of NMs and cells that lead to toxicological responses, but also of the mechanisms involved in NP-mediated toxicity. NPs cause oxidative stress and may have relevant cytotoxic, genotoxic, and carcinogenic potential (Jain *et al.*, 2018). After entering the human body, metal-based NPs (e.g. AgNPs, AuNPs, and CuNPs) disrupt the body's antioxidant mechanism by interacting with cell antioxidant enzymes and proteins and generate ROS, triggering inflammatory reactions and sometimes even destroying the mitochondria, which in turn causes cell apoptosis and necrosis (Schrand *et al.*, 2010).

Most of the time, in nature, the human body is subjected to poly-exposure through air particles, and the health effects are combined. Entry of particles from the environment into the body is facilitated by the interaction between the particles and NPs, which changes their toxicological profile. Compounds that individually do not cause adverse effects can induce significant toxicity when they are part of a mixture, which can generate an underestimation of risk (Jiang *et al.*, 2020). Mixing NPs with environmental pollutants has revealed surprising adverse effects, even though the individual toxicities of the compounds were well characterized (Silins and Högberg, 2011), necessitating an assessment of their joint toxicity (Yang *et al.*, 2018). A holistic approach to multi-pollution (Yun *et al.*, 2015), referred to by Deville *et al.* (2016) as 'mixture toxicology,' is required for optimal and reliable management of the risk of NPs to human health, regardless of the individual pollutants. The effects of these interactions can be additive, synergistic, or even antagonistic (Liu *et al.*, 2018).

10.3 Tools for Identifying and Monitoring the Health and Environmental Risks of NMs

For judicious design of experiments to identify and monitor the risks produced by NPs and NFs used in agriculture, numerous factors must be considered, together with the use of several investigative techniques that provide complementary information (Kahru *et al.*, 2008; Wojcieszek *et al.*, 2020).

Standard methods for evaluating the ecotoxicity of a potentially polluting substance can also be applied to nanoregulations but with some adaptations (Boros and Ostafe, 2020). For the analysis and characterization of NPs and their course during the process of their interaction with organisms (e.g. absorption, bioaccumulation, and translocation), electron microscopy techniques (transmission, scanning, and atomic force microscopy), spectrometry (atomic absorption and Raman spectroscopy, Fourier-transform infrared spectroscopy), and X-ray imaging techniques and computed tomography have been used. Applied separately or in different combinations, these investigative techniques are all the more relevant as they improve

the detection limits and spatial resolution of NPs (Castillo-Michel *et al.*, 2017). The choice of the investigation method/methods aims, in addition to the relevance and correctness of the results obtained, is determined by the simplicity of the test, the ease of preparation of the samples to be investigated, and the noninvasive nature.

Applied bioassays for simple and rapid ecotoxicity testing recommend the selection of appropriate test organisms from different trophic levels (i.e. microorganisms, plants, invertebrates, and vertebrates), and from aquatic and terrestrial environments. In general, for ethical reasons, vertebrates are not considered suitable organisms, nor species with protection status or that require difficult culture methods and conditions, or have a longer reproductive cycle. Test systems have been standardized for some of these organisms, in certain clearly specified exposure situations (e.g. composition of the environment, age of organisms, exposure times, measured parameters) (OECD, 2014; Hund-Rinke *et al.*, 2016).

Rana and Kalaichelvan (2013) considered that, in the case of microorganisms (bacteria, fungi, protozoa, and algae), due to their abundant presence in diverse habitats, small sizes, rapid rate of multiplication, and their role in ecosystems, individual observations can also be taken into account through miniaturized and rapid tests. The shortcoming of many studies on the toxic effects of NPs on organisms is that pure systems have been used and little consideration has been given to possible interactions between NPs and environmental constituents that affect their bioavailability and toxicity. Cytotoxicity and genotoxicity induced by NPs can be quantified by observing the mitotic index, micronuclei, chromosomal aberrations, and lipid peroxidation, through *in vivo* and/or *in vitro* study methods (Landsiedel *et al.*, 2009; Dwivedi *et al.*, 2018).

Chromosomal aberrations and the degree of disruption of DNA repair mechanisms and the cell cycle, manifested in the case of NP toxicity, can be highlighted by a series of tests, the most common of which are the bacterial reverse mutation assay (Ames test), single-cell gel electrophoresis assay (Comet assay), micronucleus assay, and 8-hydroxy-2′-deoxyguanosine (8-OHdG) analysis. *In vivo* tests can highlight the carcinogenic potential of NMs. A compound tested by the bacterial reversal mutation assay has mutagenic potential if it is capable of causing mutations that allow the bacterium *Salmonella* Typhimurium to synthesize histidine. The disadvantages of this test are that it can generate false-positive results (Khandoudi *et al.*, 2009), some NMs are bactericidal (Landsiedel *et al.*, 2009), it is difficult to transpose the data obtained for prokaryotic cells, and it requires other tests after the initial screening.

The Comet assay can assess the extent of genotoxicity in eukaryotic cells by highlighting DNA damage. The limitation of this test is that it is sensitive and requires proper sample handling to ensure reproducibility of results. The number of micronuclei in treated cells can provide information to measure chromosome breakage, impaired DNA repair, chromosome loss, nondisjunction, necrosis, and apoptosis. The 8-OHdG assay is based on the fact that ROS can attack guanine residues in DNA forming 8-OHdG lesions with mutagenic potential and is therefore a good biomarker for carcinogenesis. *In vivo* research validates NP toxicity studies on animals. The difficulties lie in deciding the parameters that should be considered in the toxicity examination and correct interpretation of the study results, taking into account the entry route, bioavailability, and subsequent translocation of the NPs, and the organs where accumulation is significant. In the design of genotoxicity research, in order to obtain results comparable to those of other researchers and on other NMs, the specificity of previously tested NMs must be taken into account; standardized methods should be uses, the mechanisms of genotoxic effects of NMs should be known in order to consider their absorption and distribution, and *in vivo* studies shod be used to confirm *in vitro* results in order to extrapolate to human risk.

To approach genotoxicity investigations for new NMs as realistically as possible, the use of several standard test methods covering a wide range of mechanisms is required (Oesch and Landsiedel, 2012). In the context of management of the effects generated by the exposure of different professional categories to NPs, evaluation tools have been developed for specific risks, and criteria for classification into risk classes have been established, associated with measures at the personal and organizational or technical level, in order to reduce possible risks (Spruit, 2015; Liguori *et al.*, 2016).

10.4 Considerations Regarding the Risk Assessment of Nanomaterials

As mentioned earlier, NMs are widely used in agriculture and the food industry as fertilizers, pesticides, and additives, and for detection of plant diseases and toxins in food and agricultural products. Many studies have certified the benefits of NMs in human and veterinary medicine, soil and waste-water treatment, and energy conservation. Although the advantages of nanoproducts are well known, there are many ecotoxicological implications of their use. For example, a review by Hashim *et al.* (2022) highlighted the potential use of graphene NMs for food industry as well as their limitations. Cebadero-Domínguez *et al.* (2022) demonstrated that the studies in this field do not respect the recommendations of the European Food Safety Authority (EFSA) and that graphene NMs can lead to genotoxicity, so toxicological evaluation is required for consumer safety.

The use of engineered NMs (ENMs) in different fields can be sustainable and safe, both through a better understanding of the associations between ENM characteristics and living organisms, and through permanent improvement of the approaches for assessment and mitigation of risks of ENMs during their life cycle (Fadeel *et al.*, 2018). Although many of the benefits of NMs for humans and the environment are recognized, there remain many disadvantages due to their physico-chemical properties and interactivity with plant/animal cells, soil, water, and air. According to Resnik (2019), "Most of the information concerning the risks of ENMs has come from *in vitro* cell studies, *in vivo* animal experiments, or computer modelling of physical, chemical, and biological processes related to exposure to and distribution, excretion, aggregation, and toxicity of engineered NMs.' Thus, it is necessary to assess the risk of nanoproducts in each step of their life cycle: production, utilization, distribution, and recycling (Linkov *et al.*, 2009; Naqvi *et al.*, 2018). Oksel *et al.* (2017) considered that "risk assessment is an action to identify, assess and prioritize potential risks that are most likely to occur as a result of a given exposure to a particular substance.'

Risk assessment involves the identification, evaluation, and characterization of the potential risks of a substance for human safety and environmental protection by adoption of specific risk-management measures by the decision makers (Center for International Environmental Law, 2016). One of the most important risk-assessment strategies has been described by Oomen *et al.* (2015) and Bos *et al.* (2015), as part of the European Union (EU) FP7 MARINA (Managing Risks of Nanoparticles) project. This strategy has integrated the grouping and read-across approaches and includes two stages: (i) problem framing; and (ii) risk assessment. In the first stage, based on the physico-chemical properties, behaviour, and reactivity of NMs, the Relevant Exposure Scenarios for the entire life cycle of a NM can be identify. Also, in this stage, we can investigate whether there are some negative health or environment effects of NMs. Four steps are important for the second stage: (i) risk characterization (is there sufficient quantitative and qualitative information to provide conclusions with an adequate degree of certainty?); (ii) defining data needs; (iii) data gathering regarding fate/kinetics and hazards adapted to a specific exposure scenario; and, finally, (iv) data evaluation in close connection with those already available.

There are many risk-assessment frameworks of NMs in the literature proposed by different authors and organizations (National Research Council, 2009; Kuempel *et al.*, 2012; Center for International Environmental Law, 2016; Oksel *et al.*, 2017; Mattsson and Simkó, 2017). An exhaustive analysis on this subject revealed that the risk assessment involves going through several stages: risk hypothesis, hazard assessment, exposure assessment, dose–response assessment (hazard characterization), risk characterization, risk management, and risk communication. The first step for an adequate risk assessment of a particular NM consists of a suitable problem formulation: "A robust risk assessment relies on a testable (falsifiable) risk hypothesis based on a precise identification of relevant sources of a given substance as well as suspected hazards to specific 'targets' such as human beings, animals or the environment'

(Center for International Environmental Law, 2016).

An important part of the risk assessment is to point out the potential hazards of nano-products on health and the environment, and then to evaluate biological effects in terms of type and magnitude using toxicology testing methods. The qualitative and quantitative exposure assessment to NMs estimates the extent to which living things and environment have been exposed to them. Exposure assessment involves calculation of predicted environmental concentrations (PECs) of a substance over its entire life cycle starting with production and continuing with processing, application, and dispersion in the environment. Dose–response assessment provides quantitative data, based on mathematical models, regarding the interconnection between dose of a chemical substance (i.e. the NP) and its negative effects on health and environment.

Risk characterization is based on the proportion between PECs and predicted no-effect concentrations (PNECs), known as the risk characterization ratio (RCR). When this ratio is greater than 1, the substance can cause serious problems and measures are necessary for risk reduction; if the ratio is less than 1, the substance is of low concern and the risk is considered to be sufficiently controlled (Amiard and Amiard-Triquet, 2015; European Chemicals Agency, 2016; Wigger and Nowack, 2019). The environmental risks of three ENMs (AgNPs, TiO$_2$NPs, and ZnONPs) in European freshwater were evaluated by Hong *et al.* (2021) using the RCR. They concluded that for each form (pristine, dissolved, and transformed) of ENM released into the environment, a specific risk assessment is required.

Due to the development of nanotechnologies, the exposure rate of organisms to NPs has increased, so risk-management strategies are needed for human and environment safety. All data obtained in the previous stages of risks assessment are essential for risk-management decision making and risk communication (Kuempel *et al.*, 2012). Several strategies of risk management for ENMs have been developed over time. The majority take into account a hierarchy of the hazard control strategies, from most control to least control, beginning with elimination/reduction of all hazards, and continuing with utilization of engineering and administrative controls (Oksel *et al.*, 2017).

With the continuous development of nanotechnologies, it is essential that the regulators and risk communicators disseminate information regarding the risk of NMs through all common types of media, so that policies can be developed based on risk science (Bostrom and Löfstedt, 2010). A comprehensive online survey was carried out over a period of 17 months using SurveyMonkey software as part of the project 'Nan-O-Style' by Joubert *et al.* (2020). The data produced illustrated a generally optimistic attitude toward NMs but with some concerns on safety and potential risks in dietary products. In addition, the authors highlight the necessity to expand scientific education and communication, so that risk governance of NMs could be achieved through active involvement of society.

The latest recommendations regarding the risk assessment of the use of nanotechnologies in agriculture and food industry were published by the EFSA in 2021. According to this document, the physico-chemical properties, exposure, and toxicological effects are the main elements for risk characterization of a NM (More *et al.*, 2021). Lenz e Silva and Hurt (2016) (cited by Silva *et al.*, 2019) emphasized the considerable importance of the physico-chemical properties (chemical composition, surface area, morphology, shape, solubility, size, and distribution concentration) for risk assessment of carbon NMs, Naqvi *et al.* (2018) suggested that the behaviour and toxicity data of the NMs and an estimation of the predicted amounts in the environment are essential in the development of risk-assessment strategies. In their opinion, there are three steps for risk assessment of NMs. First, hazard identification, meaning the possible toxicity of NMs for biological systems based on their physico-chemical properties. Second, the hazard characterization should be taken into consideration, meaning the dose–response relationship, mechanism of toxicity, exposure pathways, and immune response. Lastly, it is important to evaluate the exposure, taking into account the carcinogenic, immunotoxic, mutagenic, and genotoxic potentials.

Rao (2014) proposed a framework for a life cycle analysis to evaluate the possible risk of NMs, taking into account all stages of manufacturing and marketing of nanoproducts, as well as their use and end-of-life disposal. Some strategic plans regarding NM risk assessment on human health have been elaborated over time. One of these was presented by Dekkers *et al.* (2016) and takes into account six important elements: exposure potential, NM disintegration, NM metamorphosis, accumulation, toxicity potential on genetic information, and the immune system. Afantitis *et al.* (2018) used protein corona fingerprinting to create a model, available through the Enalos Cloud Platform, for evaluation of the biological response of surface-modified AuNPs, which can improve the risk assessment based on experimental information.

A comprehensive study on frameworks and instruments for NM risk assessment was carried out by Hristozov *et al.* (2016). In their review, they took into account both information from literature (scientific articles and research projects) and online questionnaires to organizations engaged in elaboration of tools for identification of environment and health risk. In order to evaluate 48 tools distributed over eight categories, 15 criteria were established. Each of these frameworks and tools has strengths and weaknesses. For example, the authors selected those frameworks that fulfil most of the criteria considered: European Chemicals Agency guidance on risk assessment of NMs under REACH (Registration, Evaluation, Authorization and Restriction of Chemicals), EFSA guidance on risk assessment in the food and feed chain, the MARINA framework, the Nano Risk Framework, NMs under REACH, Comprehensive Environmental Assessment, Nano Life Cycle Risk Analysis (Nano-LCRA), and others.

Control banding is a tool for qualitative risk assessment, used for the first time in the pharmaceutical domain, with applicable in small enterprises, on the basis of which security measures can be taken for the workers who handle products for which little information is available (i.e. toxicological and exposure data) (Riediker *et al.*, 2012). Several control banding models in the nanoindustry field have been proposed: CB Nanotool; IVAM Guidance; Stoffenmanager Nano, NanoSafer, and ANSES; and Precautionary Matrix, Good Nano Guide, and EC Guidance (Liguori *et al.*, 2016; Dunn *et al.*, 2018). Eight control banding tools for risk assessment of nano-Fe_2O_3, nano-Al_2O_3 and nano-$CaCO_3$ were analysed comparatively by

Gao *et al.* (2019), highlighting the differences between them in seven exposure scenarios. They concluded that the CB Nanotool and Stoffenmanager Nano have a number of advantages in terms of quantitative and qualitative evaluation, but recommended choosing more than one tool to evaluate alternatives and make better decisions. In order to evaluate the risks of NMs, studied an application named NanoSerpa for the insurance sector taking into account different exposure scenarios, integrating hazard-related data and optimized exposure models. For example, in the case of an accident associated with NMs, they described five input parameters that must be introduced: the type and quantity of NM, the technological process type, the type of accident, and other observations related to the accident. After this, based on probabilistic models, this app can give data regarding the health hazard values and risk indices.

Species sensitivity distributions are a quantitative model used for estimation of the ecological risk of a chemical substance. For example, Coll *et al.* (2015) used PNECs, obtained from probabilistic species sensitivity distribution, to estimate the RCR for five ENMs. They considered that fullerenes, carbon nanotubes and nano-Ag were the ENMs of lowest concern, while nano-ZnO and nano-TiO_2 have an important risk for the environment, especially in the fresh waters and in soil. Two projects financed by the EU have developed complementary instruments for risk assessment of ENMs: SUNDS (SUN Decision Support) software system for the estimation of professional, environmental, and consumer risks of ENMs in industrial products, and the GUIDEnano tool, which can be useful in designing a strategy for risk assessment suitable for a specific product (Fadeel *et al.*, 2018). Isigonis *et al.* (2019) carried out an exhaustive study covering over 25 years of research in the field of risk of nanotechnologies in which 60 instruments for risk evaluation, mitigation, and communication were identified. They took into consideration the most important criteria for improving the decision-support tools for the risk governance, as follows: "uncertainty analysis, structured decision making, fair and knowledge communication process, understandable, quantitative information, documented applications, transparency application, comprehension, influence on final policy.' In their opinion, the

SUNDS, GUIDEnano and NanoSafer tools satisfy most of the evaluation criteria, and can be used for risk concern/safety assessment, risk evaluation, and risk management and decision making. In 2008, the TÜV SÜD Industry Service and The Innovation Society (St. Gallen, Switzerland) inaugurated CENARIOS, the first certified risk management and monitoring system, making an important contribution to development of the NM industry (The Innovation Society, 2008). This system has three modules: risk evaluation, risk monitoring and issues of management/communication. Although this system does not include all types of risk, it can be used for supporting decision making integrated in SUNDS, using the CENARIOS stand-alone module (Isigonis *et al.*, 2019).

An analysis regarding the involvement of 34 companies that produce/use EMNs (especially based on metal) in development of risk assessment and management measures was carried out by Iavicoli *et al.* (2019). According to their study, half of these companies carried out engineering exposure controls and less than half undertook environmental monitoring. The majority were involved with information and training programs, and unfortunately, less than 10% achieved the required levels of health monitoring for ENM workers. They concluded that all efforts should focus on both the development of standardized industrial frameworks for assessment of ENM risks, and the establishment of the risk-management policies for human health protection.

Although worldwide there are many well-known frameworks that approach risk assessment over the entire life cycle of nanostructured products, in order to obtain the full picture concerning their safety and sustainability, further studies on organisms and the environment are needed to cover the gaps in terms of toxicity and bioaccumulation (Weyell *et al.*, 2020).

10.5 Perspectives on Nanotechnology in Agriculture

One of the most significant roles in the global economy belongs to agriculture, due to its importance in ensuring food for both animals and humans. The latest statistics reveal that the human population has increased to 8 billion individuals (National Geographic, 2022), and is expected to reach at approximately 10 billion by 2050 (Niazian *et al.*, 2021). There are many aspects that farmers have to deal with, such as climate change, desertification, loss of crop fields due to urbanization, decreasing soil fertility, and increasing input of chemical fertilizers and pesticides (Basavegowda and Baek, 2021). Currently, one of the greatest challenges for humanity is to find solutions to increase productivity and food supplies, taking into account that 828 million people are affected by hunger (UN Food and Agriculture Organization, 2022). Related to this, nanotechnology has a tremendous contribution to make to sustainable development of agriculture and the food industry in terms of water and land protection, productivity increases, crop fortification, improving pest and pathogen control, increasing plant stress tolerance, improving animal health and reproduction, increasing the nutritional value of food, and promoting NMs for packaging to increase food quality and stability (Prasad *et al.*, 2017; Thakur *et al.*, 2018; Ali *et al.*, 2021b; Niazian *et al.*, 2021). For example, fluorescent silica NPs, AuNPs, and CuONPs, nanolayers, and nanowire biosensors are used in plant pathogen diagnosis (Alghuthaymi *et al.*, 2021). These authors highlighted the efficiency of nanocomposites consisting of graphene oxide (GO) and iron oxide (Fe_3O_4), GO–AgNPs, Ag–titanate nanotubes, and Ag–chitosan for crop protection, and for mycotoxin diminution, as an eco-safe and rapid solution for crop control in the future. Over time, the importance of NFs and NPs has been demonstrated in the field of pest control. According to Khandelwal *et al.* (2016), aspects that should be studied further include: natural compounds that can be use as biopesticides; specific NMs or NFs, preferably biocompatible, biodegradable, and nontoxic; and the consequences of NF usage on plant physiology. Dimkpa *et al.* (2020b) highlighted the role of nanofilms, nanopolymers, and nano-additives for efficient absorption of nitrogen and increasing plant productivity. In addition, the authors underlined the idea that research should focus on the development of enhanced-efficiency fertilizers, elucidation of nutrient mobilization mechanisms in soil, improving methods of

detecting nutrient deficiency in the soil and in plants, filling the gaps in plant nutrient requirements, and the diversity of routes to access nutrients. Additionally, knowing the modes of function of nanofertilizers means that specific regulatory frameworks can be promoted for their safe use by farmers (Raliya *et al.*, 2018).

Ali *et al.* (2021b) undertook a comprehensive study concerning the implications of NM technology in the agri-food sector, and presented a SWOT (strengths, weaknesses, opportunities and threats) analysis. In their opinion, nanotoxicologists need to evaluate the safety of NMs at each step of their life cycle, and must find answers to the problems related to surface-area dosimetry, the correlation between laboratory results and data field, and tracking the fate of NPs and nanocompounds in the environment. In their review on perspectives of nanotechnology and nanoplastics, Allan *et al.* (2021) summarized the aims concluded by the Global Summit on Regulatory Sciences in 2019 in Italy, as follows: improving the regulatory science using efficient tools and methodologies, concordance between methodologies for risk evaluation of nanoproducts, and regulatory guidance for scientific studies. According to the latest findings in the food industry, special attention should be given to the following aspects: development of functional foods using NPs to resolve some health problem, improvement of active packaging using NP composites to increase the validity period of food products, improvements in the nutritional quality of food products, identification of toxic doses of NPs for food safety, continuation of studies regarding the antioxidants and anti-tumoral properties of NPs as well as the toxicity of new composite NPs, using environmentally safe methods for production of NPs, and upgrading the physicochemical characteristics of NMs (Thakur *et al.*, 2018; Thiruvengadam *et al.*, 2018).

Efforts are currently concentrated on finding solutions to satisfy the food requirements of the human race. However, we also have to take into consideration the qualitative aspects of the food. In this regard, Khan *et al.* (2021) indicated the importance of increasing the nutritional quality of wheat grain using NMs (e.g. fullerol, TiO_2, CeO_2, carbon nanotubes, Ag, Au, platinum (Pt), selenium, cobalt), and also the necessity of evaluating their biosafety. They consider that biologically synthesized NMs should also be considered as well as chemically synthesized NMs. Hanafy (2018) also highlighted the advantages of nanotechnology in the veterinary sector for improving the drug delivery systems, breeding, diseases diagnosis, and animal treatment. In his opinion, more attention should be paid to the fungal biomass due to its high potential for NP synthesis (e.g. Au, Ag, Si, zirconium, Pt, Ti) vs other biological systems.

The qualitative and quantitative value of modified genetic plants can be improved using nano-enabled molecules. Niazian *et al.* (2021) described types of NP (calcium phosphate, polymeric-based, carbon-based, mesoporous silica, and magnetic-based) that can be mixed with DNA, RNA, and proteins and are useful for genetic plant transformation. According to these authors, more studies are needed in the field of bionanotechnology to: determine the best NPs that can be use as nanocarriers; optimize the use of mesoporous silica NPs for stable and efficient plant transformation; analyse the fine-tuning of NM properties and their stability; and increase the quantity of experimental data from the field. The advantages of nanotechnology for the agri-food sector are well known, but the scientific community is agreed that in the future the studies should be continued and focused on: obtaining NMs that are safe for humans and the environment, physico-chemical characterization of NPs and NMs, clarifying the routes of NPs and NMs along the trophic chains and their interactions with the metabolic pathways of organisms, establishing the safety dose of NPs, development of experimental models directly in the field under natural conditions, development of new instruments for risk evaluation and risk management of products obtained using nanotechnology, and improving the regulatory policies and standards for manipulation of NPs and NMs. Due to the potential risks of NMs and nanoproducts for human health and the environment, a safe-by-design concept has been implemented. In this regard, a good relationship between different stakeholders (scientists, representatives from industry, the media, environmental organizations, government, and community) is required.

10.6   Worldwide Regulations Concerning Nanomaterials

Although on the one hand nanotechnology offers timely huge improvements in quality of life through agricultural applications and food, on the other hand, equally legitimate are questions looking at their potential negative effects, which, according to Pandey and Jain (2020), are detected at three levels different: ethical, social (dangers for the environment, risks for health, and economic effects), and legal.

A brief review of the worldwide regulations regarding NMs has shown that there is no uniformity in the approach to this problem. Even at the level of the EU, a satisfactory coherence is not observed, although there are trends in this sense, but these have not gone toward all-encompassing regulations. There is a need for product sheets that contain data about the security of NMs (characteristics of their morphology, size, solubility, and even process descriptions) and investigation of the effects for every potential route of absorption (e.g. respiratory, dermal, intestinal) (Groso *et al.*, 2010). The idea that it would be very necessary to draw up a European register of products containing NMs arose a long time ago, and materializing, for example in Germany, as early as 2012 (www.umweltbundesamt.de; accessed 17 March 2023).

Several developed and emerging countries around the world are active in solving the problems raised by the regulation of nanotechnologies, but they have different approaches regarding the products that use NMs in the agri-food field, and often these regulations are not convergent. The European countries of the EU, together with Switzerland, represent a region of the world where specific provisions regarding the use of nanotechnologies have been incorporated into existing legislation, while in other regions the use of NMs is regulated based mainly on industrial applications (Amenta *et al.*, 2015). This area has progressed slowly, despite the obvious need to regulate this aspect of NMs.

Some of the explanations for this slow development are objective in nature. Thus, some would argue that the use of NMs is recent and there has not been enough time to develop a framework to regulate their uses, but this is not true. It is more than 25 years since NMs started being used on a large scale and they have since become even more widely used. Innovation in nanotechnology takes place very quickly and cannot be followed at the same speed by the regulatory frameworks; however, it is important for regulators to keep up with developments, because some NMs could have a large potential risk (Allan *et al.*, 2021). The extraordinary variety of the nature of NMs, their high complexity, and their multitude of uses make the regulations difficult to develop and harder to implement. Currently, it is almost utopian to pretend that there are unitary regulations that cover all aspects of the problems, that can be applied in all fields in which NMs are used, and that are accepted by all countries. Even a clear but all-encompassing definition of what NM means that can be accepted by all parties involved is difficult to give. We can note that the European Commission Recommendation for the definition of NMs, which was adopted in 2011, has been a relative success (Rauscher *et al.*, 2019). This one definition is also consistent with the definition by the International Organization for Standardization (Allan *et al.*, 2021).

One characteristic common to all NMs is a particle outer size on a nanometric scale (1–100 nm), which is related only to NM solids (Rauscher *et al.*, 2019). A difficulty in finding a unanimously accepted definition of NMs is caused by the complexity of NM structures and compositions; thus, sophisticated techniques are needed to adequately characterize each new NM, from as many points of view as possible, but practically each technique provides only a partial definition of the NM (Iavicoli *et al.*, 2017). Another reason why it is difficult to develop appropriate regulations is the risks of using these NMs in different fields; the side effects are not fully known or are still insufficiently determined, and in some pioneering situations when these NPs are used for the first time, some of the consequences of their use cannot be foreseen. The procedures for assessment of risks are very complex, and the implementation legislation to regulate the use of NMs is still very fragmented at a country level (Sodano, 2018). At the global level, the use of NMs has been implemented in a differentiated and uneven way, with the level of inconsistency depending on the geographical area and the degree of development of the countries, which makes it difficult to develop unitary regulations.

As far as emerging countries are concerned, an increasing number of governmental and scientific institutions have come to the conclusion that through the use of nanotechnology in agriculture, poverty alleviation and stability in terms of food security can be achieved, but there are potential risks for these countries regarding the use of these NMs (Mukhopadhyay, 2014).

The question should thus be asked as to whether the use of NMs in different fields is risky without prior rigorous tests, without regulation of all aspects that may arise, and without having determined the side effects and consequences of their use. If there are risks and if there have been cases in which the use of NMs has led to health problems for people, or adverse effects on animals, plants, or the environment, then the question arises as to why there has not been a rapid development of a system of rules, standards, and methodologies to regulate these sensitive issues.

There remain delays in this process of regulating the application of nanotechnologies in the agri-food sector, some of the causes of which are of a subjective nature: for example, the different people and organizations involved in these regulations (e.g. consumers, organizations, the community scientific, companies active in the agricultural field, experts in the field of safety and health at work, government structures, insurance companies) may have different interests and opinions, some of which may be contradictory; and different motivations often mean that the various entities who are interested take different attitudes, sometimes even of nonintervention, declining the responsibility and betraying their professional Codex ethics (Sodano, 2018). However, many nongovernmental organizations with objectives of environment and human rights issues have tried to identify the applications and real consequences of food nanotechnology and to provide public information regarding the real risks of NMs for health and environment (Sodano, 2018).

On a positive note, in 2013, the Global Coalition for Regulatory Science Research was established under the direction of the US Food and Drug Administration. This institution is composed of regulatory bodies from ten countries, some from the EU. This global partnership aims to improve scientific research in the regulation of the safety and efficacy of consumer products, in particular food and medicine (Allan et al., 2021). The use of NMs in agriculture requires tests and in-depth studies, considering the diversity of plants that are cultivated worldwide. One problem lies in the fact that the use of NMs in agriculture should first be evaluated under controlled conditions in greenhouses and indoor grow rooms, and then, if the results are promising, tests should be done on crops in the field in mass production (Pérez-de-Luque, 2017). This requires considerable effort, but also approval and regulation for the transition from the use of NMs in crops under controlled conditions to crops grown over large areas, after evaluating the potential risks. The precautionary principle could be applied here, if a preliminary assessment indicates that there are reasonable grounds for concern. The European Commission has published a Commission Communication on the Precautionary Principle (EU Resolution on the Precautionary Principle) in 2000, which provides a general framework for its use in EU policy.

A scientific contribution is required to continue to complete the gaps in our knowledge while continuing judicious application of the precautionary caution. The appropriate goal of these efforts must be to protect the environment and health human, at the same time ensuring sustainability in long term of the nanotechnology. The precautionary principle is at the forefront of scientific and regulatory discussions about nanotechnology and NMs (Saldívar-Tanaka and Hansen, 2021), and the application of this principle in Europe is supported by numerous authors (Saldívar-Tanaka and Hansen, 2021; Pivato et al., 2022). In fact, general safe practices for working with NMs in research laboratories or in experimental areas are evolving rapidly, and their constant development and updating is necessary for each field (e.g. https://ehs.cornell.edu; accessed 17 March 2023), especially in the area of agri-food. In the context of 8 billion people in the world and this set to increase, with an obvious need for growth in food production, an agriculture based on the overloading of ecosystems and on excessive use of fertilizers, fungicides, and herbicides is not sustainable and is seriously damaging the environment. The results of research on the toxicity of various NFs are still largely limited to laboratory studies, and the rules and regulations in question are not well documented to convey a common message to the scientific community, authorities, or the general public regarding this subject.

References

Adisa, I.O., Pullagurala, V.L.R., Peralta-Videa, J.R., Dimkpa, C.O., Elmer, W.H. *et al.* (2019) Recent advances in nano-enabled fertilizers and pesticides: a critical review of mechanisms of action. *Environmental Science: Nano* 6(7), 2002–2030. DOI: 10.1039/C9EN00265K.

Adrees, M., Khan, Z.S., Ali, S., Hafeez, M., Khalid, S. *et al.* (2020) Simultaneous mitigation of cadmium and drought stress in wheat by soil application of iron nanoparticles. *Chemosphere* 238, 124681. DOI: 10.1016/j.chemosphere.2019.124681.

Afantitis, A., Melagraki, G., Tsoumanis, A., Valsami-Jones, E. and Lynch, I. (2018) A nanoinformatics decision support tool for the virtual screening of gold nanoparticle cellular association using protein corona fingerprints. *Nanotoxicology* 12(10), 1148–1165. DOI: 10.1080/17435390.2018.1504998.

Afsharinejad, A., Davy, A., Jennings, B. and Brennan, C. (2015) Performance Analysis of Plant Monitoring Nanosensor Networks at THz Frequencies. *IEEE Internet of Things Journal* 3(1), 59–69. DOI: 10.1109/JIOT.2015.2463685.

Ahamed, M., Akhtar, M.J., Alaizeri, Z.M. and Alhadlaq, H.A. (2020) TiO$_2$ nanoparticles potentiated the cytotoxicity, oxidative stress and apoptosis response of cadmium in two different human cells. *Environmental Science and Pollution Research International* 27(10), 10425–10435. DOI: 10.1007/s11356-019-07130-6.

Alghuthaymi, M.A., C., R., P., R., Kalia, A., Bhardwaj, K. *et al.* (2021) Nanohybrid antifungals for control of plant diseases: current status and future perspectives. *Journal of Fungi* 7(1), 48. DOI: 10.3390/jof7010048.

Ali, M.A., Rehman, I., Iqbal, A., Din, S., Rao, A.Q. *et al.* (2014) Nanotechnology, a new frontier in agriculture. *Advancements in Life Sciences* 1, 129–138.

Ali, N.M., Altaey, D.K.A. and Altaee, N.H. (2021a) The impact of selenium, nano (SiO$_2$) and organic fertilization on growth and yield of potato *Solanum tuberosum* L. under salt stress conditions. IOP Conference Series: Earth and Environmental Science 735(1), 012042. DOI: 10.1088/1755-1315/735/1/012042.

Ali, S.S., Al-Tohamy, R., Koutra, E., Moawad, M.S., Kornaros, M. *et al.* (2021b) Nanobiotechnological advancements in agriculture and food industry: applications, nanotoxicity, and future perspectives. *The Science of the Total Environment* 792, 148359. DOI: 10.1016/j.scitotenv.2021.148359.

Allan, J., Belz, S., Hoeveler, A., Hugas, M., Okuda, H. *et al.* (2021) Regulatory landscape of nanotechnology and nanoplastics from a global perspective. *Regulatory Toxicology and Pharmacology* 122, 104885. DOI: 10.1016/j.yrtph.2021.104885.

Alsamhary, K.I. (2020) Eco-friendly synthesis of silver nanoparticles by *Bacillus subtilis* and their antibacterial activity. *Saudi Journal of Biological Sciences* 27(8), 2185–2191. DOI: 10.1016/j.sjbs.2020.04.026.

Amenta, V., Aschberger, K., Arena, M., Bouwmeester, H., Botelho Moniz, F. *et al.* (2015) Regulatory aspects of nanotechnology in the agri/feed/food sector in EU and non-EU countries. *Regulatory Toxicology and Pharmacology* 73(1), 463–476. DOI: 10.1016/j.yrtph.2015.06.016.

Amiard, J.-C. and Amiard-Triquet, C. (2015) Conventional risk assessment of environmental contaminants. In: Amiard-Triquet, C., Amiard, J.-C., and Mouneyrac, C. (eds.) *Aquatic Ecotoxicology*. Academic Press, London, pp. 25–49.

Arora, S., Murmu, G., Mukherjee, K., Saha, S. and Maity, D. (2022) A comprehensive overview of nanotechnology in sustainable agriculture. *Journal of Biotechnology* 355, 21–41. DOI: 10.1016/j.jbiotec.2022.06.007.

Asadishad, B., Chahal, S., Akbari, A., Cianciarelli, V., Azodi, M. *et al.* (2018) Amendment of agricultural soil with metal nanoparticles: effects on soil enzyme activity and microbial community composition. *Environmental Science & Technology* 52(4), 1908–1918. DOI: 10.1021/acs.est.7b05389.

Asgari, F., Majd, A., Jonoubi, P. and Najafi, F. (2018) Effects of silicon nanoparticles on molecular, chemical, structural and ultrastructural characteristics of oat (*Avena sativa* L.). *Plant Physiology and Biochemistry* 127, 152–160. DOI: 10.1016/j.plaphy.2018.03.021.

Askary, M., Talebi, S.M., Amini, F. and Bangan, A.D.B. (2017) Effects of iron nanoparticles on *Mentha piperita* L. under salinity stress. *Biologija* 63(1), 65–75. DOI: 10.6001/biologija.v63i1.3476.

Athanassiou, C.G., Kavallieratos, N.G., Benelli, G., Losic, D., Usha Rani, P. *et al.* (2018) Nanoparticles for pest control: current status and future perspectives. *Journal of Pest Science* 91(1), 1–15. DOI: 10.1007/s10340-017-0898-0.

Aziz, N., Fatma, T., Varma, A. and Prasad, R. (2014) Biogenic synthesis of silver nanoparticles using *Scenedesmus Abundans* and evaluation of their antibacterial activity. *Journal of Nanoparticles* 2014, 1–6. DOI: 10.1155/2014/689419.

Aziz, N., Pandey, R., Barman, I. and Prasad, R. (2016) Leveraging the attributes of *Mucor hiemalis*-derived silver nanoparticles for a synergistic broad-spectrum antimicrobial platform. *Frontiers in Microbiology* 7, 1984. DOI: 10.3389/fmicb.2016.01984.

Aznar Mollá, F., Fito-López, C., Heredia Alvaro, J.A. and Huertas-López, F. (2021) New tools to support the risk assessment process of nanomaterials in the insurance sector. *International Journal of Environmental Research and Public Health* 18(13), 6985. DOI: 10.3390/ijerph18136985.

Badawy, A.A., Abdelfattah, N.A.H., Salem, S.S., Awad, M.F. and Fouda, A. (2021) Efficacy assessment of biosynthesized copper oxide nanoparticles (CuO-NPs) on stored grain insects and their impacts on morphological and physiological traits of wheat (*Triticum aestivum* L.) plant. *Biology* 10(3), 233. DOI: 10.3390/biology10030233.

Balogun, D.A., Oke, M.A., Rocha-Meneses, L., Fawole, O.B. and Omojasola, P.F. (2022) Phosphate solubilization potential of indigenous rhizosphere fungi and their biofertilizer formulations. *Agronomy Research* 20, 40–55. DOI: 10.15159/AR.21.163.

Basavegowda, N. and Baek, K.H. (2021) Current and future perspectives on the use of nanofertilizers for sustainable agriculture: the case of phosphorus nanofertilizer. *3 Biotech* 11(7), 357. DOI: 10.1007/s13205-021-02907-4.

Basit, F., Asghar, S., Ahmed, T., Ijaz, U., Noman, M. *et al.* (2022) Facile synthesis of nanomaterials as nanofertilizers: a novel way for sustainable crop production. *Environmental Science and Pollution Research* 29(34), 51281–51297. DOI: 10.1007/s11356-022-20950-3.

Behboudi, F., Tahmasebi-Sarvestani, Z., Kassaee, M.Z., Modarres-Sanavy, S.A.M., Sorooshzadeh, A. *et al.* (2019) Evaluation of chitosan nanoparticles effects with two application methods on wheat under drought stress. *Journal of Plant Nutrition* 42(13), 1439–1451. DOI: 10.1080/01904167.2019.1617308.

Boros, B.V. and Ostafe, V. (2020) Evaluation of Ecotoxicology assessment methods of nanomaterials and their effects. *Nanomaterials* 10(4), 610. DOI: 10.3390/nano10040610.

Bos, P., Gottardo, S., Scott-Fordsmand, J., van Tongeren, M., Semenzin, E. *et al.* (2015) The MARINA risk assessment strategy: a flexible strategy for efficient information collection and risk assessment of nanomaterials. *International Journal of Environmental Research and Public Health* 12(12), 15007–15021. DOI: 10.3390/ijerph121214961.

Bostrom, A. and Löfstedt, R.E. (2010) Nanotechnology risk communication past and prologue. *Risk Analysis* 30(11), 1645–1662. DOI: 10.1111/j.1539-6924.2010.01521.x.

Cao, X., Khare, S., DeLoid, G.M., Gokulan, K. and Demokritou, P. (2021) Co-exposure to boscalid and TiO2 (E171) or SiO2 (E551) downregulates cell junction gene expression in small intestinal epithelium cellular model and increases pesticide translocation. *NanoImpact* 22, 100306. DOI: 10.1016/j.impact.2021.100306.

Castillo-Michel, H.A., Larue, C., Pradas Del Real, A.E., Cotte, M. and Sarret, G. (2017) Practical review on the use of synchrotron based micro- and nano- X-ray fluorescence mapping and X-ray absorption spectroscopy to investigate the interactions between plants and engineered nanomaterials. *Plant Physiology and Biochemistry* 110, 13–32. DOI: 10.1016/j.plaphy.2016.07.018.

Cataldo, F. (2002) A study on the thermal stability to 1000°C of various carbon allotropes and carbonaceous matter both under nitrogen and in air. *Fullerenes, Nanotubes and Carbon Nanostructures* 10(4), 293–311. DOI: 10.1081/FST-120016451.

Cebadero-Domínguez, Ó., Jos, A., Cameán, A.M. and Cătunescu, G.M. (2022) Hazard characterization of graphene nanomaterials in the frame of their food risk assessment: a review. *Food and Chemical Toxicology* 164, 113014. DOI: 10.1016/j.fct.2022.113014.

Center for International Environmental Law (2016) Factsheet: risk assessment of nanomaterials in a regulatory context. Center for International Environmental Law. Available at: www.ciel.org/reports/factsheet-risk-assessment-nanomaterials-regulatory-context/ (accessed 20 March 2023).

Chaud, M., Souto, E.B., Zielinska, A., Severino, P., Batain, F. *et al.* (2021) Nanopesticides in agriculture: benefits and challenge in agricultural productivity, toxicological risks to human health and environment. *Toxics* 9(6), 131. DOI: 10.3390/toxics9060131.

Chaurasia, U., Kumar, A., Maurya, K., Yadav, V., Hussain, T, *et al.* (2021) Role of nano-biotechnology in agriculture and allied sciences. In: Mallick, M., Solanki, M., Baby Kumari, B. and Verma, S. (eds) *Nanotechnology in Sustainable Agriculture*. CRC Press, Boca Raton, FL, pp. 91–118.

Coll, C., Notter, D., Gottschalk, F., Sun, T., Som, C. *et al.* (2015) Probabilistic environmental risk assessment of five nanomaterials (nano-TiO2, nano-Ag, nano-ZnO, CNT, and fullerenes). *Nanotoxicology* 10(4), 436–444. DOI: 10.3109/17435390.2015.1073812.

Cox, A., Venkatachalam, P., Sahi, S. and Sharma, N. (2016) Reprint of: Silver and titanium dioxide nanoparticle toxicity in plants: a review of current research. *Plant Physiology and Biochemistry* 107, 147–163. DOI: 10.1016/j.plaphy.2016.05.022.

Danish, M. and Hussain, T. (2019) Nanobiofertilizers in crop production. In: Panpatte, D. and Jhala, Y. (eds) *Nanotechnology for Agriculture: Crop Production & Protection*. Springer, Singapore, pp. 107–118. DOI: 10.1007/978-981-32-9374-8.

Dasgupta, N., Ranjan, S., Rajendran, B., Manickam, V., Ramalingam, C. *et al.* (2016) Thermal co-reduction approach to vary size of silver nanoparticle: its microbial and cellular toxicology. *Environmental Science and Pollution Research International* 23(5), 4149–4163. DOI: 10.1007/s11356-015-4570-z.

Das, B., Yonzone, R., Saha, S., Murmu, D.K. and Kundu, S. (2022) Comprehensive assessment of Zno, P and TiO2 nanoparticles sustaining environment in response to seed germination, antioxidants activity, nutritional quality and yield of spinach beet (*beta Vulgaris* VAR. *Bengalensis*). *Research Square, preprint*. DOI: 10.21203/rs.3.rs-1315549/v1.

Dekkers, S., Oomen, A.G., Bleeker, E.A.J., Vandebriel, R.J., Micheletti, C. *et al.* (2016) Towards a nano-specific approach for risk assessment. *Regulatory Toxicology and Pharmacology* 80, 46–59. DOI: 10.1016/j.yrtph.2016.05.037.

Deville, S., Baré, B., Piella, J., Tirez, K., Hoet, P. *et al.* (2016) Interaction of gold nanoparticles and nickel(II) sulfate affects dendritic cell maturation. *Nanotoxicology* 10(10), 1395–1403. DOI: 10.1080/17435390.2016.1221476.

Dimkpa, C.O., Singh, U., Bindraban, P.S., Elmer, W.H., Gardea-Torresdey, J.L. *et al.* (2019) Zinc oxide nanoparticles alleviate drought-induced alterations in sorghum performance, nutrient acquisition, and grain fortification. *Science of The Total Environment* 688, 926–934. DOI: 10.1016/j.scitotenv.2019.06.392.

Dimkpa, C.O., Andrews, J., Sanabria, J., Bindraban, P.S., Singh, U. *et al.* (2020a) Interactive effects of drought, organic fertilizer, and zinc oxide nanoscale and bulk particles on wheat performance and grain nutrient accumulation. *Science of The Total Environment* 722, 137808. DOI: 10.1016/j.scitotenv.2020.137808.

Dimkpa, Ch.O., Fugice, J., Singh, U. and Lewis, T.D. (2020b) Development of fertilizers for enhanced nitrogen use efficiency – Trends and perspectives. *Science of The Total Environment* 731, 139113. DOI: 10.1016/j.scitotenv.2020.139113.

Dorobantu, L.S., Fallone, C., Noble, A.J., Veinot, J., Ma, G. *et al.* (2015) Toxicity of silver nanoparticles against bacteria, yeast, and algae. *Journal of Nanoparticle Research* 17(4), 172. DOI: 10.1007/s11051-015-2984-7.

Drozdov, A.S., Nikitin, P.I. and Rozenberg, J.M. (2021) Systematic review of cancer targeting by nanoparticles revealed a global association between accumulation in tumors and spleen. *International Journal of Molecular Sciences* 22(23), 13011. DOI: 10.3390/ijms222313011.

Du, W., Gardea-Torresdey, J.L., Ji, R., Yin, Y., Zhu, J. *et al.* (2015) Physiological and biochemical changes imposed by CeO$_2$ nanoparticles on wheat: a life cycle field study . *Environmental Science & Technology* 49(19), 11884–11893. DOI: 10.1021/acs.est.5b03055.

Dunn, K.H., Eastlake, A.C., Story, M. and Kuempel, E.D. (2018) Control banding tools for engineered nanoparticles: what the practitioner needs to know. *Annals of Work Exposures and Health* 62(3), 362–388. DOI: 10.1093/annweh/wxy002.

Dussert, F., Arthaud, P.-A., Arnal, M.-E., Dalzon, B., Torres, A. *et al.* (2020) Toxicity to RAW264.7 macrophages of silica nanoparticles and the E551 food additive, in combination with genotoxic agents. *Nanomaterials* 10(7), 1418. DOI: 10.3390/nano10071418.

Dwivedi, S., Saquib, Q., Ahmad, B., Ansari, S.M., Azam, A. *et al.* (2018) Toxicogenomics: a new paradigm for nanotoxicity evaluation. *Advances in Experimental Medicine and Biology* 1048, 143–161. DOI: 10.1007/978-3-319-72041-8_9.

Elizabath, A., Bahadur, V., Misra, P., Prasad, V.M. and Thomas, T. (2017) Effect of different concentrations of iron oxide and zinc oxide nanoparticles on growth and yield of carrot (*Daucus Carota* L). *Journal of Pharmacognosy and Phytochemistry* 6, 1266–1269.

Elsheery, N.I., Sunoj, V.S.J., Wen, Y., Zhu, J.J., Muralidharan, G. *et al.* (2020) Foliar application of nanoparticles mitigates the chilling effect on photosynthesis and photoprotection in sugarcane. *Plant Physiology and Biochemistry* 149, 50–60. DOI: 10.1016/j.plaphy.2020.01.035.

Elmer, W., Ma, C. and White, J. (2018) Nanoparticles for plant disease management. *Current Opinion in Environmental Science & Health* 6, 66–70. DOI: 10.1016/j.coesh.2018.08.002.

Elmer, W.H. and White, J.C. (2016) The use of metallic oxide nanoparticles to enhance growth of tomatoes and eggplants in disease infested soil or soilless medium. *Environmental Science: Nano* 3(5), 1072–1079. DOI: 10.1039/C6EN00146G.

European Chemicals Agency (2016) Guidance on information requirements and chemical safety assessment. Available at: https://echa.europa.eu/guidance-documents/guidance-on-information-requirements-and-chemical-safety-assessment (accessed 20 March 2023).

Fadeel, B., Farcal, L., Hardy, B., Vázquez-Campos, S., Hristozov, D. *et al.* (2018) Advanced tools for the safety assessment of nanomaterials. *Nature Nanotechnology* 13(7), 537–543. DOI: 10.1038/s41565-018-0185-0.

Feng, L., Yang, X., Asweto, C.O., Wu, J., Zhang, Y. *et al.* (2017) Low-dose combined exposure of nanoparticles and heavy metal compared with PM2.5 in human myocardial AC16 cells. *Environmental Science and Pollution Research* 24(36), 27767–27777. DOI: 10.1007/s11356-017-0228-3.

Feng, L., Yang, X., Shi, Y., Liang, S., Zhao, T. *et al.* (2018) Co-exposure subacute toxicity of silica nanoparticles and lead acetate on cardiovascular system. *International Journal of Nanomedicine* 13, 7819–7834. DOI: 10.2147/IJN.S185259.

Fernández-Bertólez, N., Costa, C., Bessa, M.J., Park, M., Carriere, M. *et al.* (2019) Assessment of oxidative damage induced by iron oxide nanoparticles on different nervous system cells. *Mutation Research/Genetic Toxicology and Environmental Mutagenesis* 845, 402989. DOI: 10.1016/j.mrgentox.2018.11.013.

Forest, V. (2021) Combined effects of nanoparticles and other environmental contaminants on human health - an issue often overlooked. *NanoImpact* 23, 100344. DOI: 10.1016/j.impact.2021.100344.

Gao, X., Li, L., Cai, X., Huang, Q., Xiao, J. *et al.* (2021) Targeting nanoparticles for diagnosis and therapy of bone tumors: opportunities and challenges. *Biomaterials* 265, 120404. DOI: 10.1016/j. biomaterials.2020.120404.

Gao, X., Zou, H., Zhou, Z., Yuan, W., Quan, C, *et al.* (2019) Qualitative and quantitative differences between common control banding tools for Nanomaterials in workplaces. *RSC Advances* 9(59), 34512–34528. DOI: 10.1039/c9ra06823f.

Ghorbanpour, M., Mohammadi, H. and Kariman, K. (2020) Nanosilicon-based recovery of barley (*Hordeum Vulgare*) plants subjected to drought stress. *Environmental Science: Nano* 7(2), 443–461. DOI: 10.1039/C9EN00973F.

Gonzalez, C., Rosas-Hernandez, H., Ramirez-Lee, M.A., Salazar-García, S. and Ali, S.F. (2016) Role of silver nanoparticles (AgNPs) on the cardiovascular system. *Archives of Toxicology* 90(3), 493–511. DOI: 10.1007/s00204-014-1447-8.

Gordillo-Delgado, F., Zuluaga-Acosta, J. and Restrepo-Guerrero, G. (2020) Effect of the suspension of Ag-incorporated TiO2 nanoparticles (Ag-TiO2 NPs) on certain growth, physiology and phytotoxicity parameters in spinach seedlings. *PloS One* 15(12), e0244511. DOI: 10.1371/journal.pone.0244511.

Groso, A., Petri-Fink, A., Magrez, A., Riediker, M. and Meyer, T. (2010) Management of nanomaterials safety in research environment. *Particle and Fibre Toxicology* 7, 40. DOI: 10.1186/1743-8977-7-40.

Guan, G., Zhang, L., Zhu, J., Wu, H., Li, W. *et al.* (2021) Antibacterial properties and mechanism of biopolymer-based films functionalized by CuO/ZnO nanoparticles against *Escherichia coli* and *Staphylococcus aureus*. *Journal of Hazardous Materials* 402, 123542. DOI: 10.1016/j. jhazmat.2020.123542.

Guo, C., Liu, Y. and Li, Y. (2021) Adverse effects of amorphous silica nanoparticles: focus on human cardio-vascular health. *Journal of Hazardous Materials* 406, 124626. DOI: 10.1016/j.jhazmat.2020.124626.

Gupta, A. and Senthil-Kumar, M. (2017) Concurrent stresses are perceived as new state of stress by the plants: overview of impact of abiotic and biotic stress combinations. In: Senthil-Kumar, M. (ed.) *Plant Tolerance to Individual and Concurrent Stresses*. Springer, New Delhi, pp. 1–31. DOI: 10.1007/978-81-322-3706-8.

Hanafy, M.H. (2018) Myconanotechnology in veterinary sector: status quo and future perspectives. *International Journal of Veterinary Science and Medicine* 6(2), 270–273. DOI: 10.1016/j. ijvsm.2018.11.003.

Handy, R.D., von der Kammer, F., Lead, J.R., Hassellöv, M., Owen, R. *et al.* (2008) The ecotoxicology and chemistry of manufactured nanoparticles. *Ecotoxicology* 17(4), 287–314. DOI: 10.1007/ s10646-008-0199-8.

Haq, I.U. and Ijaz, S. (2019) Use of metallic nanoparticles and nanoformulations as nanofungicides for sustainable disease management in plants. In: Prasad, R., Kumar, V., Kumar, M. and D, C. (eds) *Nanobiotechnology in Bioformulations*. Springer, Cham, pp. 289–316. DOI: 10.1007/978-3-030-17061-5_12.

Hashim, N., Abdullah, S. and Yusoh, K. (2022) Graphene nanomaterials in the food industries: quality control in promising food safety to consumers. *Graphene and 2D Materials Technologies* 7(1–2), 1–29. DOI: 10.1007/s41127-021-00045-5.

Hashimoto, T., Mustafa, G., Nishiuchi, T. and Komatsu, S. (2020) Comparative analysis of the effect of inorganic and organic chemicals with silver nanoparticles on soybean under flooding stress. *International Journal of Molecular Sciences* 21(4), 1300. DOI: 10.3390/ijms21041300.

Hassan, N.S., Salah El Din, T.A., Hendawey, M.H., Borai, I.H. and Mahdi, A.A. (2018) Magnetite and zinc oxide nanoparticles alleviated heat stress in wheat plants. *Current Nanomaterials* 3(1), 32–43. DOI: 10.2174/2405461503666180619160923.

Hegde, K., Goswami, R., Sarma, S.J., Veeranki, V.D., and Brar, S.K. (2015) Nano-ecotoxicology of natural and engineered nanomaterials for different ecosystems. In: Brar, S.K., Zhang, T.C., Verma, P.E., Surampalli, R.Y., and Tyagi,R.D. (eds.) *Nanomaterials in the Environment*. American Society of Civil Engineers, Reston, VA, pp. 487–511.

Hong, H., Adam, V. and Nowack, B. (2021) Form-specific and probabilistic environmental risk assessment of 3 engineered nanomaterials (Nano-Ag, Nano-TiO$_2$, and Nano-ZnO) in European freshwaters. *Environmental Toxicology and Chemistry* 40(9), 2629–2639. DOI: 10.1002/etc.5146.

Hristozov, D., Gottardo, S., Semenzin, E., Oomen, A., Bos, P. *et al.* (2016) Frameworks and tools for risk assessment of manufactured nanomaterials. *Environment International* 95, 36–53. DOI: 10.1016/j.envint.2016.07.016.

Hund-Rinke, K., Baun, A., Cupi, D., Fernandes, T.F., Handy, R. *et al.* (2016) Regulatory ecotoxicity testing of nanomaterials - proposed modifications of OECD test guidelines based on laboratory experience with silver and titanium dioxide nanoparticles. *Nanotoxicology* 10(10), 1442–1447. DOI: 10.1080/17435390.2016.1229517.

Hussain, A., Rizwan, M., Ali, S., Rehman, M.Z. ur, Qayyum, M.F. *et al.* (2021) Combined use of different nanoparticles effectively decreased cadmium (Cd) concentration in grains of wheat grown in a field contaminated with Cd. *Ecotoxicology and Environmental Safety* 215, 112139. DOI: 10.1016/j.ecoenv.2021.112139.

Iavicoli, I., Leso, V., Beezhold, D.H. and Shvedova, A.A. (2017) Nanotechnology in agriculture: opportunities, toxicological implications, and occupational risks. *Toxicology and Applied Pharmacology* 329, 96–111. DOI: 10.1016/j.taap.2017.05.025.

Iavicoli, Ivo, Leso, V., Piacci, M., Cioffi, D.L., Guseva Canu, I. *et al.* (2019) An exploratory assessment of applying risk management practices to engineered nanomaterials. *International Journal of Environmental Research and Public Health* 16(18), 3290. DOI: 10.3390/ijerph16183290.

Iqbal, M.A. (2019) Nano-fertilizers for sustainable crop production under changing climate: a global perspective. In: Hasanuzzaman, M., Filho, M.C.M.T., Fujita, M. and Nogueira, T.A.R. (eds) *Sustainable Crop Production*. IntechOpen. DOI: 10.5772/intechopen.89089.

Iqbal, M., Raja, N.I., Mashwani, Z.-U.-R., Wattoo, F.H., Hussain, M. *et al.* (2019) Assessment of AgNPs exposure on physiological and biochemical changes and antioxidative defence system in wheat (*Triticum aestivum* L) under heat stress. *IET Nanobiotechnology* 13(2), 230–236. DOI: 10.1049/iet-nbt.2018.5041.

Irshad, M.A., Rehman, M.Z.U., Anwar-Ul-Haq, M., Rizwan, M., Nawaz, R. *et al.* (2021) Effect of green and chemically synthesized titanium dioxide nanoparticles on cadmium accumulation in wheat grains and potential dietary health risk: a field investigation. *Journal of Hazardous Materials* 415, 125585. DOI: 10.1016/j.jhazmat.2021.125585.

Isigonis, P., Hristozov, D., Benighaus, C., Giubilato, E., Grieger, K. *et al.* (2019) Risk governance of nanomaterials: review of criteria and tools for risk communication, evaluation, and mitigation. *Nanomaterials* 9(5), 696. DOI: 10.3390/nano9050696.

Jain, R., Matassa, S., Singh, S., van Hullebusch, E.D., Esposito, G. *et al.* (2016) Reduction of selenite to elemental selenium nanoparticles by activated sludge. *Environmental Science and Pollution Research* 23(2), 1193–1202. DOI: 10.1007/s11356-015-5138-7.

Jain, A., Ranjan, S., Dasgupta, N. and Ramalingam, C. (2018) Nanomaterials in food and agriculture: An overview on their safety concerns and regulatory issues. *Critical Reviews in Food Science and Nutrition* 58(2), 297–317. DOI: 10.1080/10408398.2016.1160363.

Jampílek, J. and Králʼová, K. (2021) Nanoparticles for improving and augmenting plant functions. In: Jogaiah, S., Singh, H.B., Fraceto, L.F. and de Lima, R. (eds) *Advances in Nano-Fertilizers and Nano-Pesticides in Agriculture*. Woodhead Publishing, Sawston, UK, pp. 171–227.

Jha, S. and Pudake, R.N. (2016) Molecular mechanism of plant–nanoparticle interactions. In: Kole, C., Kumar, D., Khodakovskaya, M. (eds.) *Plant Nanotechnology*. Springer, Cham, Switzerland, pp. 155–181.

Jiang, N., Wen, H., Zhou, M., Lei, T., Shen, J. *et al.* (2020) Low-dose combined exposure of carboxylated black carbon and heavy metal lead induced potentiation of oxidative stress, DNA damage, inflammation, and apoptosis in BEAS-2B cells. *Ecotoxicology and Environmental Safety* 206, 111388. DOI: 10.1016/j.ecoenv.2020.111388.

Johnson, A.C., Bowes, M.J., Crossley, A., Jarvie, H.P., Jurkschat, K. *et al.* (2011) An assessment of the fate, behaviour and environmental risk associated with sunscreen TiO$_2$ nanoparticles in UK field scenarios. *The Science of the Total Environment* 409(13), 2503–2510. DOI: 10.1016/j.scitotenv.2011.03.040.

Joubert, I.A., Geppert, M., Ess, S., Nestelbacher, R., Gadermaier, G. *et al.* (2020) Public perception and knowledge on nanotechnology: a study based on a citizen science approach. *NanoImpact* 17, 100201. DOI: 10.1016/j.impact.2019.100201.

Kahru, A., Dubourguier, H.C., Blinova, I., Ivask, A. and Kasemets, K. (2008) Biotests and biosensors for ecotoxicology of metal oxide nanoparticles: a minireview. *Sensors* 8(8), 5153–5170. DOI: 10.3390/s8085153.

Khan, M.K., Pandey, A., Hamurcu, M., Gezgin, S., Athar, T. *et al.* (2021) Insight into the prospects for nanotechnology in wheat biofortification. *Biology* 10(11), 1123. DOI: 10.3390/biology10111123.

Khan, M.R. and Rizvi, T.F. (2014) Nanotechnology: scope and application in plant disease management. *Plant Pathology Journal* 13(3), 214–231. DOI: 10.3923/ppj.2014.214.231.

Khan, Z. and Ansari, M.Y.K. (2018) Impact of engineered Si nanoparticles on seed germination, vigour index and genotoxicity assessment via DNA damage of root tip cells in *Lens culinaris*. *Journal of Plant Biochemistry & Physiology* 06(03), 219. DOI: 10.4172/2329-9029.1000218.

Khan, Z., Shahwar, D., Yunus Ansari, M.K. and Chandel, R. (2019) Toxicity assessment of anatase (TiO$_2$) nanoparticles: a pilot study on stress response alterations and DNA damage studies in *Lens culinaris* Medik. *Heliyon* 5(7), e02069. DOI: 10.1016/j.heliyon.2019.e02069.

Khandelwal, N., Barbole, R.S., Banerjee, S.S., Chate, G.P., Biradar, A.V. *et al.* (2016) Budding trends in integrated pest management using advanced micro- and nano-materials: challenges and perspectives. *Journal of Environmental Management* 184(Pt 2), 157–169. DOI: 10.1016/j.jenvman.2016.09.071.

Khandoudi, N., Porte, P., Chtourou, S., Nesslany, F., Marzin, D. *et al.* (2009) The presence of arginine may be a source of false positive results in the Ames test. *Mutation Research/Genetic Toxicology and Environmental Mutagenesis* 679(1–2), 65–71. DOI: 10.1016/j.mrgentox.2009.03.010.

Kolbert, Z., Szőllősi, R., Rónavári, A., Molnár, Á. and Cuypers, A. (2022) Nanoforms of essential metals: from hormetic phytoeffects to agricultural potential. *Journal of Experimental Botany* 73(6), 1825–1840. DOI: 10.1093/jxb/erab547.

Kolenčík, M., Ernst, D., Komár, M., Urík, M., Šebesta, M. *et al.* (2019) Effect of foliar spray application of zinc oxide nanoparticles on quantitative, nutritional, and physiological parameters of foxtail millet (*Setaria italica* L.) under field conditions. *Nanomaterials* 9(11), 1559. DOI: 10.3390/nano9111559.

Kolenčík, M., Ernst, D., Urík, M., Ďurišová, Ľ., Bujdoš, M. *et al.* (2020) Foliar application of low concentrations of titanium dioxide and zinc oxide nanoparticles to the common sunflower under field conditions. *Nanomaterials* 10(8), 1619. DOI: 10.3390/nano10081619.

Kuempel, E.D., Geraci, C.L. and Schulte, P.A. (2012) Risk assessment and risk management of nanomaterials in the workplace: translating research to practice. *The Annals of Occupational Hygiene* 56(5), 491–505. DOI: 10.1093/annhyg/mes040.

Kumar, M., Shamsi, T.N., Parveen, R., and Fatima, S. (2017) Application of nanotechnology in enhancement of crop productivity and integrated pest management. In: Prasad, R., Kumar, M., Kumar, V. (eds.) *Nanotechnology*. Springer, Singapore, pp. 361–371.

Kumari, R. and Singh, D.P. (2020) Nano-biofertilizer: an emerging eco-friendly approach for sustainable agriculture. *Proceedings of the National Academy of Sciences, India Section B* 90(4), 733–741. DOI: 10.1007/s40011-019-01133-6.

Lam, C.W., James, J.T., McCluskey, R., Arepalli, S. and Hunter, R.L. (2006) A review of carbon nanotube toxicity and assessment of potential occupational and environmental health risks. *Critical Reviews in Toxicology* 36(3), 189–217. DOI: 10.1080/10408440600570233.

Landsiedel, R., Kapp, M.D., Schulz, M., Wiench, K. and Oesch, F. (2009) Genotoxicity investigations on nanomaterials: methods, preparation and characterization of test material, potential artifacts and limitations--many questions, some answers. *Mutation Research* 681(2–3), 241–258. DOI: 10.1016/j.mrrev.2008.10.002.

Lenz e Silva, G.F.B. and Hurt, R. (2016) Risk assessment of nanocarbon materials using a multi-criteria method based on analytical hierarchy process. *Current Bionanotechnology* 2, 106–111.

Liguori, B., Foss Hansen, S., Baun, A. and Jensen, K.A. (2016) Control banding tools for occupational exposure assessment of nanomaterials — Ready for use in a regulatory context? *NanoImpact* 2, 1–17. DOI: 10.1016/j.impact.2016.04.002.

Linkov, I., Steevens, J., Adlakha-Hutcheon, G., Bennett, E., Chappell, M. *et al.* (2009) Emerging methods and tools for environmental risk assessment, decision-making, and policy for nanomaterials: summary of NATO advanced research workshop. *Journal of Nanoparticle Research* 11(3), 513–527. DOI: 10.1007/s11051-008-9514-9.

Liu, J., Dhungana, B. and Cobb, G.P. (2018) Environmental behavior, potential phytotoxicity, and accumulation of copper oxide nanoparticles and arsenic in rice plants. *Environmental Toxicology and Chemistry* 37(1), 11–20. DOI: 10.1002/etc.3945.

Lyon, D.Y., Thill, A., Rose, J. and Alvarez, P.J.J. (2007) Ecotoxicological impacts of nanomaterials. In: Wiesner, M.R. and Bottero, J.Y. (eds) *Environmental Nanotechnology. Applications and Impacts of Nanomaterials*. McGraw Hill, New York, pp. 445–479.

Malejko, J., Godlewska-Żyłkiewicz, B., Vanek, T., Landa, P., Nath, J. *et al*. (2021) Uptake, translocation, weathering and speciation of gold nanoparticles in potato, radish, carrot and lettuce crops. *Journal of Hazardous Materials* 418, 126219. DOI: 10.1016/j.jhazmat.2021.126219.

Mattsson, M.-O. and Simkó, M. (2017) The changing face of nanomaterials: risk assessment challenges along the value chain. *Regulatory Toxicology and Pharmacology* 84, 105–115. DOI: 10.1016/j.yrtph.2016.12.008.

Mehnaz, S. (2016) An overview of globally available Bioformulations. In: Arora, N., Mehnaz, S. and Balestrini, R. (eds) *Bioformulations: For Sustainable Agriculture*. Springer, New Delhi, pp. 267–281. DOI: 10.1007/978-81-322-2779-3.

Mittal, D., Kaur, G., Singh, P., Yadav, K. and Ali, S.A. (2020) Nanoparticle-based sustainable agriculture and food science: recent advances and future outlook. *Frontiers in Nanotechnology* 2, 10. DOI: 10.3389/fnano.2020.579954.

Moaveni, P., Farahani, HA., and Maroufi, K. (2011) Effect of TiO_2 nanoparticles spraying on barley (*Hordeum vulgare* L.) under field condition. *Advances in Environmental Biology* 5, 2220–2223.

More, S., Bampidis, V., Benford, D., Bragard, C., Halldorsson, T, *et al*. (2021) Guidance on risk assessment of Nanomaterials to be applied in the food and feed chain: Human and animal health. *EFSA Journal* 19(8), e06768. DOI: 10.2903/j.efsa.2021.6768.

Mukhopadhyay, SS. (2014) Nanotechnology in agriculture: prospects and constraints. *Nanotechnology, Science and Applications* 7, 63–71. DOI: 10.2147/NSA.S39409.

Naqvi, S., Kumar, V., and Gopinath, P. (2018) Nanomaterial toxicity: a challenge to end users. In: Bhagyaraj, S.M., Oluwafemi, O.S., Kalarikkal, N., and Thomas, S. (eds.) *Applications of Nanomaterials. Advances and Key Technologies*. Elsevier, pp. 315–343.

National Geographic (2022) Earth now has 8 billion people—and counting. where do we go from here? National Geographic, Washington, DC. Available at: www.nationalgeographic.com/environment/article/the-world-now-has-8-billion-people (accessed 17 March 2023).

National Research Council (2009) Science and decisions: Advancing risk assessment committee. The National Academies Press, Washington, DC. Available at: https://nap.nationalacademies.org/catalog/12209/science-and-decisions-advancing-risk-assessment (accessed 30 March 2023).

Nho, R. (2020) Pathological effects of nano-sized particles on the respiratory system. *Nanomedicine* 29, 102242. DOI: 10.1016/j.nano.2020.102242.

Niazian, M., Molaahmad Nalousi, A., Azadi, P., Ma'mani, L. and Chandler, S.F. (2021) Perspectives on new opportunities for nano-enabled strategies for gene delivery to plants using nanoporous materials. *Planta* 254(4), 83. DOI: 10.1007/s00425-021-03734-w.

OECD (Organisation for Economic Co-operation and Development) (2014) *ENV/JM/Mono1: Ecotoxicology and Environmental Fate of Manufactured Nanomaterials: Test Guidelines. Environment Directorate Joint Meeting of the Chemicals Committee and the Working Party on Chemicals, Pesticides and Biotechnology*. OECD, Paris.

Oesch, F. and Landsiedel, R. (2012) Genotoxicity investigations on nanomaterials. *Archives of Toxicology* 86(7), 985–994. DOI: 10.1007/s00204-012-0838-y.

Oomen, A.G., Bleeker, E.A.J., Bos, P.M.J., van Broekhuizen, F., Gottardo, S. *et al*. (2015) Grouping and read-across approaches for risk assessment of nanomaterials. *International Journal of Environmental Research and Public Health* 12(10), 13415–13434. DOI: 10.3390/ijerph121013415.

Oksel, C., Hunt, N., Wilkins, T. and Wang, X.Z. (2017) *Guidelines for the Safe Manufacture and Use of Nanomaterials*. Sustainable Nanotechnologies Project 52841713.

Pandey, G. and Jain, P. (2020) Assessing the nanotechnology on the grounds of costs, benefits, and risks. *Beni-Suef University Journal of Basic and Applied Sciences* 9(1), 63. DOI: 10.1186/s43088-020-00085-5.

Panpatte, D.G., Jhala, Y.K., Shelat, H.N. and Vyas, R.V. (2016) Nanoparticles: the next generation technology for sustainable Agriculture. In: Singh, D., Singh, H. and Prabha, R. (eds) *Microbial Inoculants in Sustainable Agricultural Productivity*. Springer, New Delhi, pp. 289–300. DOI: 10.1007/978-81-322-2644-4_18.

Paramo, L.A., Feregrino-Pérez, A.A., Guevara, R., Mendoza, S. and Esquivel, K. (2020) Nanoparticles in agroindustry: applications, toxicity, challenges, and trends. *Nanomaterials* 10(9), 1654. DOI: 10.3390/nano10091654.

Patel, A., Patel, B., Shah, P., and Jayvadan, P. (2022) Risk assessment and management of occupational exposure to nanopesticides in agriculture. In: Yashwant, P., Govindan, P., and Patel, J. (eds.) *Sustainable Nanotechnology: Strategies, Products, and Applications*. John Wiley & Sons, Hoboken, NJ, pp. 249–264.

Patil, H.J. and Solanki, M.K. (2016) Microbial inoculant: modern era of fertilizers and pesticides. In: Singh, D., Singh, H., and Prabha, R. (eds.) *Microbial Inoculants in Sustainable Agricultural Productivity.* Springer, New Delhi, pp. 319–343.

Pérez-de-Luque, A. (2017) Interaction of nanomaterials with plants: what do we need for real applications in agriculture? *Frontiers in Environmental Science* 5, 12. DOI: 10.3389/fenvs.2017.00012.

Pestovsky, Y.S. and Martínez-Antonio, A. (2017) The use of nanoparticles and nanoformulations in agriculture. *Journal of Nanoscience and Nanotechnology* 17(12), 8699–8730. DOI: 10.1166/jnn.2017.15041.

Prasad, T.N.V.K.V., Sudhakar, P., Sreenivasulu, Y., Latha, P., Munaswamy, V. *et al.* (2012) Effect of nanoscale zinc oxide particles on the germination, growth and yield of peanut. *Journal of Plant Nutrition* 35(6), 905–927. DOI: 10.1080/01904167.2012.663443.

Phillips, D.J., Davies, G.L. and Gibson, M.I. (2015) Siderophore-inspired nanoparticle-based biosensor for the selective detection of Fe^{3+} *Journal of Materials Chemistry B* 3(2), 270–275. DOI: 10.1039/c4tb01501k.

Pivato, A., Beggio, G., Bonato, T., Butti, L., Cavani, L, *et al.* (2022) The role of the precautionary principle in the agricultural reuse of sewage sludge from urban wastewater treatment plants. *Detritus* 19, V–XII. DOI: 10.31025/2611-4135/2022.15202.

Prasad, R., Bhattacharyya, A. and Nguyen, Q.D. (2017) Nanotechnology in sustainable agriculture: recent developments, challenges, and perspectives. *Frontiers in Microbiology* 8, 1014. DOI: 10.3389/fmicb.2017.01014.

Prażak, R., Święciło, A., Krzepiłko, A., Michałek, S. and Arczewska, M. (2020) Impact of Ag nanoparticles on seed germination and seedling growth of green beans in normal and chill temperatures. *Agriculture* 10(8), 312. DOI: 10.3390/agriculture10080312.

Rafińska, K., Pomastowski, P. and Buszewski, B. (2019) Study of *Bacillus subtilis* response to different forms of silver. *The Science of the Total Environment* 661, 120–129. DOI: 10.1016/j.scitotenv.2018.12.139.

Rafique, R., Zahra, Z., Virk, N., Shahid, M., Pinelli, E. *et al.* (2018) Dose-dependent physiological responses of *Triticum aestivum* L. to soil applied TiO_2 nanoparticles: alterations in chlorophyll content, H_2O_2 production, and genotoxicity. *Agriculture, Ecosystems & Environment* 255, 95–101. DOI: 10.1016/j.agee.2017.12.010.

Rai, M. and Kratosova, G. (2015) Management of phytopathogens by application of green nanobiotechnology: emerging trends and challenges. *Acta Agraria Debreceniensis* 66(66), 15–22. DOI: 10.34101/actaagrar/66/1884.

Rajput, V., Minkina, T., Ahmed, B., Sushkova, S., Singh, R, *et al.* (2019) Interaction of copper-based nanoparticles to soil, terrestrial, and aquatic systems: critical review of the state of the science and future perspectives. In: de Voogt, P. (ed.) *Reviews of Environmental Contamination and Toxicology*, Vol. 252. Springer, Cham, pp. 51–96. Available at: https://doi.org/10.1007/398_2019_34

Rajput, V.D., Minkina, T., Fedorenko, A., Tsitsuashvili, V., Mandzhieva, S. *et al.* (2018a) Metal oxide nanoparticles: Applications and effects on soil Ecosystems. In: Lund, J.E. (ed.) *Soil Contamination: Sources, Assessment and Remediation*. Nova Science Publishers, New York, pp. 81–106.

Rajput, V.D., Minkina, T., Fedorenko, A., Mandzhieva, S., Sushkova, S, *et al.* (2018b) Destructive effect of copper oxide nanoparticles on ultrastructure of chloroplast, plastoglobules and starch grains in spring barley (*Hordeum Sativum Distichum*). *International Journal of Agriculture and Biology* 21, 171–174. DOI: 10.17957/IJAB/15.0877.

Rajput, V.D., Minkina, T., Kumari, A., Singh, V.K., Verma, K.K. *et al.* (2021) Coping with the challenges of abiotic stress in plants: new dimensions in the field application of nanoparticles. *Plants* 10(6), 1221. DOI: 10.3390/plants10061221.

Raliya, R., Saharan, V., Dimkpa, Ch. and Biswas, P. (2018) Nanofertilizer for precision and sustainable agriculture: current state and future perspectives. *Journal of Agricultural and Food Chemistry* 66(26), 6487–6503. DOI: 10.1021/acs.jafc.7b02178.

Rana, S. and Kalaichelvan, P.T. (2013) Ecotoxicity of nanoparticles. *International Scholarly Research Notices* 2013, 1–11. DOI: 10.1155/2013/574648.

Rao, V. (2014) Systems approach to biosafety and risk assessment of engineered nanomaterials. *Applied Biosafety* 19(1), 11–19. DOI: 10.1177/153567601401900102.

Rauscher, H., Roebben, G., Mech, A., Gibson, P., Kestens, V, *et al.* (2019) An overview of concepts and terms used in the European Commission's definition of nanomaterial. EUR 29647 EN.Publications Office of the European Union, Luxembourg. 10.2760/459136

Resnik, D.B. (2019) How should engineered nanomaterials be regulated for public and environmental health? *AMA Journal of Ethics* 21(4), E363–369. DOI: 10.1001/amajethics.2019.363.

Rico, C.M., Majumdar, S., Duarte-Gardea, M., Peralta-Videa, J.R. and Gardea-Torresdey, J.L. (2011) Interaction of nanoparticles with edible plants and their possible implications in the food chain. *Journal of Agricultural and Food Chemistry* 59(8), 3485–3498. DOI: 10.1021/jf104517j.

Riediker, M., Ostiguy, C., Triolet, J., Troisfontaine, P., Vernez, D. *et al.* (2012) Development of a control banding tool for nanomaterials. *Journal of Nanomaterials* 2012, 1–8. DOI: 10.1155/2012/879671.

Rizwan, M., Ali, S., Rehman, M.Z.U. and Maqbool, A. (2019a) A critical review on the effects of zinc at toxic levels of cadmium in plants. *Environmental Science and Pollution Research International* 26(7), 6279–6289. DOI: 10.1007/s11356-019-04174-6.

Rizwan, M., Ali, S., Zia ur Rehman, M., Adrees, M., Arshad, M. *et al.* (2019b) Alleviation of cadmium accumulation in maize (*Zea mays* L.) by foliar spray of zinc oxide nanoparticles and biochar to contaminated soil. *Environmental Pollution* 248, 358–367. DOI: 10.1016/j.envpol.2019.02.031.

Robichaud, C.O., Tanzil, D., and Wiesner, M.R. (2007) Assessing life-cycle risks of nanomaterials. In: Wiesner, M.R. and Bottero, J.Y. (eds.) *Environmental Nanotechnology. Applications and Impacts of Nanomaterials*. McGraw Hill, New York, pp. 481–524.

Sadak, M.S. and Bakry, B.A. (2020) Zinc-oxide and nano ZnO oxide effects on growth, some biochemical aspects, yield quantity, and quality of flax (*Linum uitatissimum* L.) in absence and presence of compost under sandy soil. *Bulletin of the National Research Centre* 44(1), 98. DOI: 10.1186/s42269-020-00348-2.

Saldívar-Tanaka, L. and Hansen, S.F. (2021) Should the precautionary principle be implemented in Europe with regard to nanomaterials? Expert interviews. *Journal of Nanoparticle Research* 23(3), 70. DOI: 10.1007/s11051-021-05173-w.

Sarkar, J., Chakraborty, N., Chatterjee, A., Bhattacharjee, A., Dasgupta, D. *et al.* (2020) Green synthesized copper oxide nanoparticles ameliorate defence and antioxidant enzymes in *Lens culinaris*. *Nanomaterials* 10(2), 312. DOI: 10.3390/nano10020312.

Sarraf, M., Vishwakarma, K., Kumar, V., Arif, N., Das, S. *et al.* (2022) Metal/metalloid-based nanomaterials for plant abiotic stress tolerance: an overview of the mechanisms. *Plants* 11(3), 316. DOI: 10.3390/plants11030316.

Schrand, A.M., Rahman, M.F., Hussain, S.M., Schlager, J.J., Smith, D.A. *et al.* (2010) Metal-based nanoparticles and their toxicity assessment. *Nanomedicine and Nanobiotechnology* 2(5), 544–568. DOI: 10.1002/wnan.103.

Servin, A., Elmer, W., Mukherjee, A., De la Torre-Roche, R., Hamdi, H. *et al.* (2015) A review of the use of engineered nanomaterials to suppress plant disease and enhance crop yield. *Journal of Nanoparticle Research* 17(2), 92. DOI: 10.1007/s11051-015-2907-7.

Shalaby, T., Bayoumi, Y., Eid, Y., Elbasiouny, H., Elbehiry, F. *et al.* (2022) Can nanofertilizers mitigate multiple environmental stresses for higher crop productivity? *Sustainability* 14(6), 3480. DOI: 10.3390/su14063480.

Silins, I. and Högberg, J. (2011) Combined toxic exposures and human health: biomarkers of exposure and effect. *International Journal of Environmental Research and Public Health* 8(3), 629–647. DOI: 10.3390/ijerph8030629.

Silva, G.L., Viana, C., Domingues, D. and Vieira, F. (2019) Risk assessment and health, safety, and environmental management of carbon nanomaterials. In: Clichici, S., Filip, A. and do Nascimento, G.M. (eds) *Nanomaterials — Toxicity, Human Health and Environment*. IntechOpen. DOI: 10.5772/intechopen.85485.

Slavin, Y.N., Asnis, J., Häfeli, U.O. and Bach, H. (2017) Metal nanoparticles: understanding the mechanisms behind antibacterial activity. *Journal of Nanobiotechnology* 15(1), 65. DOI: 10.1186/s12951-017-0308-z.

Smita, S., Gupta, S.K., Bartonova, A., Dusinska, M., Gutleb, A.C. *et al.* (2012) Nanoparticles in the environment: assessment using the causal diagram approach. *Environmental Health* 11(Suppl 1), S13. DOI: 10.1186/1476-069X-11-S1-S13.

Sodano, V. (2018) Nano-food regulatory issues in the European Union. *AIP Conference Preceedings* 1990, 020018. DOI: 10.1063/1.5047772.

Spruit, S.L. (2015) Choosing between precautions for nanoparticles in the workplace: complementing the precautionary principle with caring. *Journal of Risk Research* 20(3), 326–346. DOI: 10.1080/13669877.2015.1043574.

Sushil, A., Kamla, M., Nisha, K., Karmal, M. and Sandeep, A. (2021) Nano-enabled pesticides in agriculture: budding opportunities and challenges. *Journal of Nanoscience and Nanotechnology* 21(6), 3337–3350. DOI: 10.1166/jnn.2021.19018.

Teszlák, P., Kocsis, M., Scarpellini, A., Jakab, G. and Kőrösi, L. (2018) Foliar exposure of grapevine (*Vitis vinifera* L.) to TiO$_2$ nanoparticles under field conditions: photosynthetic response and flavonol profile. *Photosynthetica* 56(4), 1378–1386. DOI: 10.1007/s11099-018-0832-6.

Thakur, S., Thakur, S. and Kumar, R. (2018) Bio-nanotechnology and its role in agriculture and food industry. *Journal of Molecular and Genetic Medicine* 12, 1. DOI: 10.4172/1747-0862.1000324.

The Innovation Society (2008) Certification Standard CENARIOS. Available at: https://innovationsgesellsc haft.ch/en/competences/risk-management/cenarios/ (accessed 30 March 2023).

Thiruvengadam, M., Rajakumar, G. and Chung, I.M. (2018) Nanotechnology: current uses and future applications in the food industry. *3 Biotech* 8(1), 74. DOI: 10.1007/s13205-018-1104-7.

Torabian, S., Zahedi, M., and Khoshgoftarmanesh, A. (2016) Effect of foliar spray of zinc oxide on some antioxidant enzymes activity of sunflower under salt stress. *Journal of Agriculture, Science and Technology* 18, 1013–1025.

Tripathi, D.K., Singh, S., Singh, V.P., Prasad, S.M., Dubey, N.K. *et al.* (2017) Silicon nanoparticles more effectively alleviated UV-B stress than silicon in wheat (*Triticum aestivum*) seedlings. *Plant Physiology and Biochemistry* 110, 70–81. DOI: 10.1016/j.plaphy.2016.06.026.

Tymoszuk, A. (2021) Silver nanoparticles effects on in vitro germination, growth, and biochemical activity of tomato, radish, and kale seedlings. *Materials* 14(18), 5340. DOI: 10.3390/ma14185340.

Umar, W., Hameed, M.K., Aziz, T., Maqsood, M.A., Bilal, H.M. *et al.* (2021) Synthesis, characterization and application of ZnO nanoparticles for improved growth and Zn biofortification in maize. *Archives of Agronomy and Soil Science* 67(9), 1164–1176. DOI: 10.1080/03650340.2020.1782893.

UN Food and Agriculture Organization (FAO) (2022) Food security and nutrition in the world. FAO, Rome. Available at: www.fao.org/3/cc0639en/online/sofi-2022/food-security-nutrition-indicators.html (accessed 20 March 2022).

Verma, K.K., Song, X.-P., Joshi, A., Tian, D.-D., Rajput, V.D. *et al.* (2022) Recent trends in nano-fertilizers for sustainable agriculture under climate change for global food security. *Nanomaterials* 12(1), 173. DOI: 10.3390/nano12010173.

Wang, L., Hu, C. and Shao, L. (2017) The antimicrobial activity of nanoparticles: present situation and prospects for the future. *International Journal of Nanomedicine* 12, 1227. DOI: 10.2147/IJN.S121956.

Weyell, P., Kurland, H.-D., Hülser, T., Grabow, J., Müller, F.A. *et al.* (2020) Risk and life cycle assessment of nanoparticles for medical applications prepared using safe- and benign-by-design gas-phase syntheses. *Green Chemistry* 22(3), 814–827. DOI: 10.1039/C9GC02436K.

Wiemann, M., Vennemann, A., Blaske, F., Sperling, M. and Karst, U. (2017) Silver nanoparticles in the lung: toxic effects and focal accumulation of silver in remote organs. *Nanomaterials* 7(12), 441. DOI: 10.3390/nano7120441.

Wigger, H. and Nowack, B. (2019) Material-specific properties applied to an environmental risk assessment of engineered nanomaterials—implications on grouping and read-across concepts. *Nanotoxicology* 13(5), 623–643. DOI: 10.1080/17435390.2019.1568604.

Wojcieszek, J., Jiménez-Lamana, J., Ruzik, L., Szpunar, J. and Jarosz, M. (2020) To-do and not-to-do in model studies of the uptake, fate and metabolism of metal-containing nanoparticles in plants. *Nanomaterials* 10(8), 1480. DOI: 10.3390/nano10081480.

Yang, X., Feng, L., Zhang, Y., Hu, H., Shi, Y. *et al.* (2018) Co-exposure of silica nanoparticles and methylmercury induced cardiac toxicity *in vitro* and *in vivo*. *Science of The Total Environment* 631–632, 811–821. DOI: 10.1016/j.scitotenv.2018.03.107.

Yang, Z., Xiao, Y., Jiao, T., Zhang, Y., Chen, J. *et al.* (2020) Effects of copper oxide nanoparticles on the growth of rice (*Oryza Sativa* L.) seedlings and the relevant physiological responses. *International Journal of Environmental Research and Public Health* 17(4), 1260. DOI: 10.3390/ijerph17041260.

Yasmin, H., Mazher, J., Azmat, A., Nosheen, A., Naz, R. *et al.* (2021) Combined application of zinc oxide nanoparticles and biofertilizer to induce salt resistance in safflower by regulating ion homeostasis and antioxidant defence responses. *Ecotoxicology and Environmental Safety* 218, 112262. DOI: 10.1016/j.ecoenv.2021.112262.

Yun, Y., Gao, R., Yue, H., Li, G., Zhu, N. *et al.* (2015) Synergistic effects of particulate matter (PM$_{10}$) and SO$_2$ on human non-small cell lung cancer A549 via ROS-mediated NF-κB activation. *Journal of Environmental Sciences* 31, 146–153. DOI: 10.1016/j.jes.2014.09.041.

Yusefi-Tanha, E., Fallah, S., Rostamnejadi, A. and Pokhrel, L.R. (2020) Particle size and concentration dependent toxicity of copper oxide nanoparticles (CuONPs) on seed yield and antioxidant defense system in soil grown soybean (*Glycine max* cv. Kowsar). *Science of The Total Environment* 715, 136994. DOI: 10.1016/j.scitotenv.2020.136994.

Zahedi, S.M., Moharrami, F., Sarikhani, S. and Padervand, M. (2020) Selenium and silica nanostructure-based recovery of strawberry plants subjected to drought stress. *Scientific Reports* 10(1), 17672. DOI: 10.1038/s41598-020-74273-9.

Zhang, Y.-N., Poon, W., Tavares, A.J., McGilvray, I.D. and Chan, W.C.W. (2016) Nanoparticle–liver interactions: cellular uptake and hepatobiliary elimination. *Journal of Controlled Release* 240, 332–348. DOI: 10.1016/j.jconrel.2016.01.020.

Zhu, X., Hondroulis, E., Liu, W. and Li, C.Z. (2013) Biosensing approaches for rapid genotoxicity and cytotoxicity assays upon nanomaterial exposure. *Small* 9(9–10), 1821–1830. DOI: 10.1002/smll.201201593.

Zulfiqar, F., Navarro, M., Ashraf, M., Akram, N.A. and Munné-Bosch, S. (2019) Nanofertilizer use for sustainable agriculture: advantages and limitations. *Plant Science* 289, 110270. DOI: 10.1016/j.plantsci.2019.110270.

Index